Electronic Waste

Electronic Waste

Recycling and Reprocessing for a Sustainable Future

Edited by
Maria E. Holuszko
Amit Kumar
Denise C. R. Espinosa

WILEY-VCH

Editors

Dr. Maria E. Holuszko
The University of British Columbia
NBK Institue of Mining Engineering
517-6350 Stores Road
V6T 1Z4 Vancouver, BC
Canada

Dr. Amit Kumar
The University of British Columbia
NBK Institue of Mining Engineering
517-6350 Stores Road
V6T 1Z4 Vancouver, BC
Canada

Dr. Denise C. R. Espinosa
University of Sao Paulo
Polytechnic School, Department of Chemical Engineering
Av. Prof. Luciano Gualberto
380 – Butantã, São Paulo, SP 05508-010
Brazil

Cover Design: Wiley
Cover Image: © Andrei Kuzmik/Shutterstock, Recyling symbol – public domain

Library of Congress Card No.: applied for

British Library Cataloguing-in-Publication Data
A catalogue record for this book is available from the British Library.

Bibliographic information published by the Deutsche Nationalbibliothek
The Deutsche Nationalbibliothek lists this publication in the Deutsche Nationalbibliografie; detailed bibliographic data are available on the Internet at <http://dnb.d-nb.de>.

Print ISBN: 978-3-527-34490-1
ePDF ISBN: 978-3-527-81642-2
ePub ISBN: 978-3-527-81640-8
oBook ISBN: 978-3-527-81639-2

Typesetting Straive, Chennai, India
Printing and Binding CPI Group (UK) Ltd, Croydon, CR0 4YY

Printed on acid-free paper

C085811_021121

Contents

Preface

Digitalization has played an essential role in social and technological development globally, while electronic and electrical equipment has become integral to our everyday lives. Digital devices provide broad access to education, instant information, continuous entertainment and contribute to mass communication, thus improving the overall quality of our lives. During the COVID-19 pandemic, the internet allowed us to function and remain a productive society worldwide.

Meanwhile, the life expectancy of most electronic devices, specifically small devices such as cellphones, tablets, and laptops, is getting shorter and shorter, resulting in alarming amounts of e-waste generation. Many discarded electronics are being improperly disposed of, hence posing a significant risk to the environment and human health. With an estimated annual growth of 3–4%, electronic waste is the fastest growing waste stream worldwide, exceeding 50 Mt annually in 2019, while only 20% of the e-waste is collected and recycled globally. The electronic devices have been reported to contain gold and copper grades, significantly exceeding the grades of many operating mines. The existence of precious metals in e-waste provides an economic incentive for recycling. On the other hand, the presence of hazardous substances in e-waste calls for complex reprocessing to decontaminate before its final disposal.

The development of efficient e-waste recycling methods and the recovery of precious metals and critical materials from e-waste are interesting and technically challenging. Furthermore, the informal urban mining of e-waste creates significant social and public health issues. Therefore, there was a need for a comprehensive overview of the current situation with e-waste generation, disposal, regulations, recycling technologies while providing a global perspective.

This book aims to overview the current global situation regarding e-waste, including technological issues with e-waste recycling and recovery of value from e-waste streams. The chapters in this book outline the definition of electronic waste, explore methods for e-waste estimation, identify challenges related to the timely information on e-waste collection and management, and elaborate on the practices in developed and developing countries. The book delivers information on currently used recycling technologies, including physical separation technologies, pyrometallurgy, hydrometallurgy, and biohydrometallurgy, and reviews materials used in the manufacturing of electronics as well as the development of new materials for green-ecological and biodegradable electronics. Additionally, methods and ideas for new practices to facilitate sustainability in the electronics industry are proposed to “close the loop” in industrial production to minimize waste generation and possibly to promote a zero-waste scenario. The book concludes with a chapter on the circular economy in electronics and provides some perspective on the future of electronic waste.

This book was made possible through collaboration between international experts in the field of e-waste recycling. It collates academic and industrial expertise to provide a comprehensive overview of the scope of the problem with electronics worldwide, specifically on their fate as e-waste and the recycling efforts to shed light on the current e-waste paradigm.

Vancouver, July 2021

Maria E. Holuszko
Amit Kumar
Denise C. R. Espinosa

1

Introduction, Vision, and Opportunities

Maria E. Holuszko[1], Denise C. R. Espinosa[2], Tatiana Scarazzato[3], and Amit Kumar[1]

[1] *NBK Institute of Mining Engineering, University of British Columbia, 6350, Stores Road, Vancouver, BC V6T 1Z4, Canada*
[2] *University of São Paulo, Polytechnic School, Department of Chemical Engineering, Av. Prof. Luciano Gualberto, 380 – Butantã, São Paulo – SP 05508-010, Brazil*
[3] *Federal University of Rio Grande do Sul, Department of Materials, 9500, Av. Bento Gonçalves, Porto Alegre – RS, 91509-900, Brazil*

1.1 Background

The concept of sustainability defined by The United Nations Organization in 1987, which is valid even today, is based on the idea of "meeting the needs of the present without compromising the ability of future generations to meet their own needs" (Nations 2019). Such a concept was complemented in The Johannesburg Declaration on Sustainable Development in which the three pillars of sustainability were defined: economic, environmental, and social development (Comission 2002).

Notwithstanding, the world currently faces a transition between the third and the fourth industrial revolutions, which began about five decades ago and has transformed our way of living. Also known as the Information Revolution, this period has been marked by swift advances in computer technologies, massive popularization of high-technology devices, and the growth of artificial intelligence (Carvalho et al. 2018; Rai and Lal 2000). The technological revolution brought up the creation of lithium-ion batteries, touchscreen devices, supercomputers, photovoltaic panels, and nanocomposites, and practically revolutionized the way the society interacts, the way energy is stored, and the advanced materials field for all industrial sectors.

Electrical and electronic equipment is one of the major consumers of metals such as copper, gold, silver, and iron. Namias (2013) suggested that electronic devices can contain up to 60 different elements that could be valuable or hazardous. Natural Resources Canada (2019) showed that globally 18% of aluminum, 31% of copper, 9.5% of gold, 9% of platinum group metals, and 24% of rare-earth elements were used in electrical and electronic equipment manufacturing in 2017. In the United States of America, 9% of total aluminum, 21% of beryllium, 19% of copper, 40%

Electronic Waste: Recycling and Reprocessing for a Sustainable Future, First Edition.
Edited by Maria E. Holuszko, Amit Kumar, and Denise C. R. Espinosa.

of gold, and 26% of silver were used in the electrical and electronic equipment manufacturing industry in 2019 (U.S. Geological Survey 2019). BullionStreet (2012) showed that approximately 290 tonnes of gold and 6800 tonnes of silver are consumed by the electronic industry every year. In the current scenario, the new manufacturing industry became dependent on less-known raw materials and increased the extraction of common metals from ores simultaneously. Indium, for example, despite being discovered in 1863, was found to be industrially applicable only in 1934. The use of indium in thin-film coatings, mainly as indium-tin-oxide compound (ITO) in liquid crystal display screens, increased its world consumption over 1000% since 1993 (Alfantazi and Moskalyk 2003).

Rare-earth elements (REEs) are also widely used in digital technologies such as disc drivers and communication systems but also in batteries and fuel cells for hydrogen storage, catalysts, light-emitting diodes (LEDs), and fluorescent lighting. Back in 1950, the applications of REE in magnets of electric and electronic equipment were already known. Nevertheless, until 2010 their recycling rate was lower than 1% due to their relatively low prices (Gunn 2013). Between 2010 and 2015, the demand for REE surpassed its supply and continuously increased. As the production is almost totally held by few countries, the recycling of REE has become a paramount concern (Edahbi et al. 2019).

1.2 E-Waste

With the development of new technologies, especially in laptops, cellphones, and tablets, older technologies are getting obsolete, reducing the lifespan of electrical and electronics products and thus contributing to a higher rate of waste generation. As a result, close to 1 billion devices will be discarded within 4–5 years. The discarded electric and electronic equipment or their parts are considered e-waste. The European Commission Directive 2008/98/EC (2008) and the European Union Directive 2012/19/EU (2012) described e-waste as:

> any electrical or electronic equipment which is waste (substance or object which the holder discards or intends or is required to discard), including all components, sub-assemblies, and consumables which are part of the product at the time of discarding.

Based on the definition of e-waste, the electrical or electronic equipment (EEE) itself was divided into six (6) classes in the Directive 2012/19/EU (The European Union 2012). These categories with the items (not limited to) in the categories are listed as,

1) Temperature-exchange equipment: refrigerators, freezers, air conditioning equipment and, heat pumps
2) Screens, monitors, and equipment containing screens (surface > $100\,cm^2$): screens, televisions, LCD photo frames, monitors, laptops, and notebooks

3) Lamps: fluorescent lamps, high-intensity discharge lamps, including high-pressure sodium lamps and metal halide lamps, low-pressure sodium lamps and LED lamps
4) Large equipment (external dimensions > 50 cm): washing machines, dryers, dishwashers, electric stoves, musical equipment, large computer mainframes, large printing machines, copying equipment, large coin-slot machines, large medical devices, large automatic dispensers, and photovoltaic panels
5) Small equipment (external dimensions < 50 cm): vacuum cleaners, appliances for sewing, luminaires, microwaves, irons, toasters, electric kettles, clocks and watches, electric shavers, scales, radio, video cameras, electrical and electronic toys, sports equipment, smoke detectors, heating regulators, thermostats, small electrical and electronic tools, small medical devices, and small automatic dispensers
6) Small IT and telecommunication equipment (external dimension < 50 cm): mobile phones, GPS, pocket calculators, routers, personal computers, printers, and telephones

This electronic waste (discarded electronics) has been a growing concern around the world. The total e-waste generated around the globe in 2019 was 53.6 million tonnes and is expected to reach 74 million tonnes in 2030. The waste generated per capita increased from 6.1 kg per inhabitant in 2016 to 7.3 kg per inhabitant in 2019 (Forti et al. 2020). Wahlen (2019) reported that under the business-as-usual case, the total e-waste generation would increase to 120 million tonnes by 2050. The growth rate of e-waste generation has been reported to be 3–5% by Cucchiella et al. (2015), 3–4% by Baldé et al. (2017) and Aaron (2019), and as high as 8% by LeBlanc (2018). According to Transparency Market Research report (2017), the global e-waste market is anticipated to increase at a compound annual growth rate of 5.6% by volume from 2016 to 2026.

The fate of the e-waste can be described by the simplified diagram shown in Figure 1.1. The primary focus of any country or organization should be the collection and recycling of e-waste. However, not all the e-waste is collected, and a portion of the e-waste stream is disposed of in landfills. The collected materials are sent for recycling, and the high-value components such as metals and high-value plastics are fed back to the manufacturing stream, whereas low-value materials are disposed of in landfills.

The primary focus of any country or organization should be the collection and recycling of e-waste. However, not all the e-waste is collected, and a portion of the e-waste stream is disposed of in landfills. The collected materials are sent for recycling, and the high-value components such as metals and high-value plastics are fed back to the manufacturing stream, whereas low-value materials are disposed of in landfills. The e-waste collection volume must be increased to boost the circular economy in any part of the world, and the waste stream after the recycling process has to be studied simultaneously for its potential recovery and usage so that the fractions to be disposed of are minimized.

E-waste recycling decreases the amount of extracted raw materials from ores and solid waste inadequate disposal. The recycling routes must also be technically and

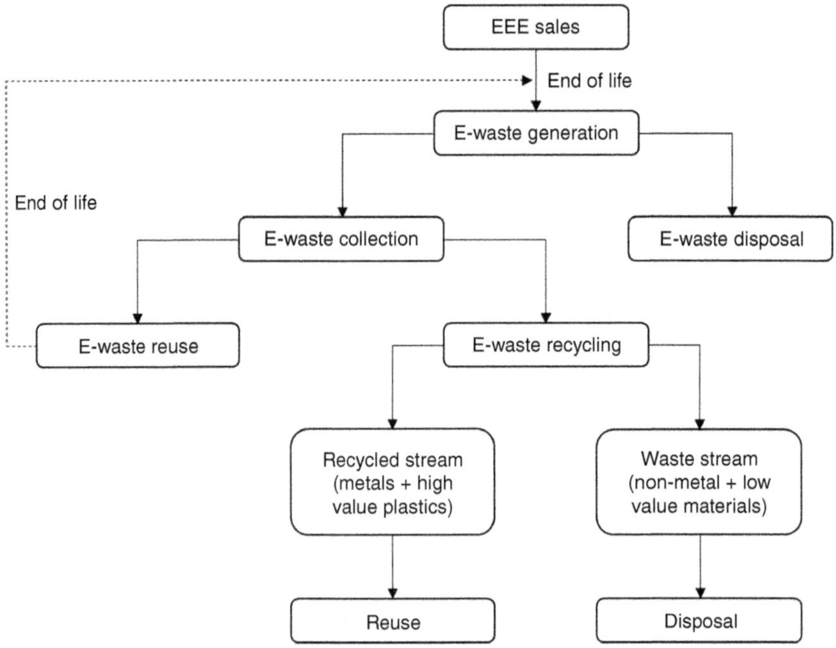

Figure 1.1 Simplified flow of EEE products.

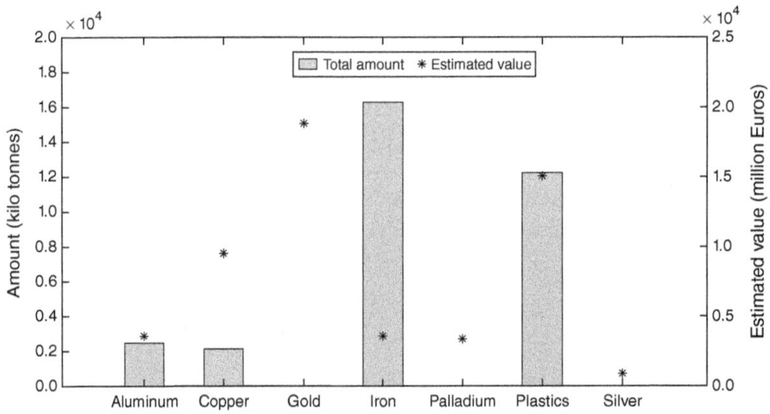

Figure 1.2 Estimated value of materials present in e-waste.

economically feasible. Given the added value of precious metals and critical metals found in the majority of e-waste, such requirements are not difficult to be fulfilled. Baldé et al. (2017) estimated the amount of various elements and materials present in e-waste. It showed that the total contained/potential value of selected metal and materials present in e-waste was US$ 57 billion in 2019 (Forti et al. 2020). Figure 1.2 shows a breakdown of the various metals and materials present in e-waste with their total amount and estimated values. It should be noted that the estimated value depicted in Figure 1.2 represents an ideal-case scenario of 100% collection and metal recovery and without accounting for costs associated with collection and recycling. It indicates the economic opportunity for e-waste recycling.

It should also be taken into consideration that the concentration of metals in the e-waste is significantly higher than that of a conventional mining operation. The global ore grade has been decreasing, and the increased global metal demand has forced mining operations to increase the plant throughput and excavate more complex and fine-grained ore deposits (Lèbre and Corder 2015).

Table 1.1 shows the concentration of various metals in different types of e-waste and an average ore body. Calvo et al. (2016) summarized that the global average copper grade for run-of-mine ore is ~0.62% and will decline in the coming years due to the exhaustion of high-grade mines. AME Research (2018) showed that the average copper grade has decreased from 0.74% in 2005 to 0.59% in 2017, with a compound annual decline rate of 1.8%. The global average gold grade of all the deposits was 1.01 g/t in 2013 (Desjardins 2014). The highest gold grade for the underground operation was 21.5 ppm in Fosterville, Australia, whereas for open pit was 7.60 ppm in Way Linggo, Indonesia (Basov 2018). The average output of top-six silver mines was 7.6 oz (215 g) per tonne in 2012 and has dropped to 4–5 oz per ton in 2017 (McLeod 2014; Money Metals Exchange 2018). The palladium grade reported in Table 1.1 is based on the average palladium grade of the Lac des Iles Mine Property in Northern Ontario. It indicates that the average metal grade present in e-waste is significantly higher than conventional mines and thus provides the opportunity for the extraction/urban mining.

Table 1.1 Average metal content in various waste EEE and typical ore.

	Copper %	Aluminum %	Iron %	Gold ppm	Silver ppm	Palladium ppm
Air conditioner	6–19	7–9	46	15	58	–
Desktop	7–20	2–4	18–47	46–240	207–570	18–25
Laptop	1–19	1–2	20	32–630	190–1100	19
Mobile phone	10–33	3	5	30–1500	2000–3800	300–1700
Printed circuit board	12–19	2–8	0–11	29–1120	100–5200	33–220
Refrigerator	3–17	1–2	48–50	44	42	–
Television	1–21	1–15	13–43	5–300	120–600	0–20
Washing machine	3–7	0–3	52–53	17	51	–
e-Waste (average)	12–35	1.5–5	5–11	30–350	80–1000	30–200
Typical ore	0.5–3	20–24	30–60	0.5–10	5–10	1–10

Sources: Based on Bizzo et al. (2014), Calvo et al. (2016), Desjardins (2014), Işıldar (2016), Bizzo et al. (2014), Calvo et al. (2016), Desjardins (2014), Fornalczyk et al. (2013), Işıldar (2016), Khaliq et al. (2014), Liu (2014), McLeod (2014), Namias (2013), North American Palladium Ltd. (2018), Shah et al. (2014), Tickner et al. (2016), Zeng et al. (2016).

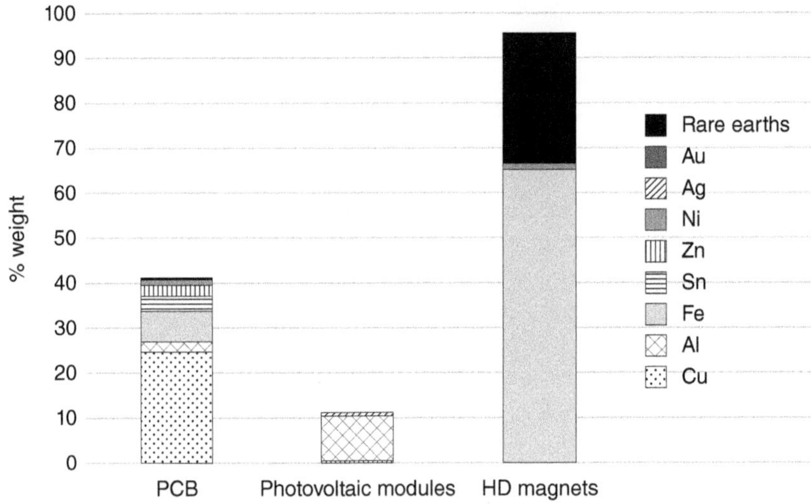

Figure 1.3 Average percent weight of some common metals found in PCB, photovoltaic modules, and HD magnets (Caldas et al. 2015; Dias et al. 2016, 2018; Kasper et al. 2011; München et al. 2018; München and Veit 2017; Petter et al. 2014; Rozas et al. 2017; Sant'ana et al. 2013; Silvas et al. 2015; Stuhlpfarrer et al. 2015; Yamane et al. 2011). Sources: Based on Caldas et al. (2015), Dias et al. (2016), and Kasper et al. (2011).

As depicted in Figure 1.3, photovoltaic modules present a high percent weight of a single element (aluminum), while printed circuit boards are composed of a mixture of different metals, mainly copper, iron, aluminum, tin, zinc, and nickel. It is estimated that printed circuit boards (PCBs) may contain an average of 18 elements from the periodic table (Caldas et al. 2015; Kasper et al. 2011; Petter et al. 2014; Rozas et al. 2017; Sant'ana et al. 2013; Silvas et al. 2015; Yamane et al. 2011). Remarkably, hard disk magnets, although they may contain high amounts of iron, also present exciting amounts of rare-earth elements, mainly neodymium, praseodymium, and dysprosium (München and Veit 2017).

Metals that are present in smaller amounts may also be economically advantageous to be recovered. Gold and silver in cell phones, for example, represent about 0.06% and 0.045% in weight, respectively (Caldas et al. 2015; Sant'ana et al. 2013). Considering the number of cell phones worldwide (Kreyenhagen 2018), it can be estimated that almost 300-ton gold is present only in cellphone devices.

The added value of such metals supports their recycling even if they are present in low percent weight. Photovoltaic modules, for example, although present less than 1% of silver in their composition (Figure 1.3), may be economically feasible to be recycled, as shown in Figure 1.4.

From Figure 1.4, it becomes evident that the establishment of feasible recycling routes may be advantageous for reclaiming metals in either high or low amounts. This is undoubtedly challenging, as e-waste from distinct origins would present different compositions and structures. In a microscopic vision, metals and other elements may be bonded with each other in many possibilities, which may require different techniques to achieve their extraction and recovery. Thus, recycling

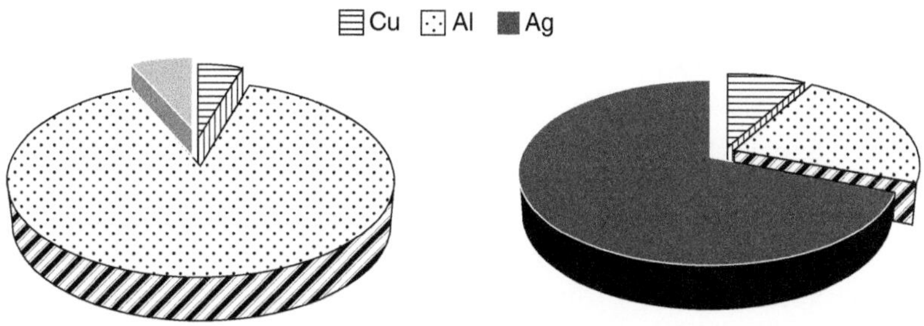

(a) Average weight composition of photovoltaic modules (according to Figure 1)

(b) Metal prices* per kilogram of photovoltaic modules

*Prices in US$/kg from London Exchange Metals on December/19

Figure 1.4 (a) Composition of photovoltaic modules illustrated as kilograms of each metal per kilogram of photovoltaic module. (b) Value of each metal per kilogram of photovoltaic modules.

routes must be versatile. Electric and electronic equipment are constantly being improved, and their chemical composition may change both the required process and the obtained materials. Lithium-ion batteries (LIBs) are a good example of continuous improvement. Cathodes from LIBs are typically composed of lithium, cobalt, manganese, and nickel oxides (Blomgren 2017; Zhao et al. 2019). Because cobalt is considered a critical metal, novel cathode materials are being developed, such as sulfur-based cathodes, which eliminate the need for cobalt (Li et al. 2018). High-efficiency anodes are also being developed to increase the performance of LIBs, such as titanium-niobium based anodes for automotive applications (Takami et al. 2018). Considering the speed of technological evolution, in a few decades from now, the composition of obsolete lithium-ion batteries may present a substantial change.

Recycling processing routes are typically hydrometallurgical, pyrometallurgical, or hydro-pyrometallurgical. Each of them presents both advantages and disadvantages, and all of them may produce toxic tailings, which must be considered. The pyrometallurgical recycling routes consist of several processes that, among others, include smelting, combustion, pyrolysis, molten salt processing, and pyro-chemical techniques. As for drawbacks to the pyrolysis process, there is usually the release of toxic gases and halogen formation arising from fire retardants and plastic mixtures that compose the waste electrical or electronic equipment (WEEE) scrap. For the efficient recovery of metals, some modern methods, such as vacuum pyrolysis, the molten salt process, and pyro-chemistry, appear to be promising candidates because of their innovative solutions for environmental issues and recovery efficiency. Nevertheless, they still need more scientific and technical contributions coupled with industrial validation (Zhang et al. 2015).

Hydrometallurgical processing involves a sequence of methods for producing metals and metal compounds from aqueous media. In general, the process is composed of leaching techniques, followed by purification and recovery of metals, resulting in the desired product. Such a product may be either metals or alloys and compounds containing the metals of interest, such as oxides (Gupta 2003). Precipitation, hydrolysis, electrochemistry, conversion, complexation, solvation, and ionic dissociation are often used in different processing steps. However, such routes often require the use of aggressive and concentrated reactants for leaching steps. The treatment of the generated waste belongs to the development of a hydrometallurgical route. In addition, some processes still present low efficiency. To solve such issues, the scientific community is currently focusing on the development of less-toxic processing routes, using, for example, ionic liquids, supercritical fluids, and organic acids, aiming at achieving optimum extraction rate using greener chemicals.

Nevertheless, another important matter is the efficient collection of the disposed of WEEE to get to recycling industries. The incitement of reversal logistics and circular economy concepts must be improved and optimized to achieve satisfactory rates of recycled WEEE. In this sense, the establishment of strict legislation and efficient management are as well important. In developing countries, for example, WEEE recycling is still incipient, there is little legislation, and incentives are low. Even in developed countries, there is still much to be improved.

These legislations and programs are essential to increase awareness and boost the collection rate. The e-waste management programs also provide collection targets, summarize, and report the collection volumes, which helps to understand the performance and provide better planning tools for the future. An efficient e-waste management program could increase the collection rates, provide a better estimation of e-waste collection, and would also provide a better understanding of the steps required to improve e-waste collection and recycling and thus promote circular economy. At the same time, a higher collection rate in tandem with an efficient recycling system that not only recovers metals but also provides a solution for the nonmetals would be necessary to increase the e-waste circularity.

Therefore, it becomes evident that e-waste recycling is an interdisciplinary and multidisciplinary theme, as depicted in Figure 1.5. Technical, economic, legislative, social, and environmental aspects are involved throughout the life cycle of all-electric and electronic equipment, including recycling after their disposal. This book seeks to provide an overview of all aspects of a sustainable future.

1.3 Outline

To fully understand the e-waste, this book presents various sections to describe different aspects of e-waste, e-waste management systems and involved technologies, and other related disciplines.

Chapter 2 presents an overview of e-waste management practices adopted in developed and developing countries. The differences in the availability of regulations in a country directly impact the fate of e-waste in that country.

In Chapter 3, the regulations related to the transboundary movement of e-waste around the world are discussed. The import and export of e-waste is another major challenge in the e-waste management system.

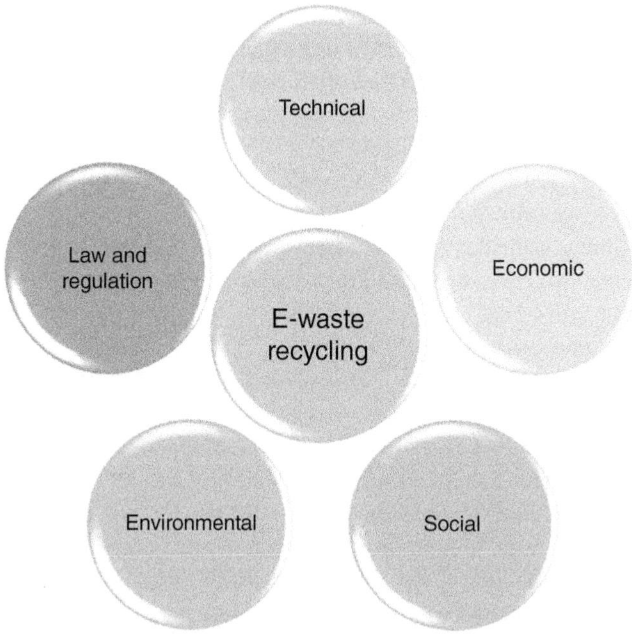

Figure 1.5 Multidisciplinary aspects of e-waste recycling.

Various models and methods used for the quantification of total e-waste globally are presented in Chapter 4. The success of an e-waste recycling or disposal facility depends on the annual throughput to plant.

Chapter 5 emphasizes the materials used in the manufacturing of electronic devices. The determination of type and quantity of the materials used in the manufacturing process define the technologies adapted for recycling.

A detailed view of recycling technologies used in e-waste processing is presented in Chapters 6–10.

Chapter 11 presents insights into the life cycle analysis of obsolete electric and electronic equipment. The life cycle analysis is an essential tool for quantifying the environmental impact of e-waste.

Finally, the future of electronic devices and e-waste is discussed in Chapters 12 and 13, focusing on innovative aspects of manufacturing electronic devices, green chemistry, and circular economy.

References

Aaron, J. (2019). Why is E-waste management important in 2019? https://get-green-now.com/e-waste-management (accessed 15 March 2021).

Alfantazi, A.M. and Moskalyk, R.R. (2003). Processing of indium: a review. *Minerals Engineering* 16 (8): 687–694. https://doi.org/10.1016/S0892-6875(03)00168-7.

AME Research (2018). Copper – declining copper ore grades. https://aus.amegroup.com/Website/FeatureArticleDetail.aspx?faId=437 (accessed 15 March 2021).

Baldé, C. P., Forti, V., Gray, V., et al. (2017). The global E-waste monitor – 2017. https://www.itu.int/en/ITU-D/Climate-Change/Documents/GEM2017/Global-E-waste Monitor 2017.pdf (accessed 15 March 2021).

Basov, V. (2018). Highest-grade gold mines in 2017. https://www.mining.com/highest-grade-gold-mines-2017 (accessed 15 March 2021).

Bizzo, W., Figueiredo, R., and de Andrade, V. (2014). Characterization of printed circuit boards for metal and energy recovery after milling and mechanical separation. *Materials* 7 (6): 4555–4566. https://doi.org/10.3390/ma7064555.

Blomgren, G.E. (2017). The development and future of lithium ion batteries. *Journal of the Electrochemical Society* 164 (1): A5019–A5025. https://doi.org/10.1149/2.0251701jes.

BullionStreet (2012). Electronics industry uses 320 tons of gold, 7500 tons of silver annually. http://www.bullionstreet.com/news/electronics-industry-uses-320-tons-of-gold7500-tons-of-silver-annually/2255 (accessed 15 March 2021).

Caldas, M.P.K., Moraes, V.T., De Junca, E. et al. (2015). Recuperação De Prata : Estudo De Uma Rota Hidrometalúrgica recycling of printed circuit boards aiming silver recovery. *Tecnologia Em Metalurgia, Materiais e Mineração* 12: 102–108.

Calvo, G., Mudd, G., Valero, A., and Valero, A. (2016). Decreasing ore grades in global metallic mining: a theoretical issue or a global reality? *Resources* 5 (4): 36. https://doi.org/10.3390/resources5040036.

Carvalho, N., Chaim, O., Cazarini, E., and Gerolamo, M. (2018). Manufacturing in the fourth industrial revolution: a positive prospect in sustainable manufacturing. *Procedia Manufacturing* 21: 671–678. https://doi.org/10.1016/j.promfg.2018.02.170.

Cucchiella, F., D'Adamo, I., Lenny Koh, S.C., and Rosa, P. (2015). Recycling of WEEEs: an economic assessment of present and future e-waste streams. *Renewable and Sustainable Energy Reviews* 51: 263–272. https://doi.org/10.1016/j.rser.2015.06.010.

Desjardins, J. (2014). Global gold mine and deposit rankings 2013. http://www.visualcapitalist.com/global-gold-mine-and-deposit-rankings-2013 (accessed 15 March 2021).

Dias, P., Javimczik, S., Benevit, M. et al. (2016). Recycling WEEE: extraction and concentration of silver from waste crystalline silicon photovoltaic modules. *Waste Management* 57: 220–225.

Dias, P., Schmidt, L., Gomes, L.B. et al. (2018). Recycling waste crystalline silicon photovoltaic modules by electrostatic separation. *Journal of Sustainable Metallurgy* 4 (2): 176–186. https://doi.org/10.1007/s40831-018-0173-5.

Edahbi, M., Plante, B., and Benzaazoua, M. (2019). Environmental challenges and identification of the knowledge gaps associated with REE mine wastes management. *Journal of Cleaner Production* 212: 1232–1241. https://doi.org/10.1016/j.jclepro.2018.11.228.

Fornalczyk, A., Willner, J., Francuz, K., and Cebulski, J. (2013). E-waste as a source of valuable metals. *Archives of Materials Science and Engineering* 63 (2): 87–92. https://www.researchgate.net/publication/260706030_E-waste_as_a_source_of_valuable_metals (accessed 15 March 2021).

Forti, V., Baldé, C. P., Kuehr, R., and Bel, G. (2020). Global E-waste Monitor 2020: Quantities, flows and the circular economy potential. https://www.itu.int/en/ITU-D/Environment/Documents/Toolbox/GEM_2020_def.pdf (accessed 15 March 2021).

Gunn, G. (2013). *Critical Metals Handbook*, 1–439. Wiley https://doi.org/10.1002/9781118755341.

Gupta, C.K. (2003). *Chemical Metallurgy Principles and Practice*. Wiley.

Işıldar, A. (2016). *Biological Versus Chemical Leaching of Electronic Waste for Copper and Gold Recovery*. Université Paris-Est Marne-la-Vallée https://tel.archives-ouvertes.fr/tel-01738056/document (accessed 15 March 2021).

Kasper, A.C., Berselli, G.B.T., Freitas, B.D. et al. (2011). Printed wiring boards for mobile phones: characterization and recycling of copper. *Waste Management* 31 (12): 2536–2545. https://doi.org/10.1016/j.wasman.2011.08.013.

Khaliq, A., Rhamdhani, M., Brooks, G., and Masood, S. (2014). Metal extraction processes for electronic waste and existing industrial routes: a review and Australian perspective. *Resources* 3 (1): 152–179. https://doi.org/10.3390/resources3010152.

Kreyenhagen, J. (2018). *Why Catering to Mobile Users is Vita for Today' S Services*. Forbes Communications Council.

LeBlanc, R. (2018). E-waste recycling facts and figures. https://www.thebalancesmb.com/e-waste-recycling-facts-and-figures-2878189 (accessed 15 March 2021).

Lèbre, É. and Corder, G. (2015). Integrating industrial ecology thinking into the management of mining waste. *Resources* 4 (4): 765–786. https://doi.org/10.3390/resources4040765.

Li, G., Chen, Z., and Lu, J. (2018). Lithium-sulfur batteries for commercial applications. *Chem* 4 (1): 3–7. https://doi.org/10.1016/j.chempr.2017.12.012.

Liu, S. (2014). Analysis of electronic waste recycling in the United States and potential application in China [Columbia University]. https://secureservercdn.net/198.71.233.185/epm.300.myftpupload.com/wp-content/uploads/2020/10/FINAL_Shumeng_Thesis_Dec-20_2014.pdf.

McLeod, C. (2014). Top primary silver miners facing declining ore grades. http://investingnews.com/daily/resource-investing/precious-metals-investing/silver-investing/top-primary-silver-miners-facing-declining-ore-grades (accessed 15 March 2021).

Money Metals Exchange. (2018). World's largest silver mines: suffer falling ore grades & rising costs/seeking alpha. https://seekingalpha.com/article/4141630-worlds-largest-silver-mines-suffer-falling-ore-grades-and-rising-costs (accessed 15 March 2021).

München, D.D., Bernardes, A.M., and Veit, H.M. (2018). Evaluation of neodymium and praseodymium leaching efficiency from post-consumer NdFeB magnets. *Journal of Sustainable Metallurgy* 4 (2): 288–294. https://doi.org/10.1007/s40831-018-0180-6.

München, D.D. and Veit, H.M. (2017). Neodymium as the main feature of permanent magnets from hard disk drives (HDDs). *Waste Management* 61: 372–376. https://doi.org/10.1016/j.wasman.2017.01.032.

Namias, J. (2013). *The Future of Electronic Waste Recycling in the United States: Obstacles and Domestic Solutions*. New York, United States: Columbia University.

Natural Resources Canada (2019). Minerals and metals facts. https://www.nrcan.gc.ca/minerals-metals-facts/20507 (accessed 15 March 2021).

North American Palladium Ltd. (2018). North American Palladium completes a new feasibility study featuring a major expansion of underground reserves. https://www.nap.com/investors/news-releases/news-releases-details/2018/North-American-Palladium-Completes-a-New-Feasibility-Study-Featuring-a-Major-Expansion-of-Underground-Reserves/default.aspx (accessed 15 March 2021).

Petter, P.M.H., Veit, H.M., and Bernardes, A.M. (2014). Evaluation of gold and silver leaching from printed circuit board of cellphones. *Waste Management* 34 (2): 475–482. https://doi.org/10.1016/j.wasman.2013.10.032.

Rai, L.P. and Lal, K. (2000). Indicators of the information revolution. *Technology in Society* 22 (2): 221–235. https://doi.org/10.1016/S0160-791X(00)00006-3.

Rozas, E.E., Mendes, M.A., Nascimento, C.A.O. et al. (2017). Bioleaching of electronic waste using bacteria isolated from the marine sponge Hymeniacidon heliophila (Porifera). *Journal of Hazardous Materials* 329: 120–130. https://doi.org/10.1016/j.jhazmat.2017.01.037.

Sant'ana, H.B.S., Moura, F.J., and Veit, H.M. (2013). Caracterização Físico-Química De Placas De Circuito Impresso De Aparelhos De Telefone Celular. *Tecnologia Em Metalurgia Materiais e Mineração* 10 (3): 231–238. https://doi.org/10.4322/tmm.2013.033.

Shah, M.B., Tipre, D.R., and Dave, S.R. (2014). Chemical and biological processes for multi-metal extraction from waste printed circuit boards of computers and mobile phones. *Waste Management & Research* 32 (11): 1134–1141. https://doi.org/10.1177/0734242X14550021.

Silvas, F.P.C., Jiménez Correa, M.M., Caldas, M.P.K. et al. (2015). Printed circuit board recycling: physical processing and copper extraction by selective leaching. *Waste Management* 46: 503–510. https://doi.org/10.1016/J.WASMAN.2015.08.030.

Stuhlpfarrer, P., Luidold, S., and Antrekowitsch, H. (2015). Recycling of $Nd_2Fe_{14}B$ magnets. *World of Metallurgy - ERZMETALL* 68 (5): 278–285.

Takami, N., Ise, K., Harada, Y. et al. (2018). High-energy, fast-charging, long-life lithium-ion batteries using $TiNb_2O_7$ anodes for automotive applications. *Journal of Power Sources* 396 (June): 429–436. https://doi.org/10.1016/j.jpowsour.2018.06.059.

The European Commission (2008). Directive 2008/98/EC of the European Parliament and of the council of 19 November 2008 on waste and repealing certain directives. *Official Journal of the European Union* L 312: 3–30. https://eur-lex.europa.eu/legal-content/EN/TXT/PDF/?uri=CELEX:32008L0098&from=EN (accessed 15 March 2021).

The European Union (2012). Directive 2012/19/EU of the European Parliament and of the Council of the 4 July 2012 on waste electrical and electronic equipment (WEEE). *Official Journal of the European Union* L (June) https://eur-lex.europa.eu/legal-content/EN/TXT/PDF/?uri=CELEX:32012L0019&from=EN (accessed 15 March 2021).

Tickner, J., Rajarao, R., Lovric, B. et al. (2016). Measurement of gold and other metals in electronic and automotive waste using gamma activation analysis. *Journal of Sustainable Metallurgy* 2 (4): 296–303. https://doi.org/10.1007/s40831-016-0051-y.

Transparency Market Research. (2017). E-scrap and printed circuit board (PCB) E-scrap market. https://www.transparencymarketresearch.com/e-scrap-printed-circuit-board-e-scrap-market.html (accessed 15 March 2021).

U.S. Geological Survey (2019). *Mineral Commodity Summaries*. U.S. Geological Survey https://doi.org/10.3133/70202434.

UN Department of Economic and Social Affairs (2002). The Johannesburg Declaration on Sustainable Development. https://www.un.org/esa/sustdev/documents/WSSD_POI_PD/English/POI_PD.htm (accessed 10 January 2021).

UNESCO (2019). *Sustainable Development*. https://en.unesco.org/themes/education-sustainable-development/what-is-esd/sd#:~:text=Sustainable%20development%20is%20the%20overarching,to%20meet%20their%20own%20needs.%E2%80%9D (accessed 10 January 2021).

Wahlen, C. B. (2019). UN report highlights environmental, health risks from E-waste, urges circular economy shift. https://sdg.iisd.org/news/un-report-highlights-environmental-health-risks-from-e-waste-urges-circular-economy-shift (accessed 15 March 2021).

Yamane, L.H., Moraes, V.T., De Crocce, D. et al. (2011). Recycling of WEEE : characterization of spent printed circuit boards from mobile phones and computers. *Waste Management* 31 (12): 2553–2558. https://doi.org/10.1016/j.wasman.2011.07.006.

Zeng, X., Gong, R., Chen, W.-Q., and Li, J. (2016). Uncovering the recycling potential of "new" WEEE in China. *Environmental Science & Technology* 50 (3): 1347–1358. https://doi.org/10.1021/acs.est.5b05446.

Zhang, S., Ding, Y., Liu, B. et al. (2015). Challenges in legislation , recycling system and technical system of waste electrical and electronic equipment in China. *Waste Management* 45: 361–373. https://doi.org/10.1016/j.wasman.2015.05.015.

Zhao, T., Ji, R., Yang, H. et al. (2019). Distinctive electrochemical performance of novel Fe-based Li-rich cathode material prepared by molten salt method for lithium-ion batteries. *Journal of Energy Chemistry* 33: 37–45. https://doi.org/10.1016/j.jechem.2018.08.005.

2

e-Waste Management and Practices in Developed and Developing Countries*

Pablo Dias[1,2], *Andréa M. Bernardes*[2], *and Nazmul Huda*[3]

[1] *University of New South Wales, School of Photovoltaic and Renewable Energy Engineering, Faculty of Science and Engineering, 229 Anzac Parade, Kensington, Sydney, NSW 2052, Australia*
[2] *Federal University of Rio Grande do Sul, Engineering School, Department of Materials, Av. Bento Gonçalves, 9500, Porto Alegre – RS 91509-900, Brazil*
[3] *Macquarie University, School of Engineering, Faculty of Science and Engineering, Balaclava Rd, North Ryde, NSW 2109, Australia*

2.1 Introduction

Waste electrical and electronic equipment (WEEE or e-waste) is classified as a solid waste within the hazardous waste category (Goel 2017). E-waste consists in end-of-life electronic and electrical equipment, it comprehends – but is not limited to – obsolete, broken, or used computers, televisions, stereos, photocopiers, printers, faxes, monitors, and mobile phones (Westcott 2012). It also comprehends the less notable equipment such as radios, washing machines, microwave ovens, hair dryers, and photovoltaic panels (EU Directive 2012; Robinson 2009). Moreover, the WEEE definition also includes the components, subset of parts, peripheral accessories and materials used in the manufacturing of these equipment (EU Directive 2012).

The generation of e-waste appears to be higher in developed countries than that in developing economies (Goel 2017), but the WEEE generation has been increasing in both realities (Schluep et al. 2009). Furthermore, a positive correlation between gross domestic product (GDP) and e-waste generated in a given country was confirmed in a recent research. Interestingly, no correlation was found between e-waste generation and population (Kumar et al. 2017).

The current e-waste generation pattern poses one of the world's greatest pollution problems. On top of the growing generation pattern, e-waste is a particularly important waste stream because of its potential to be pollutants that pose a risk to the environment and to sustainable economic growth; and the potential to be resources, given the significant concentration of precious metals and high-demand materials it contains (Babu et al. 2007; Goosey 2012; Sugimura and Murakami 2016).

* Received 20 December 2019. Revised 01 December 2020.

Electronic Waste: Recycling and Reprocessing for a Sustainable Future, First Edition.
Edited by Maria E. Holuszko, Amit Kumar, and Denise C. R. Espinosa.

2.2 Overview on WEEE Management and Practices

WEEE management is a global challenge especially given many countries have no structured system of reverse logistics, and most WEEE is still disposed of in landfills or open places exposed to the inclement weather (Veit and Bernardes 2015). Tools such as the life cycle analysis (LCA), material flow analysis (MFA), and tailored policies such as extended producer responsibility (EPR) created to assist in waste management are also being applied to the WEEE challenge. These, however, are generally seen in operation only in developed countries (Kiddee et al. 2013). Developed countries tend to have laws and regulation to process WEEE safely. The compliance with these regulations is difficult to assure, given sound processing frequently runs against economic interests (Sthiannopkao and Wong 2013). These take-back systems and end-of-life processing legislation for the electronics industry were originally proposed because of environmental motives (Stevels 2007). A schematic of the management of e-waste from consumption to disposal is illustrated in Figure 2.1.

A different management approach for the global WEEE challenge was proposed recently and named best-of-two-worlds philosophy (Bo2W). It seeks to achieve the most sustainable solution for developing countries under the current international panorama. In summary, the philosophy claims developing countries should take advantage of the low labor cost to employ manual dismantling to liberate e-waste components. These separated and sorted components would then be exported (sold) to developed economies, where technology and infra-structure are available for sound downstream processing. This theoretically ensures labor and revenue for developing nations while ensuring state-of-the-art and environmentally safe end-processing (Goel 2017; Wang et al. 2012).

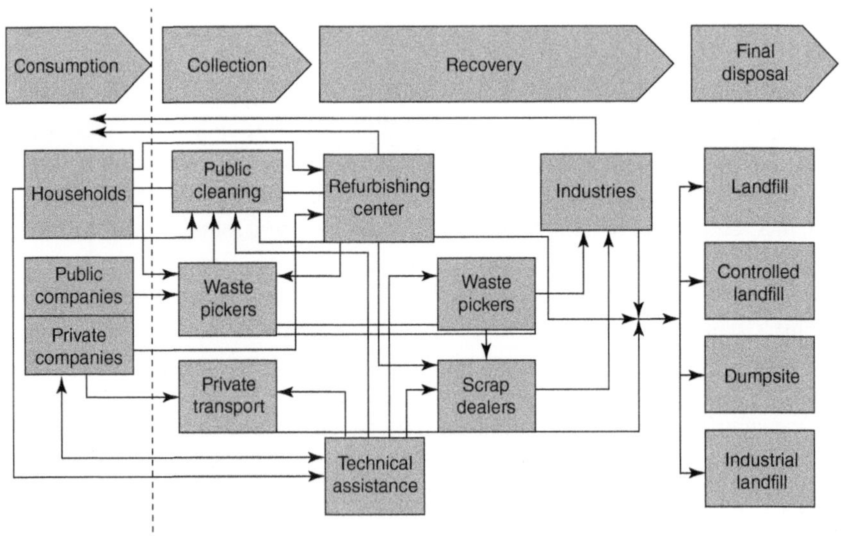

Figure 2.1 WEEE management flowchart. Source: Caiado et al. (2017).

Published research also describes measures to achieve better waste management practices. An important component identified is community awareness. Public awareness, outreach campaigns, and educational measures that show the negative impacts of incorrect e-waste disposal and their effective disposal value are particularly important. These campaigns should inform the roles and responsibilities of the agents involved in the e-waste management, including their rights as citizens to access waste management services (Rao et al. 2017; Schumacher and Agbemabiese 2019). To discourage the international e-waste transfer and enhance proper device collection, country studies on the size and destination of the complementary streams should be performed and used to create specific collection targets per WEEE category to specific countries (Huisman 2012).

The world is still searching for an ideal WEEE management model, even if that model is only fitted for a specific country or region. Currently, different country/states have different kinds of regulations and take-back systems. Europe is perhaps the best example to illustrate this great variety, as Great Britain alone holds 44 distinct take-back systems (Figure 2.2).

Regulations can allow or prohibit take-back systems to coexist and/or to compete. In some countries, the collection and recycling operational costs are distributed to the take-back systems according to the producers they represent. The verdict of whether a system run by a monopoly or a system run by companies in competition is more effective, however, is unclear at present. Moreover, the competent authorities hold the essential role of regulating the WEEE management systems to allow them to compete in a fair manner (Toffolet 2016).

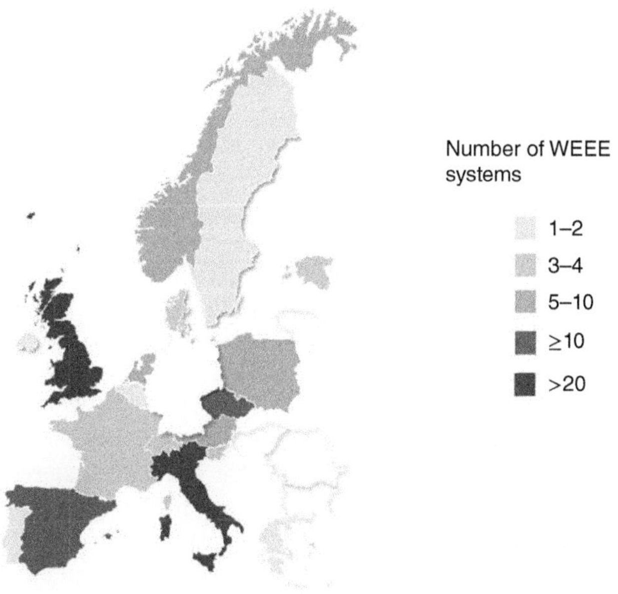

Figure 2.2 Distinct WEEE take-back systems in Europe. Source: Toffolet (2016).

2.3 International WEEE Management and Transboundary Movement

The WEEE rising waste volumes and peculiar characteristics aforementioned created a global export trend, where developed nations sent unwanted WEEE to developing nations. This has been reported in scientific papers to have started from the beginning of the twenty-first century due to large volumes of obsolete electrical and electronic equipment (EEE) in developed countries, and justified as an attempt to bridge the "digital divide" between developed and developing economies (Nnorom and Osibanjo 2008; Yang 2019). Herat and Agamuthu (2012) cited that large volumes were being sent to developing countries for reuse, refurbishment, recycling, and recovery of precious metals, and that some of the main countries receiving e-waste are India, China, Philippines, Hong Kong, Indonesia, Sri Lanka, Pakistan, Bangladesh, Malaysia, Vietnam, and Nigeria. In Nigeria, for instance, it was found that in the period between 2000 and 2010, the majority of its EEE/WEEE was coming from the USA, the UK, Germany, and China (for TVs, cathode ray tubes [CRTs], and personal computers [PCs]), and the imports have increased considerably from 2003 (Babayemi et al. 2015). The total transboundary shipment of hazardous wastes has increased since 2000 for most Organization for Economic Co-operation and Development (OECD) members and two-thirds of the EU countries, regardless of their trading positions (Yang 2019).

Most developing countries do not have a program for the storage, separation, collection, transport, or disposal of waste, nor adequate legislation and/or monitoring over the waste treatment procedures and the risks associated with incorrect disposal/treatment; this is especially true for e-waste (Nnorom and Osibanjo 2008; Sindiku et al. 2015). Several studies address the consequences of poor end-of-life treatment that happens in developing countries, the main ones involve severe environmental damage and negative impacts on human health (Egeonu and Herat 2016; Li et al. 2019; Schluep 2014; Zhang et al. 2018). These issues should be tackled by increasing the responsibility of the manufacturers (the EPR) and through the technology exchange between countries that export and import e-waste (Li et al. 2013). Nevertheless, the current global panorama remains the same as in previous decades, with large WEEE volumes being transported (legally or illegally) to developing economies (Figure 2.3) (Awasthi and Li 2017).

An important paradigm shift, however, was observed in Brazil, Mexico, South Africa, Nigeria, Indonesia, and Australia: the export of high-end components to countries with established downstream recycling industry. This involves a domestic industry setup capable of executing first stage recycling (i.e. separation of components) and organizing the logistics associated with collection and international shipping, which is achieved either by local companies that work as "material concentrators" and then sell these high-value components abroad or by large foreign downstream recycling enterprises that install sister companies abroad to act as collector, concentrator, and exporter of high-value components. In developing countries, the establishment of this industry is natural because of the relative low cost of labor, whereas in Australia it occurs due to the regulations in place.

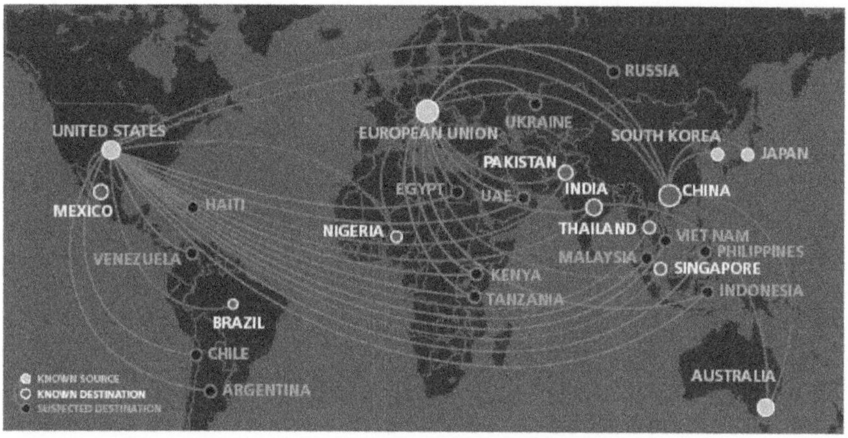

Figure 2.3 Known sources, destinations, and suspected destinations of WEEE transboundary movement worldwide. Source: Kumar et al. (2017).

This shift creates a controversial situation in which destination countries (such as Brazil, Nigeria, and Mexico) receive unwanted e-waste components with little value while exporting the e-waste components with high value (printed circuit boards (PCBs), hard drives, processors) (Dias et al. 2019; Dias et al. 2018b; Dias et al. 2018c; Iwenwanne 2019; Lydall et al. 2017; Snyman et al. 2017). Moreover, this export pattern contributes to maintain the downstream recycling industry of these destination countries stagnant (Dias et al. 2018a). Recent research even suggests the growing re-export of e-waste from the developing world back to advanced countries creates an offset by which countries importing high-quality used electronics send back an equal volume of e-waste. The (documented) transactions tend to occur between trade partners where the importer has a lower GDP per capita than the exporter. The same authors claim that while there is a movement of e-waste from developed to developing countries, there is also a substantial e-waste trade between developing countries. (Larmer 2018; Lepawsky 2015; Lepawsky and Mcnabb 2010).

2.4 WEEE Management and Practices – Developed and Developing Countries

Solid waste management generally involves (i) identifying and categorizing the source and nature of waste, (ii) separation, storage, and collection, (iii) waste transport, (iv) processing, and (v) ultimate waste disposal. This linear economy approach has been widely applied and is still a management model in many countries (Rao et al. 2017). Furthermore, solid waste management aims to minimize waste, maximize recycling and reuse, and ensure safe and environmentally sound disposal of waste. These objectives should be achieved in a sustainable manner employing and developing the capacity of the community, private enterprises, workers, and government (Rao et al. 2017). The ultimate goal of any waste management system is to increase the resource efficiency and reach the circular economy target (Nowakowski and Mrówczyńska 2018). Currently, this can be achieved by using resources more efficiently in the provision of an activity or product, using less resource-related services, reusing product and services, recycling the resources and materials in products (Worrell and Reuter 2014). The material recovery present in WEEE may be achieved by reusing its components, by recycling of the whole equipment (or a fraction of it), or by transforming waste into energy (energy recovery) (Nnorom and Osibanjo 2008). A sustainable management of WEEE has a significant role on the circular economy approach (D'Adamo et al. 2019; Nowakowski and Mrówczyńska 2018).

The cost of waste management activities is mainly associated with the cost of transport, facilities, operation (energy/fuel and labor), and real estate (Rao et al. 2017). For e-waste, specific factors have been claimed to influence the economic feasibility and environmental consequences of e-waste recycling (Hula et al. 2003; Nnorom and Osibanjo 2008): Product structure, materials, location of recycling facilities, applicable regulations, geography, and cultural context. All these factors combined will determine the feasibility of recycling certain products or goods. Another study uses four key aspects to evaluate the recycling potential and determine which element should be prioritized in the recycling of WEEE: the quantity of material in specific waste (e.g. gold in PCs), the toxicity of the given material, its market value and technology developed for recycling (Zeng et al. 2017). Thus, there is no single solution when deciding if and how to recycle WEEE because all these factors will vary on a case-to-case basis (Sinha-Khetriwal et al. 2005). While the waste management tends to be country-specific, there are general trends that outline developed countries to the detriment of developing countries.

As opposed to developing nations, developed nations usually have centralized waste treatment systems, which result in significant differences in relation to the former. The segregation of waste, for instance, is a voluntary exercise in most developed countries, but represents a source of income in developing nations, and allows the formation of a large informal network of people dedicated to waste collection (door-to-door) and meticulous waste segregation (Goel 2017; Hoornweg and Bhada-Tata 2012). Because of this network, countries of less-income have financial incentives and door-to-door collection, which create a convenient scenario toward material recycling. The waste sorting may occur prior to disposal (case of developed nations), prior to collection, during the collection or at the disposal site (Hoornweg and Bhada-Tata 2012). Schluep (2014) cites that "collection, manual dismantling, open burning to recover metals, and open dumping of residual fractions are normal practice in most developing and in transition countries." This also creates two different realities, a contrasting example is that of Switzerland and India: in the former, the consumers pay a recycling fee (for collection, treatment, etc.), whereas in the latter, the collectors in many cases pay the consumers for their obsolete appliances (Sinha-Khetriwal et al. 2005). The line that splits developed and developing economies, however, is not well defined when concerning e-waste. As shown in Sections 2.5 and 2.6, there are places like Taiwan, which are considered a benchmark in e-waste management while being dubbed a developing economy by the United Nations (UN/DESA et al. 2019). Furthermore, within a single country, there may be disparities in the way the WEEE management is administered and in its efficiency. As is the case for China (whose mainland management system varies greatly from that of Taiwan and Hong Kong, for instance), it is impossible to state a single WEEE management in the whole United States, since each state has its own policies and systems in place (Li et al. 2013; Ongondo et al. 2011; Schumacher and Agbemabiese 2019). It is noteworthy that these inter-region (or inter-country) differences can add challenges in the waste management industry due to different requirements and expectations placed upon stakeholders that practice across state/region boarders (Hickle 2014; Schumacher and Agbemabiese 2019).

Currently, developed economies have a reasonably sustainable infrastructure to deal with waste, while the rest of the world still have a large amount of organizing and implementing ahead of them (Goel 2017). The disposal of waste on land is still the most common practice worldwide due to its low cost. The main difference between developed and developing economies is that in the former this generally happens in engineered landfills, while in the latter in the form of open dumping (Goel 2017; Hoornweg and Bhada-Tata 2012). The choice between landfilling and incineration, however, is not a matter of economic development, but rather a matter of land: in countries where land is plenty, landfilling has been the matter of choice. This is mainly due to the cost–incineration is at least three-fold higher than landfilling (Goel 2017). The section hereafter will briefly describe the management system for electronic waste management in a few selected developed and developing nations to illustrate the discrepancies among them and allow further discussion about the current practices.

2.5 Developed Countries

2.5.1 Switzerland

Switzerland is considered a WEEE management benchmark as it currently has one of the world's highest collection rates: more than 79% of the average of all EEE placed onto the market in the country during the three preceding years (Toffolet 2016). The system is currently managed by the producers (manufacturers and importers), which are organized in four producer responsible organizations (PROs). The PROs manage the daily operations, set the recycling fee, and license and audit recyclers. The consumers pay a fee when they purchase an EEE (advanced recycling fee), which exempts them from any fees at the time of disposal. Consumers can drop off their WEEE at specific collection points or retailers – the latter being the main collection network. The role of the authorities is controlling and monitoring the outcomes of the different stakeholders in the WEEE management system. The government oversees the process, frames the basic regulations, and licenses the recyclers (Román 2012). There is about one service provider (retailer or designated take-back site) per 444 inhabitants in the country; recyclers have to operate according to the recycling criteria of ISO 14 000, and must have the PRO license and the government authorization (Morris and Metternicht 2016).

2.5.2 Japan

While the EU countries are mechanizing their operations, Japan still relies heavily on manual disassembly, which is one of the reasons the Japanese overall recycling cost is superior to that of the EU. It is believed that Japan's high recycling cost (and high recycling fee) leads to recycling outside the formal scheme and encourages WEEE export. That's why, in general terms, items taken back free of charge go abroad, whereas items that are paid for enter the domestic used market (Yoshida and Yoshida 2012).

2.5.3 Australia

Research suggests that e-waste collection systems in Australia does not allow a feasible domestic material recovery, which leads to significant material export for processes to be undertaken abroad (Golev et al. 2016; Kumar et al. 2017). This has been further explained by Sahajwalla et al. (2016), who claimed that "while safe resource recovery from e-waste is technically possible, it is expensive and currently relies largely on access to large scale, high tech furnaces, mostly located in Europe. Many nations, including Australia, have few or no viable resource recovery processes for e-waste." Little WEEE volume is also given as a reason for not having recovery operations in the country in another study (Dias et al. 2018a; Golev and Corder 2017). Moreover, the metal downstream recycling industry (end-processing) of the country has been reported to be well established only for iron (steel scrap). Concerning the nonferrous metals, there are only separation and smelting facilities for aluminum (Corder et al. 2015). Yet another report, this time from an industry player, agrees with the difficulties of responsibly recycling e-waste in Australia because of the high cost of labor, the low volume of e-waste recycling undertaken, the maturity of the market (low investment in infrastructure), and the availability of appropriate downstream processing (ANZRP 2015).

A study from 2015 reports that most material recovery (for PCs and TVs) occurs overseas and the role of the domestic e-waste recyclers is restricted to the collection and basic separation (Lane et al. 2015). Corder et al. (2015) showed that approximately half the scrap metal collected in Australia is currently being transported overseas. Research shows that the current economic model used in the country contributes to significant illegal exports of e-waste, given the struggle to maintain a positive economic balance and the constant approach of agents wanting to buy e-waste for unauthorized export (Lane et al. 2015). A recent study showed that these current conditions should lead to even greater WEEE material export unless incentives for domestic downstream processing are implemented (Dias et al. 2018a). The main countries to officially receive the WEEE components were China, Indonesia, and Japan (Dias et al. 2018a), but a recent report from BAN (2018) tracked down nondismantled WEEE in primitive acid-stripping operation in Thailand.

2.6 Developing Countries

2.6.1 Brazil

The WEEE generation in Brazil is estimated to be about 1.5 million tons yearly (Balde et al. 2017), and the expected increase between 2013 and 2020 is of 20% (Isıldar et al. 2018). Research suggests Brazil is still learning how to put reverse logistics of WEEE into practice, particularly with the environmental licensing of generators, transporters, and end-of-life EEE receivers (de Araujo et al. 2015). The attitudes of Brazilians toward e-waste also seem to be still behind in comparison to developed countries, given 18% of the population disposes of their end-of-life cellphone along with general waste (Moura et al. 2017). Furthermore, while the majority of Brazilians seem to hold a positive intention toward recycling e-waste, only a minority carries forward the attitude and adopt adequate recycling practices (Echegaray and Hansstein 2017). Recent studies also indicate that the lifespan of EEE used in Brazil is decreasing, in particular for cellular phones. (Moura et al. 2017).

Brazil lacks an industry capable of undertaking downstream recycling processes (end-processing) (Dias et al. 2018b; de Oliveira et al. 2012). This was reinforced in a research published in 2017, in which the authors also claim installing such industry requires high investments (de Oliveira Neto et al. 2017). This can be observed in the private sector of recycling companies in Latin America reported to mainly disassemble computers and cellular phones with the aim of recovering the valuable materials contained therein (Balde et al. 2017). Furthermore, formal reverse logistics seems to be unfeasible in the country due to operational costs and logistical constraints (Caiado et al. 2017). On the other hand, the country has built an industry around the export of valuable WEEE components, with companies specializing in stockpiling specific WEEE components and selling it overseas to developed nations, which requires a well-structured collection system and sufficient profit for at least three agents within the country (Dias et al. 2018b).

Informal end-of-life practices toward e-waste are popular in Brazil. In the collection phase, for instance, the country has a significant number of waste pickers who scavenge waste to selectively sort the e-waste materials that can be later sold (Ghisolfi et al. 2017; Guarnieri and Streit 2015). This results in a recycling system based on handpicked collection (cherry-picking) of exclusively high-value components, as opposed to a sustainable collection system (Caiado et al. 2017). Recent research suggest that the informal e-waste recycling market has increased in Brazil (Moura et al. 2017).

2.6.2 India

In India, e-waste is often viewed as a commodity of value, which causes reluctance when deciding whether to dispose of it or not, and the considerable price difference between the new and used EEE in India and other developing countries renders in equipment being forwarded for second-hand use multiple times (Borthakur and Govind 2018; Wath et al. 2010). Moreover, the main collection channel is (informal) door-to-door and involves the purchase of e-waste by the so-called *kawariwalas*. This system results in a well-established informal network driven by profit and capable of undertaking 90% of the country's WEEE. The recycling processes undertaken in the informal network, however, use rudimentary techniques and, most often, e-waste ends up in landfills mixed with municipal waste (Dwivedy et al. 2015). The difference among countries like India vs. developed countries is remarkable in the sense that population from low-income countries prefer to sell their obsolete equipment in the informal sector because they expect to obtain a profit when giving away a WEEE, whereas in countries like Japan consumers are required to pay to dispose of their WEEE – as mentioned in the Section 2.5.2 (Borthakur and Govind 2018; Dwivedy et al. 2015).

2.6.3 South Africa

Measuring and characterizing the e-waste management scenario in South Africa is important because the country is viewed by other African countries as a continent leader for developing sustainable waste management practices (Snyman et al. 2017). South Africa's e-waste management industry, however, is still at its infancy, as is the case for most developing countries (Ledwaba and Sosibo 2016).

The WEEE management is virtually voluntary and dependent on individuals, organizations, and small companies. Despite this setup, there are a handful of well-established companies capable of running their business by promoting the value chain of collecting and sorting e-waste to later sell it as a commodity or as concentrated waste components. These businesses collect WEEE by making use of advertisements and word of mouth, but also benefit from the informal sector to obtain waste (Snyman et al. 2017). Informal collection is a common activity in the country as "waste pickers" collect e-waste in addition to other waste streams – it was estimated that informal collection accounts for one-fourth of the total collection volume in the country (Salhofer et al. 2017). It is also estimated that only 20% of e-waste finds its way to recyclers because of the absence of adequate take-back centers and financing mechanisms for recyclers (Snyman et al. 2017). The recycling processes in South Africa mainly employ dismantling and sorting of simple components, while complex components are shredded and sent overseas (mainly to Asia and Europe) for downstream processing (Lydall et al. 2017; Snyman et al. 2017). A different study, however, claimed there were two companies in South Africa that undertook the extraction of precious metals from complex components such as PCBs (Salhofer et al. 2017). The dismantling of WEEE is not profitable for small businesses, which end up only undertaking it as a secondary activity (Lydall et al. 2017). This information is in contrast to the significant increase in the number of companies the country has seen in recent years (Salhofer et al. 2017). While South Africa may have developed better recycling facilities with respect to the rest of the African continent, it is still lagging behind developed nations in terms of legislation, enforcement, and characterization (Ledwaba and Sosibo 2016).

2.6.4 Nigeria

Nigeria has been a major importer of WEEE/EEE; however, it still severely lacks appropriate infrastructure and facilities to sustainably recycle the vast amount of imported e-waste (Babayemi et al. 2015). As a result, most of the imported e-waste from the developed countries end up in open dumpsites in a number of African countries, especially in Nigeria and Ghana (Asante et al. 2019). The importation of WEEE increased greatly from 2003, and especially after China reduced its importations in 2005/2006. Large volumes of waste going into Asia were then diverted to countries like Nigeria and Ghana. This has reached its peak in 2009 and was later counter through policies that regulated import and stopped shipments from entering the country (Babayemi et al. 2015). This has warranted an urgent need to formulate appropriate strategies and policies that should be in place to tackle the growing demand. The understanding of Nigeria's current setup also needs to take into account the (i) lack of technology (or industry) capable of undertaking advanced recycling processes (as defined by Dias et al. 2018b), which results in formal recyclers sending dismantled components overseas for downstream processing, and (ii) the behavior of the local population, who "are not ready to give out their e-waste for proper collection and recycling because they are expecting financial value for their waste" (Iwenwanne 2019). Assessment of quality and functionality of the e-waste to be recycled is also trivial. Assessment approach was based on the conditions stipulated in the importation guidelines developed by National Environmental Standards and Regulations Enforcement Agency (NESREA) in Nigeria (Odeyingbo et al. 2019); however, a review should be conducted on the current guidelines to improve the process and incorporate currently untapped e-waste. Furthermore, strategies should be developed for the country to ensure that responsibility falls to the (former) producers and importers of articles containing toxic/damaging materials (Sindiku et al. 2015).

2.6.5 Taiwan

Taiwan sets itself apart from its developing counterparts. Taiwan is one of the main references in e-waste management in the Asia Pacific region and has had a steady increase in e-waste recycling in recent years (Fan et al. 2018). The Taiwan Environmental Protection Administration (TEPA) is the governmental agency responsible for the main activities involved in the e-waste management. It is also responsible for monitoring the flow of waste materials (Shih 2017). These activities require a system with high levels of monitoring and, therefore, high costs associated with it, which has been described as one of the weaknesses of the setup.

The idea behind the 4-in-1 recycling program is that community residents (i), private recyclers/collectors (ii), local government (iii), and the recycling fund (iv) all play a role in the program. Residents must separate and deposit their e-waste in appropriate collection points, private sector operates recycling and collection, local governments organize and sell the appropriate waste to the private companies, and the recycling fund (managed by TEPA) subsidizes the operation of the whole system

(EPA 2012). In summary, Taiwan has a competitive e-waste take-back system that is state-operated (Shih 2017).

2.7 Conclusions

The global e-waste generation increased in the past decade, as was predicted in several scientific studies, and it should continue to increase in the following years. The overall transboundary movement of e-waste remained stagnant as unwanted items or components are still shipped in large volumes from developed to developing countries, either through legislation loopholes or illegally. A new pattern of e-waste international trade is currently observed: developing countries sending valuable e-waste components to developed countries. Examples of such movement were found in Brazil, Mexico, Indonesia, South Africa, and Nigeria. Additionally, in a developed country such as Australia, this valuable e-waste components movement was also detected. In Brazil and Australia particularly, this was accompanied by the establishment of number of private organizations whose main activity revolves around organizing and exporting valuable components. This pattern, however, requires further research to confirm whether it is a new global e-waste exchange trend or just isolated cases from a few countries. Other than the specific changes pointed out in this chapter, little has changed in the global macro WEEE *status quo* as recycling processes, management, and market remains unchanged.

References

ANZRP (2015). *Response to the National Television and Computer Scheme – Operational Review*. ANZRP – Australia and New Zealand Recycling Platform Limited www.anzrp.com.au/wp/wp-content/uploads/ANZRP-Response-to-Op-Review-Feb-15.pdf (retrieved 8 June 2017).

Asante, K.A., Amoyaw-Osei, Y., and Agusa, T. (2019). E-waste recycling in Africa: risks and opportunities. *Current Opinion in Green and Sustainable Chemistry*, Special Issue on Africa – Green solvents 18: 109–117. https://doi.org/10.1016/j.cogsc.2019.04.001.

Awasthi, A.K. and Li, J. (2017). Management of electrical and electronic waste: a comparative evaluation of China and India. *Renewable and Sustainable Energy Reviews* 76: 434–447. https://doi.org/10.1016/j.rser.2017.02.067.

Babayemi, J., Sindiku, O., Osibanjo, O., and Weber, R. (2015). Substance flow analysis of polybrominated diphenyl ethers in plastic from EEE/WEEE in Nigeria in the frame of Stockholm convention as a basis for policy advice. *Environmental Science and Pollution Research* 22: 14502–14514. https://doi.org/10.1007/s11356-014-3228-6.

Babu, B.R., Parande, A.K., and Basha, C.A. (2007). Electrical and electronic waste: a global environmental problem. *Waste Management & Research* 25: 307–318. https://doi.org/10.1177/0734242X07076941.

Balde, C., Forti, V., Kuehr, R., and Stegmann, P. (2017). *The Global E-waste Monitor*. Bonn/Geneva/Vienna: United Nations University (UNU), International Telecommunication Union (ITU) & International Solid Waste Association (ISWA).

BAN (2018). *Illegal Export of e-Waste from Australia*. The Basel Action Network (BAN) http://www.ban.org/trash-transparency (retrieved 17 October 2018).

Borthakur, A. and Govind, M. (2018). Public understandings of E-waste and its disposal in urban India: from a review towards a conceptual framework. *Journal of Cleaner Production* 172: 1053–1066. https://doi.org/10.1016/j.jclepro.2017.10.218.

Caiado, N., Guarnieri, P., Xavier, L.H., and de Lorena Diniz Chaves, G. (2017). A characterization of the Brazilian market of reverse logistic credits (RLC) and an analogy with the existing carbon credit market. *Resources, Conservation and Recycling* 118: 47–59. https://doi.org/10.1016/j.resconrec.2016.11.021.

Corder, G., Golev, A., and Giurco, D. (2015). "Wealth from metal waste": translating global knowledge on industrial ecology to metals recycling in Australia. *Minerals Engineering* 76: 2–9.

D'Adamo, I., Ferella, F., Gastaldi, M. et al. (2019). Towards sustainable recycling processes: wasted printed circuit boards as a source of economic opportunities. *Resources, Conservation and Recycling* 149: 455–467. https://doi.org/10.1016/j.resconrec.2019.06.012.

de Araujo, M.V.F., de Oliveira, U.R., Marins, A.S.F., and Jorge, M. (2015). Cost assessment and benefits of using RFID in reverse logistics of waste electrical & electronic equipment (WEEE). *Procedia Computer Science*, 3rd International Conference on Information Technology and Quantitative Management, ITQM 2015 (55): 688–697. https://doi.org/10.1016/j.procs.2015.07.075.

de Oliveira, C.R., Bernardes, A.M., and Gerbase, A.E. (2012). Collection and recycling of electronic scrap: a worldwide overview and comparison with the Brazilian situation. *Waste Management* 32: 1592–1610. https://doi.org/10.1016/j.wasman.2012.04.003.

de Oliveira Neto, G.C., de Jesus Cardoso Correia, A., and Schroeder, A.M. (2017). Economic and environmental assessment of recycling and reuse of electronic waste: multiple case studies in Brazil and Switzerland. *Resources, Conservation and Recycling* 127: 42–55. https://doi.org/10.1016/j.resconrec.2017.08.011.

Dias, P., Bernardes, A.M., and Huda, N. (2018a). Waste electrical and electronic equipment (WEEE) management: an analysis on the australian e-waste recycling scheme. *Journal of Cleaner Production* 197: 750–764. https://doi.org/10.1016/j.jclepro.2018.06.161.

Dias, P., Bernardes, A. M., Huda, N. (2018c). Defining valuable waste: the problem of defining mass as a measurement unit for waste electrical and electronic equipment [Definindo Resíduo Valioso: O Problema De Definir Massa Como Unidade De Medida De Resíduos De Equipamentos Eletroeletrônicos]. 29° Encontro Técnico AESABESP, FENASAN 2018. www.fenasan.com.br/fenasan2018 (accessed 16 March 2021).

Dias, P., Bernardes, A.M., and Huda, N. (2019). Ensuring best E-waste recycling practices in developed countries: an Australian example. *Journal of Cleaner Production* 209: 846–854. https://doi.org/10.1016/j.jclepro.2018.10.306.

Dias, P., Machado, A., Huda, N., and Bernardes, A.M. (2018b). Waste electric and electronic equipment (WEEE) management: a study on the Brazilian recycling routes. *Journal of Cleaner Production* 174: 7–16. https://doi.org/10.1016/j.jclepro.2017.10.219.

Dwivedy, M., Suchde, P., and Mittal, R.K. (2015). Modeling and assessment of e-waste take-back strategies in India. *Resources, Conservation and Recycling* 96: 11–18. https://doi.org/10.1016/j.resconrec.2015.01.003.

Echegaray, F. and Hansstein, F.V. (2017). Assessing the intention-behavior gap in electronic waste recycling: the case of Brazil. *Journal of Cleaner Production* 142: 180–190. https://doi.org/10.1016/j.jclepro.2016.05.064.

Egeonu, N. and Herat, S. (2016). E-waste: a problem or an opportunity? Review of issues, challenges and solutions in African countries. *International Journal of Environment and Waste Management* 17: 318–339. https://doi.org/10.1504/IJEWM.2016.078602.

EPA (2012). U.E.P.A. Case study on Taiwan's E-waste management system. Environmental Protection Administration (EPA). https://www.epa.gov/sites/production/files/2014-08/documents/taiwan_iemn_case_study_12.7_final.pdf (retrieved 26 March 2019).

EU Directive (2012). Directive 2012/19/EU of the European Parliament and of the Council of 4 July 2012 on waste electrical and electronic equipment, WEEE. *Official Journal of the European Union* L197: 38–71.

Fan, C., Fan, S.-K.S., Wang, C.-S., and Tsai, W.-P. (2018). Modeling computer recycling in Taiwan using system dynamics. *Resources, Conservation and Recycling* 128: 167–175. https://doi.org/10.1016/j.resconrec.2016.09.006.

Ghisolfi, V., de Diniz Chaves, G.L., Ribeiro Siman, R., and Xavier, L.H. (2017). System dynamics applied to closed loop supply chains of desktops and laptops in Brazil: a perspective for social inclusion of waste pickers. *Waste Management* 60: 14–31. https://doi.org/10.1016/j.wasman.2016.12.018.

Goel, S. (2017). *Advances in Solid and Hazardous Waste Management*. New York, NY: Springer Berlin Heidelberg.

Golev, A. and Corder, G.D. (2017). Quantifying metal values in e-waste in Australia: the value chain perspective. *Minerals Engineering* 107: 81–87. https://doi.org/10.1016/j.mineng.2016.10.021.

Golev, A., Schmeda-Lopez, D.R., Smart, S.K. et al. (2016). Where next on e-waste in Australia? *Waste Management* 58: 348–358. https://doi.org/10.1016/j.wasman.2016.09.025.

Goosey, M. (2012). The materials of WEEE. In: *Waste Electrical and Electronic Equipment (WEEE) Handbook* (eds. V. Goodship, A. Stevels and J. Huisman), 123–144. Elsevier.

Guarnieri, P. and Streit, J.A.C. (2015). Implications for waste pickers of Distrito Federal, Brazil arising from the obligation of reverse logistics by the National Policy of Solid Waste. *Latin American Journal of Management for Sustainable Development* 2: 19–35. https://doi.org/10.1504/LAJMSD.2015.067468.

Herat, S. and Agamuthu, P. (2012). E-waste: a problem or an opportunity? Review of issues, challenges and solutions in Asian countries. *Waste Management & Research* 30: 1113–1129. https://doi.org/10.1177/0734242X12453378.

Hickle, G.T. (2014). Moving beyond the "patchwork:" a review of strategies to promote consistency for extended producer responsibility policy in the U.S. *Journal of Cleaner Production* 64: 266–276. https://doi.org/10.1016/j.jclepro.2013.08.013.

Hoornweg, D. and Bhada-Tata, P. (2012). *What a Waste: a Global Review of Solid Waste Management*, Urban development series; knowledge papers no. 15. Washington, DC.: World Bank.

Huisman, J. (2012). 6 – Eco-efficiency evaluation of WEEE take-back systems. In: *Waste Electrical and Electronic Equipment (WEEE) Handbook*, Woodhead Publishing Series in Electronic and Optical Materials (eds. V. Goodship and A. Stevels), 93–119. Woodhead Publishing https://doi.org/10.1533/9780857096333.1.93.

Hula, A., Jalali, K., Hamza, K. et al. (2003). Multi-criteria decision-making for optimization of product disassembly under multiple situations. *Environmental Science & Technology* 37: 5303–5313. https://doi.org/10.1021/es0345423.

Isıldar, A., Rene, E.R., van Hullebusch, E.D., and Lens, P.N.L. (2018). Electronic waste as a secondary source of critical metals: management and recovery technologies. *Resources, Conservation and Recycling*, Sustainable Resource Management and the Circular Economy 135: 296–312. https://doi.org/10.1016/j.resconrec.2017.07.031.

Iwenwanne, V. (2019). Nigeria's e-waste mountain. Resource 7.

Kiddee, P., Naidu, R., and Wong, M.H. (2013). Electronic waste management approaches: an overview. *Waste Management* 33: 1237–1250. https://doi.org/10.1016/j.wasman.2013.01.006.

Kumar, A., Holuszko, M., and Espinosa, D.C.R. (2017). E-waste: an overview on generation, collection, legislation and recycling practices. *Resources, Conservation and Recycling* 122: 32–42. https://doi.org/10.1016/j.resconrec.2017.01.018.

Lane, R., Gumley, W., and Santos, D. (2015). *Mapping, Characterising and Evaluating Collection Systems and Organisations*. Australia: Monash University.

Larmer, B. (2018). E-waste offers an economic opportunity as well as toxicity. *The New York Times*. p. 12.

Ledwaba, P. and Sosibo, N. (2016). E-waste management in South Africa: case study: cathode ray tubes recycling opportunities. https://doi.org/10.20944/preprints201608.0021.v1.

Lepawsky, J. (2015). The changing geography of global trade in electronic discards: time to rethink the e-waste problem. *The Geographical Journal* 181: 147–159. https://doi.org/10.1111/geoj.12077.

Lepawsky, J. and Mcnabb, C. (2010). Mapping international flows of electronic waste. *The Canadian Geographer/Le Géographe canadien* 54: 177–195. https://doi.org/10.1111/j.1541-0064.2009.00279.x.

Li, H., La Guardia, M.J., Liu, H. et al. (2019). Brominated and organophosphate flame retardants along a sediment transect encompassing the Guiyu, China e-waste recycling zone. *Science of The Total Environment* 646: 58–67. https://doi.org/10.1016/j.scitotenv.2018.07.276.

Li, J., Lopez, B.N., Liu, L. et al. (2013). Regional or global WEEE recycling. Where to go? *Waste Management* 33: 923–934. https://doi.org/10.1016/j.wasman.2012.11.011.

Lydall, M., Nyanjowa, W., and James, Y. (2017). Mapping South Africa's waste electrical and electronic equipment (WEEE) dismantling, pre-processing and processing technology landscape – executive summary. Waste Research Development and Innovation Roadmap Research Report 9.

Morris, A. and Metternicht, G. (2016). Assessing effectiveness of WEEE management policy in Australia. *Journal of Environmental Management* 181: 218–230. https://doi.org/10.1016/j.jenvman.2016.06.013.

Moura, J.M.B.M., Gohr Pinheiro, I., Lischeski, D., and Valle, J.A.B. (2017). Relation of Brazilian institutional users and technical assistances with electronics and their waste: what has changed? *Resources, Conservation and Recycling* 127: 68–75. https://doi.org/10.1016/j.resconrec.2017.08.022.

Nnorom, I.C. and Osibanjo, O. (2008). Electronic waste (e-waste): material flows and management practices in Nigeria. *Waste Management* 28: 1472–1479. https://doi.org/10.1016/j.wasman.2007.06.012.

Nowakowski, P. and Mrówczyńska, B. (2018). Towards sustainable WEEE collection and transportation methods in circular economy – comparative study for rural and urban settlements. *Resources, Conservation and Recycling* 135: 93–107. https://doi.org/10.1016/j.resconrec.2017.12.016.

Odeyingbo, A.O., Nnorom, I.C., and Deubzer, O.K. (2019). Used and waste electronics flows into Nigeria: assessment of the quantities, types, sources, and functionality status. *Science of The Total Environment* 666: 103–113. https://doi.org/10.1016/j.scitotenv.2019.02.102.

Ongondo, F.O., Williams, I.D., and Cherrett, T.J. (2011). How are WEEE doing? A global review of the management of electrical and electronic wastes. *Waste Management* 31: 714–730. https://doi.org/10.1016/j.wasman.2010.10.023.

Rao, M.N., Sultana, R., and Kota, S.H. (2017). *Solid and Hazardous Waste Management: Science and Engineering*. Oxford, England: India: Butterworth-Heinemann; distributed by BS Publications.

Robinson, B.H. (2009). E-waste: an assessment of global production and environmental impacts. *Science of The Total Environment* 408: 183–191. https://doi.org/10.1016/j.scitotenv.2009.09.044.

Román, E. (2012). 22 – WEEE management in Europe: learning from best practice. In: *Waste Electrical and Electronic Equipment (WEEE) Handbook*, Woodhead Publishing Series in Electronic and Optical Materials (eds. V. Goodship and A. Stevels), 493–525. Woodhead Publishing https://doi.org/10.1533/9780857096333.5.493.

Sahajwalla, V., Pahlevani, F., Maroufi, S., and Rajarao, R. (2016). Green manufacturing: a key to innovation economy. *Journal of Sustainable Metallurgy* 2: 273–275. https://doi.org/10.1007/s40831-016-0087-z.

Salhofer, S., Kopacek, B., Gericke, M., 2017. Developing a roadmap for e-waste management in South Africa. Sardinia 2017 / Sixteenth International Waste Management and Landfill Symposium / 2 – 6 October 2017.

Schluep, M. (2014). Chapter 29 – Informal waste recycling in developing countries. In: *Handbook of Recycling* (eds. E. Worrell and M.A. Reuter), 439–444. Boston: Elsevier https://doi.org/10.1016/B978-0-12-396459-5.00029-5.

Schluep, M., Hagelüken, C., Kuehr, R., Magalini, F., Maurer, C., Meskers, C., Thiébaud-Müller, E., Wang, F. (2009), United Nations Environment Programme. Recycling – from e-waste to resources.

Schumacher, K.A. and Agbemabiese, L. (2019). Towards comprehensive e-waste legislation in the United States: design considerations based on quantitative and qualitative assessments. *Resources, Conservation and Recycling* 149: 605–621. https://doi.org/10.1016/j.resconrec.2019.06.033.

Shih, H.-S. (2017). Policy analysis on recycling fund management for E-waste in Taiwan under uncertainty. *Journal of Cleaner Production* 143: 345–355. https://doi.org/10.1016/j.jclepro.2016.12.103.

Sindiku, O., Babayemi, J.O., Tysklind, M. et al. (2015). Polybrominated dibenzo-*p*-dioxins and dibenzofurans (PBDD/Fs) in e-waste plastic in Nigeria. *Environmental Science and Pollution Research* 22: 14515–14529. https://doi.org/10.1007/s11356-015-5260-6.

Sinha-Khetriwal, D., Kraeuchi, P., and Schwaninger, M. (2005). A comparison of electronic waste recycling in Switzerland and in India. *Environmental Impact Assessment Review* 25: 492–504. https://doi.org/10.1016/j.eiar.2005.04.006.

Snyman, J., Vorster, K., Jacobs, S.J. (2017). Towards sustanable e-waste management in South Africa, *5th International Conference on Sustainable Solid Waste Management*. Presented at the ATHENS 2017.

Stevels, A.L.N. (2007). Adventures in ecodesign of electronic products 1993–2007. Design for sustainability program publication, 17.

Sthiannopkao, S. and Wong, M.H. (2013). Handling e-waste in developed and developing countries: initiatives, practices, and consequences. *Science of the Total Environment* 463–464: 1147–1153. https://doi.org/10.1016/j.scitotenv.2012.06.088.

Sugimura, Y. and Murakami, S. (2016). Problems in Japan's governance system related to end-of-life electrical and electronic equipment trade. *Resources, Conservation and Recycling* 112: 93–106. https://doi.org/10.1016/j.resconrec.2016.04.009.

Toffolet, R. (2016). Chapter 1 – WEEE management. In: *WEEE Recycling* (eds. A. Chagnes, G. Cote, C. Ekberg, et al.), 1–30. Elsevier https://doi.org/10.1016/B978-0-12-803363-0.00001-8.

UN/DESA, UNCTAD, ECA, ECE, ECLAC, ESCAP, ESCWA, UNWTO (2019). *World Economic Situation and Prospects 2019*. New York: United Nations Department of Economic and Social Affairs (UN/DESA), the United Nations Conference on Trade and Development (UNCTAD) and the five United Nations regional commissions (Economic Commission for Africa (ECA), Economic Commission for Europe (ECE), Economic Commission for Latin America and the Caribbean (ECLAC), Economic and Social Commission for Asia and the Pacific (ESCAP) and Economic and Social Commission for Western Asia (ESCWA). United Nations Publication.

Veit, H.M. and Bernardes, A.M. (2015). Electronic waste: generation and management. In: *Electronic Waste* (eds. H.M. Veit and A. Moura Bernardes), 3–12. Cham: Springer International Publishing https://doi.org/10.1007/978-3-319-15714-6_2.

Wang, F., Huisman, J., Meskers, C.E.M. et al. (2012). The best-of-2-worlds philosophy: developing local dismantling and global infrastructure network for sustainable e-waste treatment in emerging economies. *Waste Management*, Special Thematic Issue: Waste Management in Developing Countries 32: 2134–2146. https://doi.org/10.1016/j.wasman.2012.03.029.

Wath, S.B., Vaidya, A.N., Dutt, P.S., and Chakrabarti, T. (2010). A roadmap for development of sustainable E-waste management system in India. *Science of The Total Environment* 409: 19–32. https://doi.org/10.1016/j.scitotenv.2010.09.030.

Westcott, M. (2012). e-Waste. Queensland Parliamentary Library and Research Service. http://www.parliament.qld.gov.au/documents/explore/researchpublications/researchbriefs/2012/rbr201206.pdf (retrieved 06 April 2017).

Worrell, E. and Reuter, M.A. (2014). Recycling. In: *Handbook of Recycling* (eds. E. Worrell and M.A. Reuter), 3–8. Elsevier https://doi.org/10.1016/B978-0-12-396459-5.00001-5.

Yang, S. (2019). Trade for the environment: transboundary hazardous waste movements after the Basel convention. *SSRN Electronic Journal* https://doi.org/10.2139/ssrn.3386820.

Yoshida, F. and Yoshida, H. (2012). 25 – WEEE management in Japan. In: *Waste Electrical and Electronic Equipment (WEEE) Handbook*, Woodhead Publishing Series in Electronic and Optical Materials (eds. V. Goodship and A. Stevels), 576–590. Woodhead Publishing https://doi.org/10.1533/9780857096333.5.576.

Zeng, X., Wang, F., Li, J., and Gong, R. (2017). A simplified method to evaluate the recycling potential of e-waste. *Journal of Cleaner Production* 168: 1518–1524. https://doi.org/10.1016/j.jclepro.2017.06.232.

Zhang, Y., Xu, X., Chen, A. et al. (2018). Maternal urinary cadmium levels during pregnancy associated with risk of sex-dependent birth outcomes from an e-waste pollution site in China. *Reproductive Toxicology* 75: 49–55. https://doi.org/10.1016/j.reprotox.2017.11.003.

3

e-Waste Transboundary Movement Regulations in Various Jurisdictions*

Pablo Dias[1,2], *Md Tasbirul Islam*[3], *Bin Lu*[4], *Nazmul Huda*[3], *and Andréa M. Bernarde*[2]

[1] *University of New South Wales, School of Photovoltaic and Renewable Energy Engineering, Faculty of Science and Engineering, 229 Anzac Parade, Kensington, Sydney, NSW 2052, Australia*
[2] *Federal University of Rio Grande do Sul, Engineering School, Department of Materials, 9500, Av. Bento Gonçalves, Porto Alegre, RS 91509-900, Brazil*
[3] *Macquarie University, School of Engineering, Faculty of Science and Engineering, Balaclava Rd, North Ryde, NSW 2109, Australia*
[4] *Chinese Academy of Sciences, Research Center for Eco-Environmental Sciences, 18 Shuangqing Road, Haidian District, Beijing 100085, People's Republic of China*

3.1 Background

Waste electrical and electronic equipment (abbreviated as waste electrical and electronic equipment (WEEE) or e-waste) is generally regarded as all end-of-life devices that use electricity (battery or cord/circuity). It thus includes televisions, computers, phones, refrigerators, washing machines, toys, kettles, etc. It encompasses almost any household or business device in addition to medical equipment (e.g. resonance tomography scanners) (Kuehr 2019). There is no precise figure determining the generation of e-waste worldwide, but it is agreed that it has been rising consistently and should continue to increase as new technologies are released and the lifespan of equipment decreases (Balde et al. 2015; Ongondo et al. 2011). Other reasons for the continuous e-waste increase are the affordable prices of electrical and electronic equipment (EEE), the rapid economic growth, urbanization and growing demand for consumer goods, which increased both the consumption of EEE and the generation of WEEE (Babu et al. 2007; Kumar et al. 2017). The generation of e-waste appears to be higher in developed countries than in developing economies (Goel 2017), but the WEEE generation has been increasing in both realities (Schluep et al. 2009). Furthermore, a positive correlation between gross domestic product (GDP) and e-waste generated in a given country was confirmed in a recent research. Interestingly, no correlation was found between e-waste generation and population (Kumar et al. 2017).

In the search for determining the scale of the e-waste problem, several studies attempted to estimate the production of e-waste globally. In 2009, an estimation calculated that WEEE generation was between 20 and 25 million tons yearly,

* Received 20 December 2019. Revised 01 December 2020.

Electronic Waste: Recycling and Reprocessing for a Sustainable Future, First Edition.
Edited by Maria E. Holuszko, Amit Kumar, and Denise C. R. Espinosa.

originating mainly from the USA, Europe, and Australasia (Robinson 2009). In the same year, it was estimated that the yearly generation of e-waste was somewhere between 20 and 50 million tons (Schluep et al. 2009) – an upper limit almost twice as high. In 2012, the yearly generated figure totaled about 46 million tons (Perkins et al. 2014). Recent published studies indicate that the current e-waste generation in the world is between 20 and 50 million tons and should increase between 3% and 4% yearly (Abdelbasir, El-Sheltawy, and Abdo 2018; Abdelbasir, Hassan, Kamel, and El-Nasr 2018; Balde et al. 2017). Moreover, merely one-fifth of all e-waste generated is reported to be collected and recycled (Balde et al. 2017). Currently (2019), e-waste is the fastest growing waste stream in the world (Abdelbasir, Hassan, Kamel, and El-Nasr 2018; Cucchiella et al. 2021), and the estimates report it should continue to grow. The latest figures indicate that the e-waste generation in 2019 was 53.6 Mt. worldwide, and the projection is for this value to achieve 74.7 Mt. by 2030 (Forti et al. 2020). This growing waste generation underpins a pressing challenge because e-waste is associated with several potentially hazardous substances (e.g. mercury, cadmium, and halogens). Moreover, its mismanagement contributes to global warming and leads to loss of valuable and scarce materials. The latter is quite significant, as estimates indicate the total raw material value contained in global e-waste equals to roughly US$ 57 billion (Forti et al. 2020; Kuehr 2019).

3.2 International Legislation and Transboundary Movement

The excuse of the "digital divide" served for the purpose of breaching international treaties that would, otherwise, impede such movement of hazardous equipment. An example of such treaty, and arguably the most important international regulation around WEEE, is the "Basel Convention on the Control of Transboundary Movements of Hazardous Wastes and Their Disposal" (Basel Convention hereafter), which was established in 1989. The Basel Convention introduced important changes in the global waste panorama. Its goal was to restrain the illegal waste traffic and to enhance international cooperation toward hazardous waste management (UNEP 1989). It established that the consent of all the parties involved was necessary before the export, import, and transport of hazardous waste. Therefore, its introduction has made the transport of waste more difficult worldwide (Li et al. 2013). From 2002, the Basel Convention, following its sixth meeting, identified e-waste as a priority waste stream in the strategic plan of the years to come (Herat and Agamuthu 2012). Because of the Basel Convention, about one-third of the countries that imported electrical and electronic waste banned such practice (Li et al. 2013).

Nevertheless, the Basel Convention impacted neither the generation nor the flux of WEEE. This led to the creation of national directives aiming to tackle the e-waste challenge with direct regulations. Two of the most important were the European WEEE Directive (Directive on WEEE) and the restriction of hazardous substances (RoHS) Directive (Directive on the restriction of the use of certain hazardous substances), both implemented in 2003, and later updated in 2012 and 2011, respectively (European Commission 2018). The latter banned manufacture and sale of EEE containing lead, mercury, cadmium, hexavalent chromium, polybrominated biphenyls, and polybrominated diphenylethers. This, in turn, obliged other countries that wanted to trade with the European Union (EU) to stop using these substances in manufacturing their goods (Veit and Bernardes 2015). The current maximum levels by weight for the substances restricted by the European RoHS are as follows (European Parliament 2011): lead (0.1%), mercury (0.1%), cadmium (0.01%), hexavalent chromium (0.1%), polybrominated biphenyls (PBB) (0.1%), and polybrominated diphenyl ethers (PBDE) (0.1%). Exemptions may be requested and these are assessed by a certified competent body (Oeko-Institut 2019).

In 2015, the EU Directive 2015/863 was published (known as RoHS-Recast); it announced the addition of four substances (phthalates) to the RoHS list and lower levels of tolerance for the restricted substances – the latest modifications came into effect in July 2019. Regulations similar to the European RoHS were also applied in several other countries with the same concept and slight variations: these include Japan in 2000; China, California (USA), and South Korea in 2007; Singapore in 2017; and Turkey in 2018 (RoHS Guide 2020). Just like the recast of the EU, China implemented the RoHS 2, which came into effect in July 2016 and directly targets EEE by restricting the use of hazardous substances in its manufacturing (RoHS Guide 2019).

The formerly mentioned European WEEE Directive sets targets for the rate of collection, recovery, and recycling, which increase over time. Other requirements include, but are not limited to (EU Directive 2012): (i) All collected WEEE must undergo proper treatment, which means preparing for reuse, and recovery or recycling operations; (ii) The quality standards for the recycling of WEEE shall reflect state of the art; (iii) WEEE can only be exported for treatment outside of the EU if the country exporting can prove that the treatment took place in conditions that are equivalent to the requirements of the WEEE Directive; (iv) Member states should increase customer awareness by providing information such as the cost of end-of-life processing (at the time of sale of an EEE), not to dispose of WEEE as unsorted municipal waste, reverse logistics systems available to them, and the potential effects the hazardous substances of WEEE have on human health and the environment.

Moreover, the Directive introduces a greater control over WEEE and its components, requiring equipment to be weighed when entering and leaving the treatment facilities (or recycling facilities). It establishes minimum treatments required prior to exportation of any WEEE, requires that member states must distinguish EEE from WEEE, and also requires several pieces of evidences attesting the exported material is fully functional – to avoid deceptive export of WEEE as used EEE (EU Directive 2012).

However, these regulations, treaties, and conventions were not enough to stop the e-waste transboundary movement. Slowly, the countries individually began creating their own set of regulations to impede or regulate such movement (Tables 3.1 and 3.2), but in 2013 a similar import/export pattern was still reported (Figure 3.1). Recently, the Basel Action Network (a nonprofit organization) proposed alternative guidelines to the Basel Convention under the claim that there are loopholes in the current regulations, which allow any export to be considered as nonwaste under the

Table 3.1 Nonexhaustive list of legislations and regulations implemented to permit or ban import and export of WEEE and used EEE in common source countries.

Country	Year	Legislation and/or regulation	Description
		Source countries	
European Union	2002 (recast in 2012)	Directive 2002/96/EC on Waste Electrical and Electronic Equipment (Directive 2012/19/EU on Waste Electrical and Electronic Equipment)	WEEE exporter should prove that the recovery, reuse, and/or recycling operation takes place under conditions that are equivalent to the requirements of the Directive
Japan	2005	Law for the Control of Export, Import and others of specified Hazardous Wastes and other Wastes (Law No. 108, 1992; Latest revision: Law No. 33)	Stipulates export needs the consent from the country of import
Singapore	2008	Import and Export of e-wastes and used Electronic Equipment	The approval for to import and export of hazardous e-waste will only be granted on a case-by-case basis
South Korea	1994	Act on the Control of Transboundary Movement of Hazardous Wastes and their Disposal (1994)	No permission for export is issued without consent from the country of import
USA	2011	HR 2284: Responsible Electronics Recycling Act	Banned the export of certain WEEE: PCs, TVs, printers, copiers, videogame systems, telephones, and similar used electronic products that contain cathode-ray tubes, batteries, switches, and other parts containing lead, cadmium, mercury, organic solvents, hexavalent chromium, beryllium, or other toxic ingredients

Source: Based on Li et al. (2015) and EU Directive (2012).

claim that it is "e-waste for repair" (BAN 2019). Therefore, despite national and international effort, e-waste trafficking (illegal movement) from developed to developing countries continued (Premalatha et al. 2014).

Recent studies have also shown this global transboundary movement (Garlapati 2016; Kumar et al. 2017) and reported additional developing countries as

Table 3.2 Nonexhaustive list of legislations and regulations implemented to permit or ban import and export of WEEE and used EEE in common destination countries.

Country	Year	Legislation and/ or regulation	Description
		Destination countries	
China	2017	Implementation Plan for Prohibiting the Entry of Foreign Garbage and Advancing the Reform of the Solid Waste Import Administration System	Implemented strict regulations to ban most solid waste import, including WEEE from other countries
	2012	Measures for the Collection, Use, and Administration of Funds for the Disposal of Waste Electrical and Electronic Products	This regulation defines the scope of payers and users of recycling fund for Electrical and Electronic Products
	2013	Administrative Measures for the Circulation of the Used Electrical and Electronic Products	This regulation was issued by Ministry of Commerce to promote reuse of Electrical and Electronic Products. The reseller should set up information profile of used products and implement a reuse label on the products.
	2009	Regulation on the Administration of the Recovery and Disposal of Waste Electrical and Electronic Products	This regulation is the core of WEEE management in China. Based on the principles of extended producer responsibility (EPR), the recycling fund mechanism, centralized processing, and licensed recycling mode are fixed.
Ghana	2016	Hazardous and Electronic Waste Control and Management Act 2016 (Act 917) Status: Approved	Prohibition of import, export, and transportation of hazardous waste in and out of the country. Written authorization is required from the Minister for an exception. A Permit needs to be obtained from the Environmental Protection Agency (EPA) after paying an Advance Eco Levy (AEL) fees. The person must be registered to get the permit willing to involve in import/export of e-waste

(continued)

Table 3.2 (Continued)

Country	Year	Legislation and/ or regulation	Description
Hong Kong	2015	No specific law exists. Promotion of Recycling and Proper Disposal (Electrical Equipment and Electronic Equipment) (Amendment) Bill, 2015 (Producer responsibility scheme (PRS) Scheme) is under discussion. e-Waste is generally considered under The Waste Disposal Ordinance (WDO) (1980).	e-Waste is subjected to permit control under the WDO.
	2011	Advice on Import and Export of Used Electrical and Electronic Equipment Having Hazardous Components or Constituents. Environmental Protection Department (EPD), third ed.	Import/export compliance checking for used EEE
Malaysia	2010	Guidelines for the Classification of Used Electrical and Electronic Equipment in Malaysia	This Guideline assists all parties concerned in identifying and classifying used EEE as WEEE. Malaysia has ratified the Basel Convention, and as a Party must follow the procedures of the Convention for the import and export of e-waste
Nigeria	2011	Guide for Importers of used EEE into Nigeria (2011)	The government bans the importation of WEEE and near-end-of-life EEE. Every importer of used EEE should register with Nigeria's national agency
	2011	National Environmental (Electrical and Electronic Sector) Regulations, S.I. 23/2011	This regulation provides guidelines for importation of the e-waste items and any e-waste importer must subscribe to EPR program and partner with National Environmental Standards and Regulations Enforcement Agency (NESREA). Administrative cost needs to be paid by the importer to NESREA.

Table 3.2 (Continued)

Country	Year	Legislation and/ or regulation	Description
Thailand	2016	WEEE Strategic Plan Phase II (2012–2016) has been taken under WEEE Strategic Plan	This plan restricts import and export of low-quality electronic and electrical products
	2014	No specific law is available. Waste Electrical and Electronic Equipment, Draft Act, November 2014 is under draft.	
	2007	Criterion for Import of used EEE considered as Hazardous Substances	It presents a list of used EEE classified as hazardous substances that will be controlled, but in general allows its importation
Vietnam	2016	Decree No. 155/2016/ND-CP	With the enforcement of the degree, penalties are applied for administrative violations against regulations on environmental protection. However, this decree is not specific to e-waste, only.
	2015	Regulations on the Recovery, Processing and Disposal of Waste Products, Decision No. 16/2015/QD-TTg	EPR system applied on the discarded products in the year 2015
	2005	Law on Environmental Protection	It encourages waste reduction and recycling, stipulates the responsibilities for waste generators to minimize waste production. Includes articles for hazardous waste management and prohibits the import and transit of all kinds of wastes

Source: Based on Chinese Government Network (2017), Li et al. (2015), Owusu (2017), Peluola (2016), and Pharino (2017), Chinese Government Network (2017). Li et al. (2015), Owusu (2017), Peluola (2016), Pharino (2017), STEP (2019a), Thi Thu Nguyen et al. (2019), Wong (2018).

destination places (importers), which implies that this pattern is still in course. In the past three years, the Basel Action Network (BAN) has published several studies on "scam recycling," in which companies from developed nations (e.g. European countries, Canada, Australia, USA) supposedly recycle WEEE, but in reality export it abroad to junkyards where the waste is dismantled using dangerous and polluting methods (BAN 2016). Developed countries, large e-waste generators, in the name of international trade – using the "reuse" excuse to save their environment and money flow – dump their e-waste in developing countries like India and China (Goel 2017).

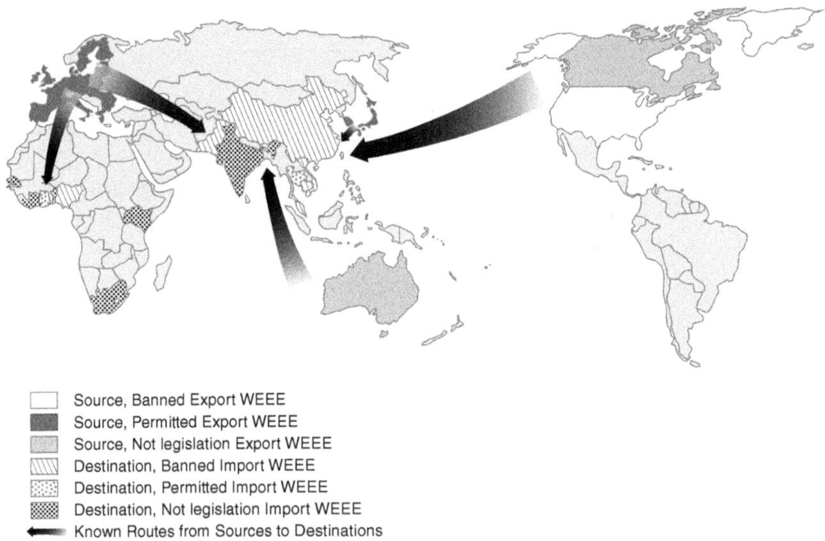

Figure 3.1 Known routes and permissions/bans for the WEEE imports/exports in 2013. Source: Li et al. (2015).

This scenario calls for a verification system capable of assessing the particularities WEEE management, recycling, and treatment. This is to ensure sound end-of-life practices, but also to track double counting of WEEE, which is suspected to happen in some countries as a means to boost results (Toffolet 2016). As can be anticipated, this need was filled by several different conformity verification systems, especially in Europe. As a result, trade between member states was made difficult because, for each country, a different set of rules applied. For this reason, the WEEELABEX (short for *WEEE Label of Excellence*) was created. Its objective is to design a set of European standards for collection, sorting, storage, transportation, preparation for reuse, treatment, and disposal of all categories of WEEE, while harmonizing the rules and procedures for conformity verification in Europe. This would facilitate trade among the member states and level the playing field for stakeholders (Leroy 2012).

In light of the transboundary movement of e-waste worldwide, important changes occurred in 2017 when China, a major waste-destination country, announced stricter monitoring regarding the quality and contamination of the scrap material received. In July 2017, the former Ministry of Environmental Protection of China announced at the World Trade Organization (WTO) that "by the end of 2017, China will forbid the import of four classes, 24 kinds of solid wastes, including plastics waste from living sources, vanadium slag, unsorted waste paper and waste textile materials" (WTO 2017). It later confirmed the announcement about the import ban on a range of polymers (highly present in e-waste), which would came into force by the end of 2018 (Resource Recycling 2018). Shipments of scrap polymers to China dropped by 99.1% from 2018 to early 2019 (Resource Recycling 2019). These changes highly impacted the international scrap trade and diverted the material flow to other

Asian countries, especially Malaysia, Vietnam, and Thailand (Staub 2018). The overload in these new destination countries was extreme to the point that, less than one year later, both the Vietnamese and the Thai governments announced they would stop receiving scrap plastics (i.e. also announced import bans) (Resource Recycling 2018). China has already proposed a total import ban, which would take place in 2020 and includes scrap fiber and every other form of solid waste. The change was implemented, however, from 01 January 2021. The country has completely banned the import of solid waste (Ministry of Ecology and Environment 2021). These measures were announced with the intent of protecting the environment, improving people's health, and favoring domestic material over international (Resource Recycling 2018; State Council 2017). The COVID pandemic has delivered yet another blow of uncertainty to this scenario, increasing the difficulty to move scrap materials internationally (Resource Recycling 2020).

3.3 Extended Producer Responsibility (EPR)

Extended producer responsibility (EPR) is a policy in which the manufacturer is held responsible for its products when they reach their end of life. It is based on the polluter-pay principle and has been widely adopted in e-waste management systems worldwide. Switzerland, Japan, European Union (e.g. UK, Holland, Germany), some regions of the USA and Canada (Table 3.3), and more recently China have adopted this policy (Kiddee et al. 2013; Wang et al. 2018). The introduction of EPR was important to achieve sustainable e-waste management, it allowed the financing of unprofitable (but necessary) processing steps, the evolution of the e-waste recycling industry, a synergy between policies and legislations, and a consumer demand for sound dispose option for e-waste (Schluep 2014). The EPR requires the producer to undertake the financial costs of the end-of-life management of WEEE (as is the case in the EU). In some cases, however, the financial burden is split with different stakeholders, including the consumers who are often required to pay a fee for the processing of its EEE once it reaches the end of its lifespan. These fees must be perceived by the payers as fair, reasonable, and based on actual costs of the end-of-life management. They should also be revised periodically as the schemes in place are better understood and audited (MS2 and Perchards 2009).

3.4 Regulations in Various Jurisdictions

As of 2019, 78 countries adopted a national e-waste policy, legislation, or regulation (Forti et al. 2020), but enforcement, investment, and other context-related factors need to be scrutinized to fully understand the regulations in place, and the extent to which they work. The particularities and characteristics of the WEEE regulation in specific countries are described in the section hereafter. Data availability, consolidated e-waste management system, and importance in the international scenario were the criteria used to choose the countries.

Table 3.3 Extended producer responsibility (EPR) in different countries.

Countries	EPR concept
European Union	The directive mainly focuses on reuse, recycle, and recovery of e-wastes and dismantling of electronic parts and recycling of materials, proper collection systems to reduce disposal and incorporate best management practices.
Switzerland	This country was the first to develop and implement methods for collection, transportation, recycling/treatment and disposal of e-waste. Three producer responsibility organizations (PROs): The Swiss Association for Information Communication and Organizational Technology (SWICO); the Stiftung Entsorgung Schweiz (SENS); and Swiss Lighting Recycling Foundation (SLRS) oversee these systems based on the concept of EPR.
United States of America	Established funding for the collection and recycling of e-waste. Consumers pay a fee called Advance Recycling Fee (ARF) at the time of purchase that goes to the state and is used to reimburse recyclers and collectors.
Japan	Manufacturers and importers are responsible for taking back end-of-life electronics for recycling and waste management. Consumers pay a fee that is directly used to meet the expenses of recycling and transportation.
South Korea	Local manufacturers, distributors, and importers of e-goods are required to achieve official recycling targets. Government keeps an account for depositing funds for recycling, which are refundable depending on the amount of waste recycled.
Australia	Importers, manufacturers, and distributors have to subscribe to mandatory, coregulatory, or voluntary schemes for managing the disposal of computers and televisions.
Singapore	Export, import, or transit waste requires a permit from the Pollution Control Department (PCD) of Singapore. If documents are available to support that the products are in good condition and can be reused, only then permission is granted.
China	(a) Pollution prevention and controls on the use, dismantling, and disposal of e-waste, under "Technical Policies for Controlling Pollution of WEEE, 2006," (b) Certificate is required for e-waste recycling systems, under "Administrative Measures for the Prevention and Control of Environmental Pollution by WEEE, 2008," and (c) All producers and importers responsible for their products, collection, and treatment funds, under "Regulation on the Administration of the Recovery and Disposal of Waste Electrical and Electronic Products, 2012."
India	According to its Ministry of Environment, Forest and Climate Change, 2011, producers are responsible for the collection of e-waste generated after the end of life of the e-products. The legislation was modified in 2015 with new responsibilities for the producers as well as consumers.
African countries	The use of EEE is very less in the African countries. No specific e-waste legislation has been implemented in those countries.

Source: Based on Goel (2017) and Premalatha et al. (2014).

3.4.1 Europe

3.4.1.1 France

In 2006, France set upon the producers, retailers (under their brand), and importers the obligation of supporting the end-of-life management of WEEE and achieving the recycling and recovery targets set by the Environmental Code. Producers could either take the responsibility individually or collectively through the so-called eco-organizations. The latter was preferred by the majority, which gave rise to the "Eco-systems," the "Eco-logic," and the "European Recycling Platform" (Vadoudi et al. 2015). To finance the operations, consumers pay an eco-fee when buying a new EEE, which is visible at the time of purchase. These vary from a few cents to a little more than tens of euros and were reported not to have changed the purchasing habits of the consumers (Toffolet 2016). However, revenue for the operation is also obtained by different taxes and by fees paid by the manufacturers, which vary according to the product (based on criteria of durability, toxicity, recycled content, and design for disassembly) (Vadoudi et al. 2015). The government participates actively in the whole system by overseeing the arrangements, charging fees, paying agents responsible for collection and recycling, approving these agents, enforcing end-of-life treatment standards, and applying penalties to uncompliant manufacturers (Vadoudi et al. 2015).

3.4.1.2 Germany

Germany uses a system dubbed "divided product responsibility." The regulations in the country require public sector recycling companies to establish WEEE recycling centers and to accept WEEE free of charge – retailers are also allowed to accept WEEE from consumers (UB 2016). The municipalities, after receiving WEEE from the public, sort the waste into groups prior to the manufacturer pickup (BMU 2015). The local municipal authorities may then undertake the recycling or forward it free of charge to the existing take-back system. This is quite different from other countries, where local authorities charge the take-back systems for such service (Toffolet 2016). All EEE manufacturers must adhere to the system, register, and assure its financing. They must also organize the treatment of WEEE and provide proof of such (BMU 2015). Moreover, they are free to provide their own recycling mechanisms (UB 2016). Unlike other countries that split "historic" WEEE (waste placed before the system was implemented) according to market share, Germany has a set of criteria to determine which producer should be responsible for it picking up (Ongondo et al. 2011). Under the system, the consumers are required by law to dispose of their WEEE in the take-back facilities (UB 2016).

The German system works on a competitive basis (BMU 2015), and there are regulations in place to limit the manufacturing clusters. Thus, each producer must collect its share of national obligations under a main take-back system that ensures the collection costs are shared equally. This main take-back system is also responsible for determining recovery and recycling targets (Ongondo et al. 2011). The targets rates are specific for the various device classes (UB 2016). The search for the lowest short-term costs sought by the system, however, resulted in scattered volumes and in the gradual desertion of the market by operators (Toffolet 2016).

3.4.1.3 Switzerland

In Switzerland, the current regulations in place oblige the EEE user to give back its end-of-life device and all retailers must take back any appliance at no cost. As opposed to many other countries, in Switzerland the legislation does not define how the industry should carry out their responsibility to manage and finance their WEEE recycling. In this regard, two producer responsibility organizations (PRO) are responsible for e-waste management under an EPR-based system, the "The Swiss Association for Information, Communication and Organizational Technology (SWICO)" and "Stiftung für Entsorgung Schweiz (SENS)" (Islam et al. 2018). Therefore, the actual industry decides how to establish and carry out the recycling, which renders a system that is voluntary and designed to be flexible and cost effective (Román 2012). Though the system operates under a voluntary condition, there are some key characteristics such as clear definition and provision of roles and responsibilities of all the actors; secure finance sourcing by implementing advanced recycling fees (ARFs); broader product coverage in the e-waste management regardless of the types of EEE, its date of purchase, and manufacturers. Other characteristics include stringent control and monitoring of the finance and material flow along with externalities such as emission and health hazards; guidelines to avoid free riders; and finally, prohibition of the illegal dumping of e-waste to non-OECD (Organization for Economic Co-operation and Development) countries (i.e. countries that belong to the Organization for Economic Co-operation and Development) (Chaudhary and Vrat 2018).

3.4.1.4 Norway

Norway is a non-EU member in spite of being tightly related to the Union politically and economically. Therefore, Norway has a WEEE management approach quite similar to other EU members, but also has its particularities and introduced the national legislation regarding e-waste in 1999, well before the EU WEEE Directive, which was officially enacted in the year 2002 (NEA 2018). The Norwegian producers and importers of WEEE are obliged to be a member of one of the authorized take-back companies, which, in turn, must have approval from the authorities. Consumer taxes ensure the financial requirements for the end treatment of WEEE, and the collection is organized either on the municipal level, by intermunicipal waste companies or by stores (Román 2012). Three major take-back companies have approval and are responsible for e-waste management: Norsirk, ERP Norway AS, and RENAS AS (NEA 2018). There is a law in force that obliges municipalities to collect WEEE, while the requirement for take-back systems is that they must ensure a free collection from enterprises, distributors, and municipalities collecting WEEE (Román 2012). It is the producers, however, who set up the take-back companies to manage WEEE from households and businesses. Under Norway's Waste Regulations, a special WEEE record is maintained to aggregate the information of all EEE product manufacturers and importers, identify free riders in the system, exchange regulatory information with importers and manufacturers complying the with the WEEE management rules, and finally, collect and collate information and data from the companies involved in the take-back system (NEA 2018).

3.4.2 Americas

3.4.2.1 United States of America

The USA does not have a proper federal regulation for electronic waste. This is mainly because legally WEEE is generally considered a nonhazardous waste. The policies are, instead, imposed by the state government (Li et al. 2015; Ongondo et al. 2011). In general, however, it is illegal for businesses in the USA to place e-waste in the trash (Namias 2013). Currently (2019), 25 states have passed laws concerning collection, recycling, or processing of e-waste – these vary considerably among states (Schumacher and Agbemabiese 2019). Some states have applied EPR programs that comprise disposal fees, deposit refund systems, and mandatory take-back systems for rechargeable nickel–cadmium batteries (Garlapati 2016); some have prohibited e-waste from being disposed of in the municipal waste stream (Namias 2013). Recent research suggests that having discrepancies in the legislation within the country obstructs the implementation of efficient waste management systems. This issue could potentially be tackled by having a federal legislation, a model state legislation, or an overall EPR policy, which clearly defines the roles and responsibilities of the stakeholders (Hickle 2014; Schumacher and Agbemabiese 2019).

In Maine, the legislation specifies that each manufacturer is individually responsible for all the collection and recycling cost of its goods in addition to a share of orphan waste, which is waste from producers that have gone out of business or no longer trading. Orphan waste must be covered by any manufacturer with more than 1% market share. Municipalities and collection points are responsible for collecting e-waste and forwarding it onto a consolidator, which counts, weighs, and identifies the brand of each product. Manufacturers can either collect a representative e-waste sample from the collector (based on a return share), pay the consolidator to recycle the e-waste on its behalf, or have their branded products separated and recycle them themselves. Any brand that is not compliant with the legislation is banned from selling its products (McCann and Wittmann 2015).

The State of North Carolina introduced an e-waste landfill ban and an EPR law comprehending PCs and TVs manufacturers, retailers, and local governments. It implemented a free take-back program for the community, required the liable parties to register (be "certified") and to pay an annual fee. The EEE producers pay the WEEE recyclers according to its selling share, which becomes their recycling target. Retailers must ensure correct EEE labeling and cross-check manufacturers/brand to make sure they are certified under the State. WEEE is mainly collected by local governments, who must provide annual reports and forward WEEE to recyclers. The role of the State government is screening and certifying the recyclers, registering the EEE producers, regulating the system, and financially supporting eligible local governments. The system discourages scavenging and encourages recyclers to improve their recycling processes and efficiency. State law also requires environmentally sound recycling (DEQ 2018).

The State of California requires retailers to charge a recycling fee from consumers who purchase certain EEE. The focus is mainly on displays (cathode ray tubes

(CRTs), liquid crystal displays (LCDs), plasma, etc.). Retailers may retain up to 3% of the fees to cover collection costs, the rest is sent to a board that reimburses recyclers and organizations, which, in turn, provide free e-waste recycling to consumers and businesses (Namias 2013).

In New York, manufacturers of certain EEE are required to collect and recycle (or reuse) their brands of products at no cost for residents and small businesses. The covered EEE include computers (desktops, laptops, tablets, e-readers), peripherals (keyboard, printer, monitors), televisions and small equipment (DVD players, video game consoles, receivers, small-scale servers) (NYSDEC 2019). Furthermore, certain WEEE are eligible for free collection through a manufacturer take-back program. The state requires manufacturers to establish collection, handling, and recycling/reuse of discarded WEEE, it also establishes annual reuse and recycling targets for all e-waste. Exceeding the imposed targets results in credit, while a shortage incurs in a surcharge. There are several take-back programs, including free postage mail-back system and local collection events (Namias 2013).

3.4.2.2 Canada

As is the case for the USA, Canada does not have a central federal regulation to deal with the e-waste. Instead, the provincial government is responsible for regulating its management (Li et al. 2013). This lack of a central government standard does not allow an even competition among recyclers from distinct provinces. By 2014, all provinces but one (New Brunswick) had a proper e-waste management program (Morawski and Millette 2014). The Canadian WEEE management systems have been designed and are managed by foundations owned by technology companies (Irani et al. 2016), while the recyclers must meet a standard established by a nonprofit entity created by the electronics industry of Canada (Morawski and Millette 2014).

Recently, the country has implemented Environmental Handling Fees (EHF) to all consumers when they purchase a new EEE. This is used to cover the cost of the end-of-life management of WEEE; the cost (and therefore the fee) varies according to EEE and location (province). The fee has been implemented in Quebec, British Columbia, Saskatchewan, Manitoba, New Brunswick, Nova Scotia, Newfoundland, and Labrador. In the Province of Quebec, recyclers that operate must be verified under the national standard. The drop-off point network is made of municipal eco centers, retailers, and other organizations and businesses. In British Columbia, EEE manufacturers and distributors are required to be part of the Stewardship Plan, which was created and is regulated by the Electronic Product Recycling Association (EPRA 2014).

In the province of Ontario, the first e-waste related regulatory program was initiated under the Waste Electrical & Electronic Equipment (WEEE) Program Plan in 2008, which was then revised by Waste Diversion Ontario (WDO) and the Ministry of Environment (MOE) on 10 July 2009. This piece of legislation delivered the compliance and enforcement procedures. Then, in 2016, the Resource Recovery and Circular Economy Act, 2016 (under schedule 1), was enacted in combination with the Waste Diversion Transition Act, 2016 (schedule 2), which focuses on achieving circular economy. The common ground of both schedules included (i) promoting the reduction, reuse, and recycling of waste; (ii) promoting competition, cooperation, and coordination among various actors involved in the resource recovery supply chain; and (iii) enhancing equity for consumers and decreasing the environmental impacts of the resource recovery activities (STEP 2019b).

3.4.2.3 Brazil

Solid waste (including e-waste) falls under the Brazilian Policy of Solid Waste (BPSW), which is the main Brazilian legal framework toward waste management in the federate level (Brasil 2010). The policy establishes the general guidelines that the country should follow concerning waste management, the responsibility of the waste generators and the government, and applicable economic mechanisms. The shared waste stewardship is among other changes imposed by the BPSW (Brasil 2010).

The framework, however, does not define specific goals to be reached through the BPWS nor holds the waste generators liable using concrete methods (evaluation, monitoring, fines, etc.). Instead, it mentions that these attributions will be addressed in a latter document called the National Solid Waste Plan (Brasil 2010). The plan was written, elaborated, and debated with the community during the years 2011 and 2012. A preliminary version was published in 2011, but it still awaits the approval of the agricultural policy committee to come into force (MMA 2012). In 2018 the plan was still in standby with no perspective of it being concluded and enforced in the near future (Grandelle 2018). Indeed, to date (2021) it is still in standby but the current expectation from the industry is that it will be approved this year (Correa and Dantas 2021). In summary, the policy was not enough to affect the Brazilian waste *status quo* and the plan – that supposedly would enforce the policy – never overcame the Brazilian bureaucracy. Explicitly targeting WEEE, the BPWS required the implementation of a specific sectoral agreement, which is composed of contracts between the government, the manufacturers, the importers, the distributors, or the vendors, with the objective of implementing the shared responsibility concept to the EEE goods. This was first drafted in 2013, but debuted only in November 2019 (Caiado et al. 2017; Ministry of Environment 2013; MMA 2020). The agreement has several stages: initial steps of the agreement mandate that companies that sell electronic products (physical stores or e-commerce) participate in a reverse logistics system. They also became obliged to accept (from consumers) and store end-of-life products temporarily. Later stages, once the reverse logistics system is in place, will include collection targets (by weight) that increase progressively (FECOMERCIOSP 2020).

3.4.3 Asia

3.4.3.1 Japan

Japan's home appliance recycling system was driven primarily because of the shortage of landfill space in the country and the need for resource recovery (Yoshida and Yoshida 2012). Under its regulations, consumers pay a recycling fee when discarding TVs, air conditioners, refrigerators, and washing machines. Retailers have the obligation of taking back items of these four categories, while producers have the obligation of recycling them (Yoshida and Yoshida 2012). PCs and copiers were later included as voluntary items (Morris and Metternicht 2016). While being responsible for collecting illegally dumped WEEE and being able to treat some designated WEEE types (Morris and Metternicht 2016), the Japanese municipalities do not perform the collection and processing recycling operations themselves. Instead, this is carried out under a producer partnership cluster that is divided between Groups A (Panasonic, Toshiba, and others) and B (Mitsubishi, Hitachi, and others) (Yoshida and Yoshida 2012). There are about 1580 inhabitants per take-back site in the country, and the EEE producers have to meet recycling standards and targets determined by legislation (Morris and Metternicht 2016). About 78% of WEEE are collected by retail stores, of which 64% are recycled by the producers and the rest is lost because of WEEE entering via channels outside the system and, therefore, without being paid for (Yoshida and Yoshida 2012). Breaching the law implies in corrective recommendations, corrective orders, or penalties (Morris and Metternicht 2016). The policy is built around recycling, which restricts product reuse (Yoshida and Yoshida 2012).

3.4.3.2 China

As the quantity of Waste Electrical and Electronic Equipment grows fast (Zeng et al. 2016), the Regulation on the Administration of the Recovery and Disposal of Waste Electrical and Electronic Products (China WEEE Regulation) was enacted in 2009 and enforced in 2012. China WEEE regulation symbolized the formal implementation of Extended Producers Responsibility theory, supported by a family of technical guidelines and policies (Zeng and Li 2019), including operation of recycling fund, etc. WEEE recycling fund was set up and the financial subsidy was distributed into certified recyclers since 2013 (Song et al. 2017).

The operation of WEEE recycling fund guaranteed and promoted development of formal WEEE recycling sector in China, with significant environmental benefit (Wang et al. 2018; Zeng and Li 2019). Until 2017, there were 109 certified WEEE recycling plants in China, distributed in 29 provinces. In 2017, 79.94 million units of WEEE were disassembled and treated in these plants, including 42.07 million units of waste televisions, 8.04 million units of waste refrigerators, 13.59 million units of waste washing machines, 3.98 million of air-conditioners, and 12.27 million units of waste PCs (MEE 2017). Further, 90% of WEEE treated in the formal plants in 2017 went through technical auditing of Ministry of Ecology and Environment of China, which means subsidy from WEEE recycling fund would be distributed to the formal plants based on the monitored quantity.

However, there are still challenges facing by WEEE management system in China. The imbalance between fund levies and subsidies led to an unsustainable WEEE funding policy, while the tedious procedures of subsidies distribution decreased the efficiency of the payment system (Zeng and Li 2019). In 2015, there was a deficit of nearly 2.6 billion CNY between the amount paid by the producers (2.8 billion CNY) and the 5.4 billion CNY subsidy distributed to the certified plants (Chen et al. 2018).

3.4.3.3 Taiwan

Taiwan's WEEE handling structure is known as the 4-in-1 recycling program and began to be drafted in 1988 with the Act that mandated an EPR system. By late 1990s, the regulatory framework in Taiwan already had a cash reward program for consumers who took their unwanted computers to designated collection points, which consisted primarily of retailers (the reward program is still active even now). These retailers also obtained financial rewards for receiving the used equipment (Lee et al. 2000). From the beginning, policies around the recycling processes stated that the computers had to be processed in a sound manner and included requirements such as avoiding landfilling and incineration, and removing the phosphorescent coating from CRTs (Lee et al. 2000). Under these policies and regulations, the Taiwanese society has had the obligation of recycling its EEE since 2001 (Shih 2017). The Taiwan Environmental Protection Administration (TEPA) is funded by manufacturers, who have to pay a fixed advanced recycling fee at the time of EEE sale. This fee is based on the auditing and verification that TEPA runs yearly alongside of industry representatives (Fan et al. 2018; Shih 2017). The funds are used mainly to subsidize the recycling industry who receives financial rewards depending on the volume (or number) of WEEE it processes.

3.4.3.4 India

In developing countries like India, the implementation of well-established EPR approaches used in developed countries is ineffective (Borthakur and Govind 2018; Wath et al. 2010). The country relies on informal collection and processing, which is capable of absorbing 90% of its e-waste (Dwivedy et al. 2015). Therefore, India currently has a voluntary take-back system, where there are no laws to enforce compliance and no penalties for not meeting the EPR goals, which were established in 2010 in the lines of the European EPR directive (Dwivedy et al. 2015). These circumstances create a different scenario from that found in developed countries, since Indian consumers prefer to sell their obsolete equipment in the informal sector (i.e. they expect a profit when discarding their WEEE) instead of adhering to the principles of formal EPR (Borthakur and Govind 2018; Dwivedy et al. 2015).

3.4.4 Africa

3.4.4.1 South Africa

There is currently no legislation specifically addressing WEEE, the country is deficient in e-waste recycling infrastructure and industries are not required to submit (or share) data on the e-waste they generate, making available information unreliable and contradictory. The South African government, however, has recently recognized the e-waste as a priority waste stream (Ledwaba and Sosibo 2016; Snyman et al. 2017). While the government has not passed any specific legislation or policy related to e-waste, it follows some of the international legislations connected to it, such as the Rotterdam Convention, the Stockholm Convention, and the Basel Convention, under which different types of hazardous substances and waste are considered. The National Environmental Management Act (Act 107 of 1998) (NEMA) is the primary national legislation that outlines several issues, for example, avoidance or minimization/waste reduction, the remediation of pollution, reuse, recycling, and proper disposal and "polluter pays" and "cradle-to-grave" principles. However, this policy guideline is not specific to e-waste (Bob et al. 2017).

3.4.4.2 Nigeria

In 2012, a study reported that there were no serious initiatives in Nigeria concerning WEEE management and that, in spite of some institutional framework, a number of challenges needed addressing due to the lack of control over the flow of used EEE, legislation to appropriately identify contraband items, public awareness concerning the dangers of handling WEEE, recycling facilities in the country, and corporate social responsibility. Moreover, obtaining reliable data on e-waste (generation, export, import, obsolescence rate, discard) is extremely difficult in the country, which increases the problem further (Adediran and Abdulkarim 2012). These constraints around data reliability are directly affected by the situation of the country that involves illegal imports, informal recycling, and poor formal recycling (Woggborg and Schroder 2018).

In 2016, however, the country introduced its own EPR program. The program is overseen by the government through the NESREA (National Environmental Standard Regulatory and Enforcement Agency) and utilizes PROs (producer responsibility organizations, like in Switzerland) to manage the funds collected from e-waste generators (importers, distributors, etc.) and transfer them to licensed recyclers and formal collection centers. The main drawback is that the EPR program does not extend to the informal sector, which plays a dominant role in Nigeria's e-waste management system (Woggborg and Schroder 2018). It should be noted, however, that the NESREA has no official presence in 10 out of 36 Nigerian states, which further decreases the reach of the EPR program (Iwenwanne 2019).

3.4.5 Australia

The Australian Bureau of Statistics (2006) claimed, in 2006, that "obsolete electronic waste or e-waste is one of the fastest growing waste types. Very little of the increasing amount of e-waste generated in Australia is being recycled, with most of it ending up in landfill." In 2008, a study reported only a small percentage of e-waste was recycled in Australia, particularly low percentages were reported for televisions (1%), computers (1.5%), and mobile phones (4%) (TEC 2008). This led to an "ever growing e-waste mountain" (Herat 2008), where significant amounts of e-waste were being sent to landfills (Ongondo et al. 2011). The biggest challenge for the country has been claimed to be legislation and compliance, given that by 2011 the initiatives to contest the WEEE problem were either too late or too little (Ongondo et al. 2011).

The lack of legislation, however, has been tackled in recent years, as Australia currently holds the only law on the management of e-waste in Oceania: The National Television and Computer Recycling Scheme (NTCRS). The National Waste Policy (Australia's general waste management guidelines), the Product Stewardship Act 2011 (framework to administer and accredit product stewardship schemes throughout the country), the Product Stewardship (Televisions and Computers) Regulations 2011, and the National Television and Computer Recycling Scheme 2011 provide the framework to manage the life cycle of computers, televisions, and their peripherals (Morris and Metternicht 2016). The scheme aims to collect and recycle e-waste, specifically televisions, computers, and its peripherals. The scheme is similar to the EU Directive, whereby the Member States must establish systems for users and distributors to return household WEEE to collection facilities free of charge and requires manufacturers to finance the collection and recycling of WEEE at these facilities (Lane et al. 2015).

On top of the NTCRS, there are other regulatory schemes and industry programs for e-waste in Australia. The Mobile Musters program aims to collect mobile phones, its batteries, and accessories. It is a 20 year-old scheme that has collected 1323 tons of mobile phone components since its commencement. (Mobile Muster 2017). Another arrangement is the Australian Battery Recycling Initiative, which aims to collect household batteries and commenced in 2008. It is currently a voluntary scheme, but there are negotiations in course to establish an industry-funded recycling program (ABRI 2018; Golev and Corder 2017). The Cartridges 4 Planet Ark is also a voluntary scheme established in 2003. It's a free scheme for the end user that has collected a total of 39.1 million cartridges since its commencement (C4PA 2018; Golev and Corder 2017).

There is no statutory bound that obliges the owner of an EEE to dispose of those products safely. There are minimal penalties for disposing of EEE in public, but it is considered a minor criminal offense. Most Australian States and Territories do not have bans on e-waste going into landfill, the exceptions are South Australia and the Australian Capital Territory (Lane et al. 2015). The state of Victoria, however, has implemented a ban on e-waste in landfills, which took effect in July 2019. The ban is enforced by the Environment Protection Authority (EPA) (Victoria State Government 2019).

3.5 Conclusions

There is a worldwide struggle to find a solution to the WEEE challenge. Different countries are trying different approaches and there is still little information about the material flow and management systems in many developing nations. While each individual country has its particularities, the exchange of experiences and comparison of different approaches toward WEEE management among countries yields important insights toward best waste management practice. Once identified, qualities and shortfalls of a given system can be adapted for a different country in a process of continuous enhancement through regulatory framework and management system setup. Examples of such exchange can be verified by looking at the foundation the European regulations gave to several other non-European countries, or the implementation of the Australian scheme, which took principles from schemes overseas and adapted to a country with large territorial size. In light of this, it becomes necessary to have more scientific outputs analyzing the e-waste management setup of different countries. This is especially important in developing countries, where details and studies are still insipid.

Regulatory initiatives should be more consumer-oriented, without scarifying the strict implementation of the EPR strategy. Specific guidelines on collection and recovery strategies should be reflected directly in the e-waste management regulation – such as financial and material flow accounting found in the Swiss e-waste management system. Critical and emerging aspect (e.g. circular economy and urban mining) should be considered when reforming and developing e-waste management-related regulations. This shall enhance the technical capabilities of the policymakers tackling the e-waste problem and waste can be converted to resources creating jobs and entrepreneurship. Additionally, it is important to have unified or collaborative legislations that assist each other in the task of achieving greater resource efficiency from e-waste, as opposed to legislations that create difficulties to stakeholder acting across different jurisdictions/regions, as is the case for the USA.

Finally, the recent waste ban implemented in China showed the magnitude of e-waste volume shipped to the country from all over the world. It also revealed how the current global setup is dependent on the Chinese recycling industry and how little infrastructure exists elsewhere. The unfolding of the ban over the next few years should render significant changes in the transboundary movement of e-waste and the regulations associated with this waste stream worldwide.

References

Abdelbasir, S.M., El-Sheltawy, C.T., and Abdo, D.M. (2018). Green processes for electronic waste recycling: a review. *Journal of Sustainable Metallurgy* 4: 295–311. https://doi.org/10.1007/s40831-018-0175-3.

Abdelbasir, S.M., Hassan, S.S.M., Kamel, A.H., and El-Nasr, R.S. (2018). Status of electronic waste recycling techniques: a review. *Environmental Science and Pollution Research* 25: 16533–16547. https://doi.org/10.1007/s11356-018-2136-6.

ABRI. (2018). ABRI – Australian Battery Recycling Initiative. www.batteryrecycling.org .au. (retrieved 13 August 2018)

Adediran, Y.A. and Abdulkarim, A. (2012). Challenges of electronic waste management in Nigeria. *International Journal of Advances in Engineering & Technology* 4: 640–648.

Australian Bureau of Statistics. (2006). 4613.0 – Australia's Environment: Issues and Trends, 2006.

Babu, B.R., Parande, A.K., and Basha, C.A. (2007). Electrical and electronic waste: a global environmental problem. *Waste Management & Research* 25: 307–318. https://doi.org/10.1177/0734242X07076941.

Balde, C., Forti, V., Kuehr, R., and Stegmann, P. (2017). *The Global E-waste Monitor*. Bonn/Geneva/Vienna: United Nations University (UNU), International Telecommunication Union (ITU) & International Solid Waste Association (ISWA).

Balde, C., Kuehr, R., Blumenthal, K. et al. (2015). *E-waste Statistics: Guidelines on Classifications, Reporting and Indicators*. Bonn, Germany: United Nations University, IAS-SCYCLE.

BAN (2016). *Scam Recycling: e-Dumping on Asia by US Recyclers*. Seattle: Basel Action Network (BAN) http://www.ban.org/trash-transparency (retrieved 03 August 2018).

BAN (2019). *BAN Intervention on the Technical Guideline on Electronic Waste*. Seattle: Basel Action Network (BAN) https://myemail.constantcontact.com/BAN-Intervention-on-the-Technical-Guideline-on-Electronic-Waste.html?soid=1114999858498&aid=nT9EqVLY1Hw (retrieved 19 December 2019).

BMU. (2015). Objectives and content of the Law on Marketing, Return and environmentally sound disposal of electrical and electronic equipment (Electrical and Electronic Equipment Act – ElektroG) (Kurzinfo zum ElektroG: Ziele und Inhal). BMU – Bundesministerium für umwelt, naturschutz und nukleare sicherheit (Federal Ministry for the Environment, Nature Conservation and Nuclear Safety). http://www.bmu.de/themen/wasser-abfall-boden/abfallwirtschaft/abfallpolitik/elektrog (retrieved 19 August 2018).

Bob, U., Padayachee, A., Gordon, M., and Moutlana, I. (2017). Enhancing innovation and technological capabilities in the management of E-waste: case study of South African government sector. *Science, Technology and Society* 22: 332–349. https://doi.org/10.1177/0971721817702293.

Borthakur, A. and Govind, M. (2018). Public understandings of E-waste and its disposal in urban India: from a review towards a conceptual framework. *Journal of Cleaner Production* 172: 1053–1066. https://doi.org/10.1016/j.jclepro.2017.10.218.

Brasil. (2010). Brazilian Policy of Solid Waste (Política Nacional de Resíduos – PNRS) Lei n. 12.305 Câmara dos Deputados, Edições Câmara. 2. ed. 73 p. 2002.

C4PA. (2018). C4PA – Cartridges 4 Planet Ark. https://cartridges.planetark.org (retrieved 13 August 2018).

Caiado, N., Guarnieri, P., Xavier, L.H., and de Lorena Diniz Chaves, G. (2017). A characterization of the Brazilian market of reverse logistic credits (RLC) and an analogy with the existing carbon credit market. *Resources, Conservation and Recycling* 118: 47–59. https://doi.org/10.1016/j.resconrec.2016.11.021.

Chaudhary, K. and Vrat, P. (2018). Case study analysis of e-waste management systems in Germany, Switzerland, Japan and India: a RADAR chart approach. *Benchmarking: An International Journal* 25: 3519–3540. https://doi.org/10.1108/BIJ-07-2017-0168.

Chen, M., Ogunseitan, O.A., Duan, H. et al. (2018). China E-waste management: struggling for future success. *Resources, Conservation and Recycling* 139: 48–49. https://doi.org/10.1016/j.resconrec.2018.08.006.

Chinese Government Network. (2017). Notice of the General Office of the State Council on Issuing the Implementation Plan for Prohibiting the Entry of Foreign Garbage and Advancing the Reform of the Solid Waste Import Administration System.

Suzana Correa and Tiago Dantas (2021). A decade after the law, most Brazilian cities still use dumps and waste production grows with pandemic (Uma década após lei, maioria das cidades brasileiras ainda usa lixões e produção de resíduos crescem com pandemia) https://oglobo.globo.com/sociedade/um-so-planeta/uma-decada-apos-lei-maioria-das-cidades-brasileiras-ainda-usa-lixoes-producao-de-residuos-cresce-com-pandemia-24988904.

Cucchiella, F., D'Adamo, I., Lenny Koh, S.C., and Rosa, P. (2015). Recycling of WEEEs: an economic assessment of present and future e-waste streams. *Renewable and Sustainable Energy Reviews* 51: 263–272. https://doi.org/10.1016/j.rser.2015.06.010.

DEQ (2018). *NC Electronics Management Program*. North Carolina: Department of Environmental Quality https://deq.nc.gov/about/divisions/waste-management/electronics (retrieved 16 August 2018).

Dwivedy, M., Suchde, P., and Mittal, R.K. (2015). Modeling and assessment of e-waste take-back strategies in India. *Resources, Conservation and Recycling* 96: 11–18. https://doi.org/10.1016/j.resconrec.2015.01.003.

EPRA. (2014). EPRA – Electronic Products Recycling Association. http://epra.ca (retrieved 20 August 2018).

EU Directive (2012). Directive 2012/19/EU of the European Parliament and of the council of 4 July 2012 on waste electrical and electronic equipment, WEEE. *Official Journal of the European Union L* 197: 38–71.

European Commission. (2018). Waste Electrical & Electronic Equipment (WEEE). http://ec.europa.eu/environment/waste/weee/index_en.htm (retrieved 03 August 2018).

European Parliament. (2011). Directive 2011/65/EU of the European Parliament and of the Council of 8 June 2011 on the restriction of the use of certain hazardous substances in electrical and electronic equipment. 174.

Fan, C., Fan, S.-K.S., Wang, C.-S., and Tsai, W.-P. (2018). Modeling computer recycling in Taiwan using system dynamics. *Resources, Conservation and Recycling* 128: 167–175. https://doi.org/10.1016/j.resconrec.2016.09.006.

FECOMERCIOSP. (2020). With a new decree, the rule provided for in the sectoral agreement is now applied by all those responsible for the life cycle of electronic products [Com novo decreto, regra prevista pelo acordo setorial passa a ser aplicada por todos os responsáveis pelo ciclo de vida de produtos eletroeletrônicos] (WWW Document). FecomercioSP. www.fecomercio.com.br/noticia/com-novo-decreto-regra-prevista-pelo-acordo-setorial-passa-a-ser-aplicada-por-todos-os-responsaveis-pelo-ciclo-de-vida-de-produtos-eletroeletronicos (accessed 12 January 2020).

Forti, V., Baldé, C.P., Kuehr, R., Bel, G. (2020). The Global E-waste Monitor 2020 120.

Garlapati, V.K. (2016). E-waste in India and developed countries: management, recycling, business and biotechnological initiatives. *Renewable and Sustainable Energy Reviews* 54: 874–881. https://doi.org/10.1016/j.rser.2015.10.106.

Goel, S. (2017). *Advances in Solid and Hazardous Waste Management*. New York, NY: Springer Berlin Heidelberg.

Golev, A. and Corder, G.D. (2017). Quantifying metal values in e-waste in Australia: the value chain perspective. *Minerals Engineering* 107: 81–87. https://doi.org/10.1016/j.mineng.2016.10.021.

Grandelle, R. (2018). Lack of plan concerning solid waste prevents the solution for 3 thousand dump sites [Falta de plano sobre resíduos sólidos impede solução para 3 mil lixões]. https://oglobo.globo.com/sociedade/falta-de-plano-sobre-residuos-solidos-impede-solucao-para-3-mil-lixoes-22857340. O Globo (retrieved 12 July 2018).

Herat, S. (2008). Australia's ever growing e-waste mountain. *Electronics World* 114: 34–36.

Herat, S. and Agamuthu, P. (2012). E-waste: a problem or an opportunity? Review of issues, challenges and solutions in Asian countries. *Waste Management & Research* 30: 1113–1129. https://doi.org/10.1177/0734242X12453378.

Hickle, G.T. (2014). Moving beyond the "patchwork": a review of strategies to promote consistency for extended producer responsibility policy in the U.S. *Journal of Cleaner Production* 64: 266–276. https://doi.org/10.1016/j.jclepro.2013.08.013.

Irani, A.K., Shirazi, F., Bener, A. (2016). A critical interrogation of e-waste management in Canada: evaluating performance of environmental management systems. 16.

Islam, M.T., Dias, P., and Huda, N. (2018). Comparison of E-waste management in Switzerland and in Australia: a qualitative content analysis. *International Journal of Environmental and Ecological Engineering, World Academy of Science, Engineering and Technology* 12. https://doi.org/doi.org/10.5281/zenodo.1474825.

Iwenwanne, V. (2019). Nigeria's e-waste mountain. *Resource* 7.

Kiddee, P., Naidu, R., and Wong, M.H. (2013). Electronic waste management approaches: an overview. *Waste Management* 33: 1237–1250. https://doi.org/10.1016/j.wasman.2013.01.006.

Kuehr, R. (2019). E-waste seen from a global perspective. In: *Waste Electrical and Electronic Equipment (WEEE) Handbook* (eds. V. Goodship, A. Stevels and J. Huisman), 1–16. Elsevier https://doi.org/10.1016/B978-0-08-102158-3.00002-1.

Kumar, A., Holuszko, M., and Espinosa, D.C.R. (2017). E-waste: an overview on generation, collection, legislation and recycling practices. *Resources, Conservation and Recycling* 122: 32–42. https://doi.org/10.1016/j.resconrec.2017.01.018.

Lane, R., Gumley, W., Santos, D. (2015). Mapping, characterising and evaluating collection systems and organisations. Characterising and Evaluating Collection Systems and Organisations, Monash University, Australia.

Ledwaba, P., Sosibo, N. (2016). E-waste management in South Africa: case study: cathode ray tubes recycling opportunities. https://doi.org/10.20944/preprints201608.0021.v1

Lee, C.-H., Chang, S.-L., Wang, K.-M., and Wen, L.-C. (2000). Management of scrap computer recycling in Taiwan. *Journal of Hazardous Materials* 73: 209–220. https://doi.org/10.1016/S0304-3894(99)00191-0.

Leroy, P. (2012). 4 - The WEEE forum and the WEEELABEX project. In: *Waste Electrical and Electronic Equipment (WEEE) Handbook, Woodhead Publishing Series in Electronic and Optical Materials* (eds. V. Goodship and A. Stevels), 66–77. Woodhead Publishing https://doi.org/10.1533/9780857096333.1.66.

Li, J., Lopez, B.N., Liu, L. et al. (2013). Regional or global WEEE recycling. Where to go? *Waste Management* 33: 923–934. https://doi.org/10.1016/j.wasman.2012.11.011.

McCann, D., Wittmann, A. (2015). E-waste prevention, take-back system design and policy approaches. Step Green Paper Series 92.

Ministry of Ecology and Environment (2021). https://www.hbzhan.com/news/detail/140248.html (accessed 05 May 2021).

Ministry of Environment. (2013). Call to elaborate the sectoral agreement to implement the reverse logistic system for WEEE and its components [Chamamento para a Elaboração de Acordo Setorial para a Implantação de Sistema de Logística Reversa de

Produtos Eletroeletrônicos e seus Componentes]. Ministry of Environment – Brazil (Ministério do Meio Ambiente).

MMA. (2012). National Solid Waste Plan [Plano Nacional de Resíduos Sólidos]. Solid waste management information system [Sistema de informações sobre a gestão dos resíduos sólidos]. Ministry of Environment [Ministério do Meio Ambiente].

MMA. (2020). SINIR – National Solid Waste Management Information System [Sistema Nacional de Informações sobre a Gestão dos Resíduos Sólidos]. Ministry of Environment [Ministério do Meio Ambiente]. https://sinir.gov.br/component/content/article/2-sem-categoria/474-acordo-setorial-de-eletroeletronicos (retrived 1 December 2020) [WWW Document]. URL https://sinir.gov.br/component/content/article/2-sem-categoria/474-acordo-setorial-de-eletroeletronicos (retrived 1 December 2020).

Mobile Muster. (2017). Annual Report-2016–2017. www.mobilemuster.com.au (retrieved 13 August 2018).

Morawski, C., Millette, S. (2014). Waste electronics management in Canada. Shortcomings and opportunities for improvement abound. Solid Waste Management. www.solidwastemag.com (retrieved 17 August 2018).

Morris, A. and Metternicht, G. (2016). Assessing effectiveness of WEEE management policy in Australia. *Journal of Environmental Management* 181: 218–230. https://doi.org/10.1016/j.jenvman.2016.06.013.

MS2 and Perchards. (2009). Product Stewardship in North America and Europe. MS2 – Martin Stewardship & Management Strategies Pty Ltd. Perchards Ltd. Final Report.

Namias, J. (2013). The future of electronic waste recycling in the United States: obstacles and domestic solutions (Master's thesis). http://www.seas.columbia.edu/earth/wtert/sofos/Namias_Thesis_07-08-13.pdf (retrieved 17 August 2018).

NEA (2018). *Waste Electrical and Electronic Equipment*. Oslo, Norway: NEA – Norwegian Environment Agency https://www.environment.no/topics/waste/avfallstyper/waste-electrical-and-electronic-equipment (retrieved 02 December 2019).

NYSDEC. (2019). "The Electronic Equipment Recycling and Reuse Act", ENVIROMENTAL CONSERVATION LAW ARTICLE 27 TITLE 26. New York State Department of Environmental Conservation. http://www.dec.ny.gov (retrieved 17 December 2019).

Oeko-Institut (2019). *RoHS Evaluations of Exemtion Requests and Other Topics*. Freiburg, Germany: Oeko-Institut https://rohs.exemptions.oeko.info/index.php?id=214 (retrieved 19 December 2019).

Ongondo, F.O., Williams, I.D., and Cherrett, T.J. (2011). How are WEEE doing? A global review of the management of electrical and electronic wastes. *Waste Management* 31: 714–730. https://doi.org/10.1016/j.wasman.2010.10.023.

Owusu, L. (2017). Developing a conducive legal framework for sustainable management of E-waste in Ghana: experience and way forward. https://www.wrforum.org/wp-content/uploads/2017/11/2017-10-24-Owusu-Ghana-legal-framework-WRF.pdf (retrieved 2 December 2019).

Peluola, A. (2016). Investigation of the implementation and effectiveness of electronic waste management in Nigeria. *Modeling Earth Systems and Environment* 2: 100. https://doi.org/10.1007/s40808-016-0155-1.

Perkins, D.N., Brune Drisse, M.-N., Nxele, T., and Sly, P.D. (2014). E-waste: a global hazard. *Annals of Global Health* 80: 286–295. https://doi.org/10.1016/j.aogh.2014.10.001.

Pharino, C. (2017). Electronic waste management in Thailand. In: *Challenges for Sustainable Solid Waste Management: Lessons from Thailand, SpringerBriefs on Case Studies of Sustainable Development* (ed. C. Pharino), 93–115. Singapore: Springer https://doi.org/10.1007/978-981-10-4631-5_7.

Premalatha, M., Abbasi, T., and Abbasi, S.A. (2014). The generation, impact, and management of E-waste: state of the art. *Critical Reviews in Environmental Science and Technology* 44: 1577–1678. https://doi.org/10.1080/10643389.2013.782171.

Resource Recycling. (2018). From Green Fence to red alert: A China timeline. https://resource-recycling.com/recycling/2018/02/13/green-fence-red-alert-china-timeline (retrieved 9 August 2018).

Resource Recycling. (2019). China: Plastic imports down 99 percent, paper down a third. https://resource-recycling.com/recycling/2019/01/29/china-plastic-imports-down-99-percent-paper-down-a-third (retrieved 20 June 2019).

Resource Recycling. (2020). Lockdowns and shipping strife snarl scrap exports. https://resource-recycling.com/recycling/2020/03/31/lockdowns-and-shipping-strife-snarl-scrap-exports (retrieved 01 December 2020).

Robinson, B.H. (2009). E-waste: an assessment of global production and environmental impacts. *Science of the Total Environment* 408: 183–191. https://doi.org/10.1016/j.scitotenv.2009.09.044.

RoHS Guide. (2019). China RoHS. https://www.rohsguide.com/china-rohs (retrieved 31 July 2019).

RoHS Guide. (2020). Other RoHS Green Initiatives Worldwide. www.rohsguide.com (retrieved 14 September 2020).

Román, E. (2012. 22). WEEE management in Europe: learning from best practice. In: *Waste Electrical and Electronic Equipment (WEEE) Handbook, Woodhead Publishing Series in Electronic and Optical Materials* (eds. V. Goodship and A. Stevels), 493–525. Woodhead Publishing https://doi.org/10.1533/9780857096333.5.493.

Schluep, M. (2014). Chapter 25 – Waste electrical and electronic equipment management. In: *Handbook of Recycling*, 397–403. Boston, MA: Elsevier https://doi.org/10.1016/B978-0-12-396459-5.00025-8.

Schluep, M., Hagelüken, C., Kuehr, R. et al. (2009). Thiébaud. In: *Recycling – from E-Waste to Resources* (eds. E. Müller and F. Wang). United Nations Environment Programme, United Nations Environmental Program (UNEP) & United Nations University (UNU) https://www.researchgate.net/publication/278849195_Recycling_-_from_e-waste_to_resources.

Schumacher, K.A. and Agbemabiese, L. (2019). Towards comprehensive e-waste legislation in the United States: design considerations based on quantitative and qualitative assessments. *Resources, Conservation and Recycling* 149: 605–621. https://doi.org/10.1016/j.resconrec.2019.06.033.

Shih, H.-S. (2017). Policy analysis on recycling fund management for E-waste in Taiwan under uncertainty. *Journal of Cleaner Production* 143: 345–355. https://doi.org/10.1016/j.jclepro.2016.12.103.

Snyman, J., Vorster, K., Jacobs, S.J. (2017). Towards sustanable e-waste management in South Africa, in: *5th International Conference on Sustainable Solid Waste Management*. Presented at the ATHENS 2017.

Song, X., Wang, J., Yang, J., and Lu, B. (2017). An updated review and conceptual model for optimizing WEEE management in China from a life cycle perspective. *Frontiers of Environmental Science & Engineering* 11: 3. https://doi.org/10.1007/s11783-017-0985-8.

State Council. (2017). Action plan to phase out waste imports. State Council. The people's republic of China. http://english.gov.cn/policies/latest_releases/2017/07/27/content_281475756814340.htm (retrieved 9 August 2018).

Staub, C. (2018). Exporter response to China: 'We are changing our whole strategy'. Resource Recycling. https://resource-recycling.com/recycling/2018/01/09/exporter-response-china-changing-whole-strategy (retrieved 9 August 2018).

STEP. (2019a). STEP – Solving the e-waste problem. E-waste Rules, Regulations, Policies and Guidance. http://www.step-initiative.org (retrieved 02 December 2019).

STEP. (2019b). Ontario (Canada): Waste-Free Ontario Act, Bill 151 Enacted, 2016. STEP – Solving the E-waste Problem. http://www.step-info.org/ontario-canada-waste-free-ontario-act-bill-151-enacted-2016.html (retrieved 02 December 2019).

TEC. (2008). Tipping point: Australia's E-waste Crisis. TEC – Total Environment Centre. www.tec.org.au/reports (retrieved 23 March 2016).

The Ministry of Ecology and Environment (MEE) (2017). *Annual Report on Environmental Pollution Prevention and Control of Solid Waste in Large and Medium Cities in 2017*. Ministry of Ecology and Environment (MME) https://www.mee.gov.cn/hjzl/sthjzk/gtfwwrfz/201912/P020191220699179334061.pdf (accessed 23 August 2019). (In Chinese).

Thi Thu Nguyen, H., Hung, R.-J., Lee, C.-H., and Thi Thu Nguyen, H. (2019). Determinants of residents' E-waste recycling behavioral intention: a case study from Vietnam. *Sustainability* 11: 164. https://doi.org/10.3390/su11010164.

Toffolet, R. (2016). Chapter 1 – WEEE management. In: *WEEE Recycling* (eds. A. Chagnes, G. Cote, C. Ekberg, et al.), 1–30. Elsevier https://doi.org/10.1016/B978-0-12-803363-0.00001-8.

UBA (2016). *Electrical and Electronic Waste*. UB – Umwelt Bundesamt (Environment Federal Office) https://www.researchgate.net/publication/278849195_Recycling_-_from_e-waste_to_resources (retrieved 19 August 2018).

UNEP (1989). 1989 Basel convention on the control of transboundary movements of hazardous wastes and their disposal. *Journal of Environmental Law* 1: 255–277. https://doi.org/10.1093/jel/1.2.255.

Vadoudi, K., Kim, J., Laratte, B. et al. (2015). E-waste management and resources recovery in France. *Waste Management & Research* 33: 919–929. https://doi.org/10.1177/0734242X15597775.

Veit, H.M. and Bernardes, A.M. (2015). Electronic waste: generation and management. In: *Electronic Waste* (eds. H.M. Veit and A. Moura Bernardes), 3–12. Cham: Springer International Publishing https://doi.org/10.1007/978-3-319-15714-6_2.

Victoria State Government. (2019). E-Waste Landfill Ban. Engage Victoria. Environment, land, water and planning. https://engage.vic.gov.au/waste/e-waste (retrieved 24 October 2018).

Wang, J., Wang, Y., liu, J. et al. (2018). Effects of fund policy incorporating extended producer responsibility for WEEE dismantling industry in China. Resources. *Conservation and Recycling* 130: 44–50. https://doi.org/10.1016/j.resconrec.2017.11.016.

Wath, S.B., Vaidya, A.N., Dutt, P.S., and Chakrabarti, T. (2010). A roadmap for development of sustainable E-waste management system in India. *Science of the Total Environment* 409: 19–32. https://doi.org/10.1016/j.scitotenv.2010.09.030.

Woggborg, A. and Schroder, P. (2018). Nigeria's E-waste management: extended producer responsibility and informal sector inclusion. *Journal of Waste Resources and Recycling* 1: 103–111.

Wong, N. (2018). Electronic waste governance under "one country, two systems": Hong Kong and mainland China. *International Journal of Environmental Research and Public Health* 15: 2347. https://doi.org/10.3390/ijerph15112347.

WTO. (2017). Notification. Committee on Technical Barriers to Trade, 17–3880. World Trade Organization. https://www.wto.org/english/news_e/news17_e/impl_03oct17_e.htm (retrieved 9 August 2018).

Yoshida, F. and Yoshida, H. (2012. 25). WEEE management in Japan. In: *Waste Electrical and Electronic Equipment (WEEE) Handbook, Woodhead Publishing Series in Electronic and Optical Materials* (eds. V. Goodship and A. Stevels), 576–590. Woodhead Publishing https://doi.org/10.1533/9780857096333.5.576.

Zeng, X., Gong, R., Chen, W.-Q., and Li, J. (2016). Uncovering the recycling potential of "new" WEEE in China. *Environmental Science & Technology* 50: 1347–1358. https://doi.org/10.1021/acs.est.5b05446.

Zeng, X. and Li, J. (2019). Chapter 19. WEEE management in China. In: *Waste Electrical and Electronic Equipment (WEEE) Handbook (Second Edition), Woodhead Publishing Series in Electronic and Optical Materials* (eds. V. Goodship, A. Stevels and J. Huisman), 521–540. Woodhead Publishing. https://doi.org/10.1016/B978-0-08-102158-3.00019-7.

4

Approach for Estimating e-Waste Generation

Amit Kumar

University of British Columbia, NBK Institute of Mining Engineering, 6350 Stores Road, Vancouver, BC V6T 1Z4, Canada

4.1 Background

Estimation of the amount of e-waste generated every year is one of the challenging processes for designing any recycling or disposal practices. The success of an e-waste recycling or disposal facility depends on the annual throughput to plant. Without a constant throughput to the facility, the plant will have to go through larger downtime due to unavailability of the input feedstock. Hence, it is important to quantify the e-waste generation statistics accurately that in turn will help the recycling/disposal facilities to size their plant capacity and equipment dimensions properly to reduce the overall downtime. The proper estimation of e-waste generation in given region also helps understand the severity of the issue and helps the concerning authorities to formulate proper action plan to tackle the e-waste problem. There have been several studies to estimate the e-waste generation. This chapter lists the most common methods for predicting the total e-waste generated in any calendar year.

4.2 Econometric Analysis

Kusch et al. (2017) studied the relation of waste generation and gross domestic product (GDP) for 50 countries of the pan-European region. It was shown that GDP per capita adjusted for purchasing power parity (PPP) has a linear dependency with e-waste generated per capita (coefficient of determination: 0.7657). Removing Luxembourg (due to its small population and high GDP per capita) improved the coefficient of determination to 0.93.

Kumar et al. (2017) also showed that a linear relationship exists between the GDP and the amount of e-waste generated in a country, whereas no significant relation was observed between the population and the amount of e-waste produced by the country. It was also concluded that the electronic waste generated per

Electronic Waste: Recycling and Reprocessing for a Sustainable Future, First Edition.
Edited by Maria E. Holuszko, Amit Kumar, and Denise C. R. Espinosa.

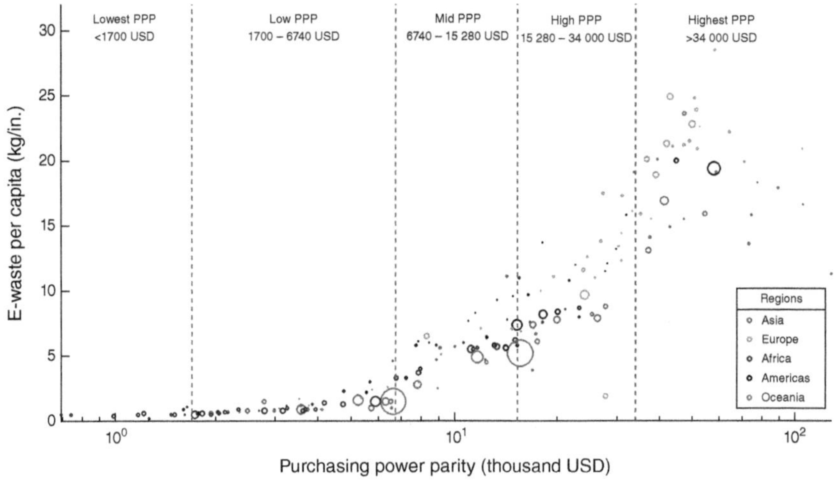

Figure 4.1 E-waste generated per capita with respect to the PPP and population. Source: Data from Baldé et al. (2017) and World Bank (2019).

inhabitant in any country was correlated with the per capita income of the inhabitants as shown in Figure 4.1. It shows that certain countries in Asia with relatively higher per capita income such as the United Arab Emirates, Qatar, and Japan produced more waste compared to other Asian countries with lower PPP such as India, Bangladesh, and Pakistan. The size of the marker in Figure 4.1 represents the population of the country. It suggests that e-waste generation is proportional to the purchasing power for any given country.

The econometric model is a regression-based model that uses various economic factors such as GDP, growth rate, population, PPP of an economy to estimate the e-waste generation (Yedla 2016). This method does not account for the consumer behavior, and the life span of electrical and electronic equipment (EEE).

4.3 Consumption and Use/Leaching/Approximation 1 Method

This method estimates an average e-waste generation based on the household stock of EEE and its average lifespan (L) assuming a saturated market where a fixed percentage of the stocked EEE will enter the waste stream every year. The stock data for a calendar year are estimated based on the number of households $H(t)$, penetration/saturation level per capita or per household $N(t)$, and the average weight (W) of the EEE. The e-waste generated in a calendar year can be calculated using Eq. (4.1) (Ikhlayel 2016).

$$\text{E-waste}(t) = \frac{H(t) \times N(t) \times W}{L} \tag{4.1}$$

However, this method also does not account for the life span distribution of EEE, rather it relies on the average lifetime of EEE.

4.4 The Sales/Approximation 2 Method

The sales method as defined by Bogar et al. (2019) or the Approximation 2 method as defined by UNEP (2007) uses sales statistics to estimate the total e-waste generated in a given year. It assumes a saturated market where e-waste in any given year is the same as the total new sales of electronics goods in the same year. Mathematically, it can be shown as Eq. (4.2).

$$\text{E-waste generated in year } t = \text{Sales in year } t \tag{4.2}$$

However, this method also does not account for the consumer behavior and the life span distribution of EEE.

4.5 Market Supply Method

The market supply method uses sales of a given EEE and its lifespan to estimate the e-waste generation. The method can be subdivided into three categories that combine the probability of reuse, storage, and disposal along with the lifespan for all these levels to predict the total e-waste generation.

4.5.1 Simple Delay

This method assumes that an EEE will enter the waste stream at the end of its lifespan. If the life span of a given product is "L" year, then that product will enter in the waste stream "n" year after the purchase. Mathematically, it can be shown as Eq. (4.3) (Bogar et al. 2019).

$$\text{E-waste } (t) = \text{Sales } (t - \text{L}) \tag{4.3}$$

However, this method also does not account for the consumer behavior and the variation of life span of EEE over the year.

4.5.2 Distribution Delay Method

The distribution delay method is a more accurate version of the simple delay method. It considers the probability of an EEE to enter the waste stream. The lifespan of a given product is not averaged but used in a probabilistic form (Miller et al. 2016; Wang et al. 2013). Weibull distribution is most widely used to determine the probabilistic obsolescence rate in any given year (Polák and Drápalová 2012) and will be discussed in Section 4.9 in detail. The amount of e-waste generated in any given year can be estimated using Eq. (4.4).

$$\text{E-waste}(t) = \sum_{i=t_0}^{t} \text{Sales}(t) \times L^{(p)}(i, t) \tag{4.4}$$

where,

t is the year for which e-waste quantity is being determined
t_o is the base year (year the product was sold)
$L^{(p)}(i, t)$ is the lifetime profile of EEE in a given year

4.5.3 Carnegie Mellon Method/Mass Balance Method

Carnegie Mellon method is a more detailed version of the simple delay method that uses different disposal levels of an EEE. It also uses different lifespans for these disposal scenarios and applies a percentage to estimate the movement of a product to one phase to the other phase and thus provide high accuracy. Different phases of products are reuse, storage, recycle, and landfill. It uses comprehensive material flows and their respective lifetime (Bogar et al. 2019; Wang et al. 2013). It requires a separate analysis for each product life phase. The e-waste generation can be determined using Eqs. (4.5)–(4.9) (Bogar et al. 2019; Miller et al. 2016).

$$O(t) = \text{Obsolete}(t) = \text{Sales}(t - L) \tag{4.5}$$

$$\text{RU}(t) = \text{Reuse}(t) = O(t) \times P_1 \tag{4.6}$$

$$S(t) = \text{Store}(t) = O(t) \times P_2 + \text{RU}(t - L_r) \times P_5 \tag{4.7}$$

$$\begin{aligned} R(t) = \text{Recycle}(t) &= O(t) \times P_3 + S(t - L_s) \times P_8 + \text{RU}(t - L_r) \times P_6 \\ &+ \text{RU}(t - L_r - L_s) \times P_5 P_8 \end{aligned} \tag{4.8}$$

$$\begin{aligned} Lf(t) = \text{Landfill}(t) &= O(t) \times P_4 + S(t - L_s) \times P_9 + \text{RU}(t - L_r) \times P_7 \\ &+ \text{RU}(t - L_r - L_s) \times P_5 P_9 \end{aligned} \tag{4.9}$$

where,

The lifespan of the product for first use is L, reuse is L_r, and store is L_s.
t is the year for which e-waste quantity is being determined.
P_1 to P_9 is the disposal percentages of obsolete reused, obsolete stored, obsolete recycled, obsolete landfilled, reused stored, reused recycled, reused landfilled, stored recycled, and stored landfilled, respectively.

4.6 Time-Step Method

The time-step method calculates the amount of e-waste generated based on the sales and stock data of EEE. E-waste amount in a calendar year is estimated by the difference between the sales of EEE in the current year and the change in EEE stock in previous two years (Bogar et al. 2019; UNEP 2007; Wang et al. 2013). The method can be shown as Eq. (4.10).

$$\text{E-waste}(t) = \text{Sales}(t) - \{\text{Stock}(t) - \text{Stock}(t - 1)\} \tag{4.10}$$

Table 4.1 Methods to estimate e-waste generation.

Method	Required data			Market type		Accuracy
	Sales	Stock	Lifespan	Saturated	Dynamic	
Econometric analysis						Low
Consumption and use		X	X	X		Low
Sales method	X			X		Low
Simple delay	X		X	X		Medium
Distribution delay	X		X	X	X	High
Carnegie Mellon	X	X	X	X	X	High
Timestep	X	X		X	X	High

Sources: Based on Chancerel (2010), Ikhlayel (2016), Polák and Drápalová (2012), UNEP (2007), and Wang et al. (2013).

4.7 Summary of Estimation Methods

There are various methods available to estimate the amount of e-waste generated in a given year. The applicability of these methods depends on the quality of available data. Most of these methods are based on the sales and stocks of EEE and consumer behavior. The reliability of these data for a given time frame and for a country will be different, and it will also define the usefulness of the method. The summary of the listed methods, the input variables, and the level of accuracy are listed in Table 4.1.

4.8 Lifespan of Electronic Products

Most of the highly accurate methods listed in Table 4.1 require the average lifespan or the distribution of the lifespan of any given EEE. The lifespan of a product is the time it takes for any product to be discarded by its owner. The estimated lifespan of a few electronics products is listed in Table 4.2. It shows that the lifespan of consumer electronics has decreased in the past few years. Ala-Kurikka (2015a,b), Ely (2014), Geere (2015), Ahmed (2016), and Umweltbundesamt (2016) have all suggested a similar trend, which is a major reason of the increasing e-waste quantities around the globe.

As mentioned in the Carnegie Mellon method, a product might go through various phases such as storage and reuse before being discarded; hence, modeling a realistic lifespan is quite challenging. Thiébaud-Müller et al. (2018) used detailed social media and emailed surveys in Swiss households to obtain service lifetime and storage lifetime of 10 electronic devices. Oguchi (2015) listed three methods to estimate the lifetime distribution of products. The methods are based on survey data to obtain the discard rate distributions for different products. The survey would be conducted at the recycling or disposal facilities to collect the discard distributions, whereas a questionnaire to the consumers will provide information about the

Table 4.2 Estimated lifespan of EEE.

Items	Average lifespan	
	Ely (2014)	**Abbondanza and Souza (2019)**
Flat-panel TV	7.4	4.3
Desktop computer	5.9	5.9
Laptop	5.5	4.0
Cellphones	4.7	2.3
Smartphones	4.6	1.8
Refrigerator	—	7.8
Washing machine	—	6.8
Basic printer	—	2.9

duration of use/reuse and domestic service lifetime. Oguchi et al. (2016) used these survey methods to estimate the lifespan of several types of EEE.

Santoso et al. (2019) suggested that the lifespan of a product can be classified into two types: average lifespan, which is often used in estimating e-waste in developing countries due to unavailability of data or unreliable data and distribution lifespan that provides more detailed lifespan but is difficult, time consuming, and expensive. A simplified method to obtain the distribution of lifespan is to use Weibull distribution with the Weibull lifespan and a distribution factor (Baldé et al. 2017; Environmental Protection Agency 2016; Miller et al. 2016; Oguchi and Fuse 2015; Polák and Drápalová 2012; Santoso et al. 2019; Sumasto et al. 2019; Tran et al. 2016).

Polák and Drápalová (2012) detailed the method to estimate the Weibull lifespan and the distribution factor using the lifespan data based on the survey results of 2008. Baldé et al. (2015a) and the Environmental Protection Agency (2016) have used this method and listed the Weibull lifespan and the distribution factor of various EEE. The values provided by Baldé et al. (2015a) have further been used to estimate the e-waste generation around the globe in 2014, 2016, and 2019 (Baldé et al. 2015b, 2017; Forti et al. 2020).

4.9 Global e-Waste Estimation

The global e-waste generation data are published as the Global E-waste Monitor. It is a collaboration between the International Telecommunication Union, the Sustainable Cycles Program that is cohosted by the United Nations University and the United Nations Institute for Training and Research, and the International Solid Waste Association. It provides the e-waste estimates around the world. The method for the e-waste estimates is published by Forti et al. (2018).

For the purpose of the measurement, the products are classified in groups labeled as UNU-KEYS with comparable average weights, material composition, end-of-life characteristics, and lifetime distributions. The full list along with their respective Harmonized System (HS) codes is presented in the document published by Forti et al. (2018). Harmonized System (HS) codes are commonly used for the purpose of import–export around the world.

The amount of total e-waste generated for each product in every category is estimated using the Distribution delay method as described in Section 5.5. It uses the amount of a product placed in the market in a given year along with the lifetime distribution of that product as shown in Eq. (4.11).

$$\text{E-waste generated}(n) = \sum_{t=t_0}^{n} \text{POM}(t) \times L^{(p)}(t, n) \tag{4.11}$$

where,

n is the year for which e-waste quantity is being determined
$POM(t)$ is the product sold (put-on-market) in a given year t
t_0 is the base year (year the product was sold)
$L^{(p)}(t, n)$ is discard-based lifetime profile of EEE sold in the year t

If the amount of a product placed/sold in the market is unknown, the POM can be estimated using Eq. (4.12) and domestic production, import and export information. If the domestic production is unknown, it is assumed to be zero. In case of imports and exports, the number of units of the product is converted to weight by using unit weight of products reported by Forti et al. (2018).

$$\text{POM}\ (t) = \text{Domestic production}\ (t) + \text{Imports}\ (t) - \text{Exports}\ (t) \tag{4.12}$$

Some corrections are needed for the outliers of the POM data. These corrections are needed if

- The POM value is too low, most likely due to unavailability of the domestic production data for a country where domestic production would be relatively large.
- The POM value is too high, mostly likely due to the misrepresentation of the reporting codes or units.
- The POM value is too high, mostly due to an obsolete product that is not manufactured anymore.

A more realistic value for POM is obtained either by using historic values or from a comparable country and then used for estimation purposes.

To account for the e-waste produced from product placed in market prior to the current year, the POM data are also estimated for past year based on available data or market trend. The future POM values are also predicted using the past, current, and future forecast of the PPP of the country provided by the world economic outlook and its ratio with the past and current POM.

The next step of estimating the e-waste generated is to estimate the lifetime distribution of the electrical and electronic products. Researchers have shown that the Weibull distribution is the most suitable probability distribution function to model the discard behavior of EEE (Baldé et al. 2017; Environmental Protection Agency

2016; Miller et al. 2016; Oguchi and Fuse 2015; Polák and Drápalová 2012; Santoso et al. 2019; Sumasto et al. 2019; Tran et al. 2016).

The simplified Weibull distribution of product discard-based lifetime profile is shown in Eq. (4.13) (Forti et al. 2018). The equation is mostly used for stable product with time independent lifetime. For these products, the variation in the shape and scale parameter is negligible overtime. The shape and scale parameters for various products are listed by Forti et al. (2018). The discard-based lifetime echoes the probability of an EEE entering the waste stream in a given year after its sale.

$$L^{(p)}(t,n) = \frac{\alpha}{\beta^{\alpha}}(n-t)^{\alpha-1}e^{-\left(\frac{n-t}{\beta}\right)^{\alpha}} \tag{4.13}$$

where,

L is the lifetime profile of an EEE product sold in a base year t
α is the distribution parameter, also known as the shape parameter
β is the average lifespan, also known as the scale parameter
n is the year for which the lifespan is being modeled

Figure 4.2 shows a sample plot for the discard probability and cumulative discard probability of various EEE using Eq. (4.13).

However, due to the technological developments (for example CRT screens), the lifetime of a EEE product could be time dependent meaning that the shape and scale parameters would change significantly overtime and the variations could not be neglected. In that case, the lifetime profile has to be modeled for each sales year using Eq. (4.14).

$$L^{(p)}(t,n) = \frac{\alpha(t)}{\beta(t)^{\alpha(t)}}(n-t)^{\alpha(t)-1}e^{-\left(\frac{n-t}{\beta(t)}\right)^{\alpha(t)}} \tag{4.14}$$

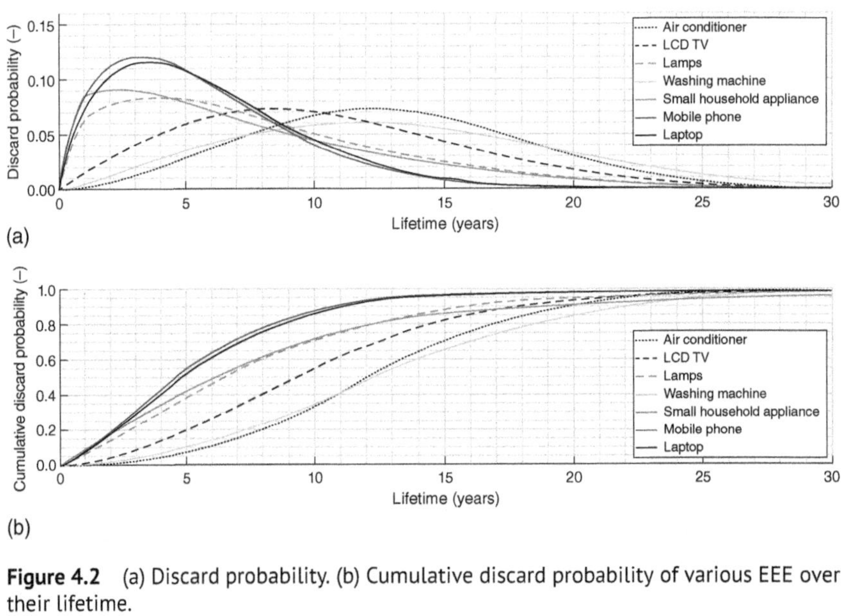

Figure 4.2 (a) Discard probability. (b) Cumulative discard probability of various EEE over their lifetime.

where,

$\alpha(t)$ is the time-varying distribution parameter
$\beta(t)$ is the time-varying scale parameter

In terms of e-waste collection, the total amount of e-waste generated is also the combination of e-waste collected through various programs and e-waste discarded can be represented by Eq. (4.15).

$$\text{E-waste generated}(n) = W_{\text{formal}} + W_{\text{other}} + W_{\text{bin}} + W_{\text{gap}} \quad (4.15)$$

where,

W_{formal} is the e-waste collected by formal system, W_{other} is the e-waste collected by other recycling streams, W_{bin} is the e-waste discarded in bins, and W_{gap} is the quantity for which the fate of e-waste is unknown.

After the understating of e-waste generation and collection, the e-waste collection rate could be estimated using Eq. (4.16).

$$\text{E-waste collection rate} = \frac{W_{\text{formal}}}{\text{Total e-waste generated}} \times 100 \quad (4.16)$$

The European Union Directive 2012/19/EU (2012) provided e-waste collection and recycling targets for member countries. They enforced an 85% collection and 80% reuse/recycling target for e-waste categories 1 and 4, 80% collection and 70% reuse/recycling target for e-waste category 2, 75% collection and 55% reuse/recycling target for e-waste categories 5 and 6, and 80% reuse/recycling target for e-waste category 3 from 15 August 2018.

References

Abbondanza, M.N.M. and Souza, R.G. (2019). Estimating the generation of household e-waste in municipalities using primary data from surveys: a case study of Sao Jose dos Campos, Brazil. *Waste Management* 85: 374–384. https://doi.org/10.1016/J.WASMAN.2018.12.040.

Ahmed, S. F. (2016). The global cost of discarded electronics waste. https://www.theatlantic.com/technology/archive/2016/09/the-global-cost-of-electronic-waste/502019 (accessed 17 March 2021).

Ala-Kurikka, S. (2015a). Electronic goods' life spans shrinking, study indicates. http://www.endseurope.com/article/39711/electronic-goods-life-spans-shrinking-study-indicates (accessed 17 March 2021).

Ala-Kurikka, S. (2015b). Lifespan of consumer electronics is getting shorter, study finds. https://www.theguardian.com/environment/2015/mar/03/lifespan-of-consumer-electronics-is-getting-shorter-study-finds (accessed 17 March 2021).

Baldé, C. P., Forti, V., Gray, V., Kuehr, R., and Stegmann, P. (2017). The global e-waste monitor – 2017. https://www.itu.int/en/ITU-D/Climate-Change/Documents/GEM2017/Global-E-waste Monitor 2017.pdf (accessed 17 March 2021).

Baldé, C. P., Kuehr, R., Blumenthal, K., et al. (2015a). E-waste statistics – Guidelines on classification, reporting and indicators. http://collections.unu.edu/view/UNU:5620 (accessed 17 March 2021).

Baldé, C.P., Wang, F., Kuehr, R., and Huisman, R. (2015b). The global e-waste monitor-2014. https://i.unu.edu/media/unu.edu/news/52624/UNU-1stGlobal-E-Waste-Monitor-2014-small.pdf (accessed 17 March 2021).

Bogar, Z.O., Capraz, O., and Güngör, A. (2019). An overview of methods used for estimating E-waste amount. In: *Electronic Waste Management and Treatment Technology* (eds. M.N.V. Prasad and M. Vithanage), 53–75. Butterworth-Heinemann, Elsevier https://doi.org/10.1016/B978-0-12-816190-6.00003-0.

Chancerel, P. (2010). *Substance Flow Analysis of the Recycling of Small Waste Electrical and Electronic Equipment: An Assessment of the Recovery of Gold and Palladium.* Berlin: Institut für Technischen Umweltschutz.

Ely, C. (2014). The life expectancy of electronics. http://www.cta.tech/Blog/Articles/2014/September/The-Life-Expectancy-of-Electronics (accessed 17 March 2021).

Environmental Protection Agency (2016). Electronic products generation and recycling in the United States, 2013 and 2014. https://www.epa.gov/smm/advancing-sustainable-materials-management-facts-and-figures-report (accessed 17 March 2021).

Forti, V., Baldé, C. P., Kuehr, R., and Bel, G. (2020). Global E-waste monitor 2020: quantities, flows and the circular economy potential. https://www.itu.int/en/ITU-D/Environment/Documents/Toolbox/GEM_2020_def.pdf (accessed 17 March 2021).

Forti, V., Baldé, K., and Kuehr, R. (2018). *E-waste Statistics: Guidelines on Classifications, Reporting and Indicators*, 2e. United Nations University http://collections.unu.edu/view/UNU:6477 (accessed 17 March 2021).

Geere, D. (2015). Electronic product lifespans are getting shorter. www.wired.co.uk/article/product-lifespans.

Ikhlayel, M. (2016). Differences of methods to estimate generation of waste electrical and electronic equipment for developing countries: Jordan as a case study. *Resources, Conservation and Recycling* 108: 134–139. https://doi.org/10.1016/J.RESCONREC.2016.01.015.

Kumar, A., Holuszko, M., and Espinosa, D.C.R. (2017). E-waste: an overview on generation, collection, legislation and recycling practices. *Resources, Conservation and Recycling* 122: 32–42. https://doi.org/10.1016/j.resconrec.2017.01.018.

Kusch, S., Hills, C.D., Kusch, S., and Hills, C.D. (2017). The link between e-waste and GDP – new insights from data from the Pan-European region. *Resources* 6 (2): 15. https://doi.org/10.3390/resources6020015.

Miller, T.R., Duan, H., Gregory, J. et al. (2016). Quantifying domestic used electronics flows using a combination of material flow methodologies: a US case study. *Environmental Science & Technology* 50 (11): 5711–5719. https://doi.org/10.1021/acs.est.6b00079.

Oguchi, M. (2015). Methodologies for estimating actual lifetime distribution of products. Product Lifetimes and the Environment, Nottingham Trent University, UK, 388–393. https://www.plateconference.org/methodologies-estimating-actual-lifetime-distribution-products.

Oguchi, M. and Fuse, M. (2015). Regional and longitudinal estimation of product lifespan distribution: a case study for automobiles and a simplified estimation method. *Environmental Science and Technology* 49 (3): 1738–1743. https://doi.org/10.1021/es505245q.

Oguchi, M., Tomohiro, T., Daigo, I., et al. (2016). Consumers' expectations for product lifetimes of consumer durables. *2016 Electronics Goes Green 2016+, EGG 2016*, 1–6. https://doi.org/10.1109/EGG.2016.7829850.

Polák, M. and Drápalová, L. (2012). Estimation of end of life mobile phones generation: the case study of the Czech Republic. *Waste Management* 32 (8): 1583–1591. https://doi.org/10.1016/J.WASMAN.2012.03.028.

Santoso, S., Yuri, M., Zagloel, T. et al. (2019). Estimating the amount of electronic waste generated in Indonesia: population balance model. *IOP Conference Series: Earth and Environmental Science* 219 (1): 012006. https://doi.org/10.1088/1755-1315/219/1/012006.

Sumasto, F., Zagloel, T.Y.M., Ardi, R., and Zulkarnain (2019). Estimation lifespan of home electronic appliances in Indonesia: the case study of Java Island. *IOP Conference Series: Earth and Environmental Science* 219 (1): 012008. https://doi.org/10.1088/1755-1315/219/1/012008.

The European Union (2012). Directive 2012/19/EU of the European Parliament and of the Council of the 4 July 2012 on waste electrical and electronic equipment (WEEE). *Official Journal of the European Union* L 197. https://eur-lex.europa.eu/legal-content/EN/TXT/PDF/?uri=CELEX:32012L0019&from=EN (accessed 17 March 2021): 38–71.

Thiébaud-Müller, E., Hilty, L.M., Schluep, M. et al. (2018). Service lifetime, storage time, and disposal pathways of electronic equipment: a Swiss case study. *Journal of Industrial Ecology* 22 (1): 196–208. https://doi.org/10.1111/jiec.12551.

Tran, H.P., Wang, F., Dewulf, J. et al. (2016). Estimation of the unregistered inflow of electrical and electronic equipment to a domestic market: a case study on televisions in Vietnam. *Environmental Science & Technology* 50 (5): 2424–2433. https://doi.org/10.1021/acs.est.5b01388.

Umweltbundesamt (2016). Lifetime of electrical appliances becoming shorter and shorter. https://www.umweltbundesamt.de/en/press/pressinformation/lifetime-of-electrical-appliances-becoming-shorter (accessed 17 March 2021).

UNEP (2007). *E-waste Volume 1: Inventory Assessment Manual*, 1e. United Nations https://wedocs.unep.org/bitstream/handle/20.500.11822/7857/EWasteManual_Vol1.pdf?sequence=3&isAllowed=y (accessed 17 March 2021).

Wang, F., Huisman, J., Stevels, A., and Baldé, C.P. (2013). Enhancing e-waste estimates: improving data quality by multivariate input–output analysis. *Waste Management* 33 (11): 2397–2407. https://doi.org/10.1016/J.WASMAN.2013.07.005.

World Bank (2019) Population, total. Available at: https://data.worldbank.org/indicator/sp.pop.totl (accessed 24 April 2019).

Yedla, S. (2016). Development of a methodology for electronic waste estimation: a material flow analysis-based SYE-waste model. *Waste Management & Research* 34 (1): 81–86. https://doi.org/10.1177/0734242X15610421.

5

Materials Used in Electronic Equipment and Manufacturing Perspectives*

Daniel D. München[1], Pablo Dias[2], and Hugo M. Veit[1]

[1] *Federal University of Rio Grande do Sul, Post Graduation Program in Materials, Mine and Metallurgy Engineering (PPGE3M), Department of Materials, 9500, Av. Bento Gonçalves, Porto Alegre – RS, 91509-900, Brazil*

[2] *University of New South Wales, Faculty of Science and Engineering, School of Photovoltaic and Renewable Energy Engineering, 229 Anzac Parade, Kensington, Sydney, NSW 2052, Australia*

5.1 Introduction

It is common ground that the invention of the transistor in 1947 by John Bardeen, Walter H. Brattain, and William B. Shockley provided the breakthrough for electronic equipment. Transistors, which had become less expensive than previous technology (e.g. vacuum tubes), did not burn out in service, and were much smaller and more reliable, had allowed electronics to enter all fields of human life and to provide functionalities that were unimaginable a few years before. Transistors then evolved to integrated circuits (ICs) that were manufactured with about 10 individual components on a 3 mm^2 silicon chip. By 1970, the number of components was up to 1000 on a chip of the same size. In the following year, the first microprocessor was introduced, which contained all the arithmetic, logic, and control circuitry required to perform the functions of a computer's central processing unit (CPU) (Scace 2016).

Then, by the mid-1980s, inexpensive microprocessors had stimulated computerization of an enormous variety of consumer products. Common examples included programmable microwave ovens and thermostats, washing and drying machines, television sets and self-focusing cameras, videocassette recorders and video games, telephones and answering machines, musical instruments, watches, and security systems. This widespread use of electronics has without a doubt generated advantages for society and individuals, since the performance of such devices had improved, not to mention the increase in functionalities. In addition, the following years allowed each new generation of products to reduce costs and advance in technology.

Consequently, for every new electrical and electronic equipment (EEE) with better features, aesthetics, and quality, a new discarded one has been created. According to StEP Initiative (StEP 2014), "*e-waste is a term used to cover items of all types of EEE*

* Received 27 December 2019.

Electronic Waste: Recycling and Reprocessing for a Sustainable Future, First Edition.
Edited by Maria E. Holuszko, Amit Kumar, and Denise C. R. Espinosa.

and its parts that have been discarded by the owner as waste without the intention of re-use." Nonetheless, final disposal, e.g. landfills or dumpsites, should not be thought as the first resource, given EEE can, in many cases, be prepared for reuse after repairs or remodeling.

Before 1990, EEE was disposed of as general waste, which corresponded to dumpsites in most countries. In reality, e-waste is chemically and physically distinct from other forms of municipal or industrial waste. It contains both valuable and hazardous materials that require special handling and recycling methods to avoid environmental contamination and detrimental effects on human health. Recycling can recover reusable components and base materials, especially copper (Cu) and precious metals. However, due to lack of infrastructure, high labor costs, and restrictive environmental regulations, rich countries tend not to recycle waste electrical and electronic equipment (WEEE). Instead, it is either landfilled, or exported from rich countries to poor countries, where it may be recycled using primitive techniques and ignoring worker safety (Robinson 2009).

Since the concept of resource conservation received considerable attention in the 1990s, European governments managed WEEE by creating laws to impose responsibility to manufacturers and importers (Fan et al. 2013). Two WEEE standard categorizations were created by the European Parliament (Parliament 2012), the first was valid from 13th August 2012 to 14th August 2018 as a transitional period, and the latter from 15th August 2018 onward. However, the most cited and used in the literature is still the classification used in the transition period. Both are listed in Table 5.1 and are based on similar characteristics, collection, function, and technology of EEE. Each category includes a whole set of examples and subcategories, which sum up more than a thousand appliances.

Over the years, the literature has been focusing on this categorization regarding WEEE characterization and recycling. Therefore, a WEEE generic composition

Table 5.1 WEEE categories according to the European Parliament.

Transitional period	After 15th August 2018
1. Large household appliances 2. Small household appliances 3. IT and telecommunications equipment 4. Consumer equipment and photovoltaic panels 5. Lighting equipment 6. Electrical and electronic tools (with the exception of large-scale stationary industrial tools) 7. Toys, leisure, and sports equipment 8. Medical devices (with the exception of all implanted and infected products) 9. Monitoring and control instruments 10. Automatic dispensers	1. Temperature exchange equipment 2. Screens, monitors, and equipment containing screens having a surface greater than 100 cm^2 3. Lamps 4. Large equipment (any external dimension more than 50 cm) 5. Small equipment (no external dimension more than 50 cm) 6. Small IT and telecommunication equipment (no external dimension more than 50 cm)

Source: Based on Parliament (2012).

Table 5.2 WEEE generic material composition (%).

Material	Composition (%)	
	Widmer et al. (2005)	De Meester et al. (2019)
Iron (Fe)/steel	47.9	20.0
Copper (Cu)	7.0	12.0
Aluminum (Al)	4.7	15.0
Nonferrous	1.0	—
Gold (Au)	—	0.05
Palladium (Pd)	—	0.01
Total metal	60.6	47.06
Flame-retardant plastics	5.3	52.94
Non-flame-retardants plastics	15.3	
Glass	5.4	—
Rubber	0.9	—
Wood and polymer	2.6	—
Ceramic	2.0	—
Printed circuit boards (PCB)	3.1	—
Other	4.6	—

Source: Based on Widmer et al. (2005).

investigation is mostly avoided due to the difficulty in sampling and the wide range of materials, which change constantly with the technology evolution. Widmer et al. (2005) have reported a WEEE generic composition estimate and it is shown in Table 5.2. This content emphasizes the massive presence of metals and plastics, whose contributions are around three-quarters of the mass of every WEEE unit generated. Another generic WEEE content is reported by de Meester et al. (2019), which not only brings a simplistic view of materials but also reveals the dominant presence of metals and plastics. However, when the investigations take into account the individual categories mentioned earlier, the composition tends to vary considerably, as seen in Figure 5.1 for six of them. As a result, Sections 5.2–5.7 explore materials and chemical elements that constitute the most common WEEE.

5.2 Large Household Appliances (LHA)

Appliances in this WEEE category subgroup appear in the literature quite often, e.g. home laundry machines, ovens and microwaves, refrigerators, stoves, and air treatment equipment, mostly in terms of volume generated. The most common materials found in these appliances are metals like Fe, Cu, Al, and steel alloys, diverse plastics and organic substances like rubber and fibers, glass and ceramics

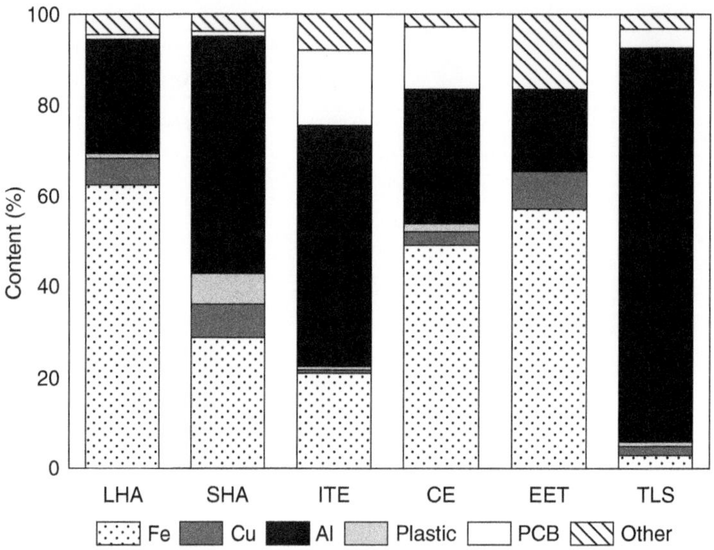

Figure 5.1 WEEE content in percentage. LHA, Large household appliances; SHA, small household appliances; ITE, IT and telecommunication equipment; CE, consumer electronics; EET, electric and electronic tools; and TLS, toys, leisure, and sport. Note: TLS content presented in the reference reaches values above 100%. Source: Adapted from Parajuly and Wenzel (2017).

mixed with low amounts of metals in printed circuit board (PCB). Figure 5.2 shows some examples of large household appliance (LHA) and their content, in which steel and plastics are dominant.

The composition of these LHA, also called as white goods, varies considerably from product to product, especially among different brands. In addition, the "greener" or "eco-friendly" these appliances become, resource efficiency increases. The critical materials are mainly found on PCB of this type of equipment, which may contain on average 20 ppm Pd, 160 ppm silver (Ag), and 38 ppm Au (Reuter et al. 2013).

5.3 Small Household Appliance (SHA)

With no more than 50 cm of size at any dimension, these appliances are the most common in terms of quantity due to the variety of features. Design is also a major factor influencing the sale of this category since consumers are often drawn into appearances. In addition, many have been produced as "not intended to be durable" items, leading to no reuse or upgrade.

By being small in size allows it to be easily disposed of in general municipal landfills or dumpsites, as opposed to, for example, a washing machine due to size and weight. Furthermore, the infrastructure worldwide that separately collects small WEEE is inadequate, and there is significant lack of expertise related to its

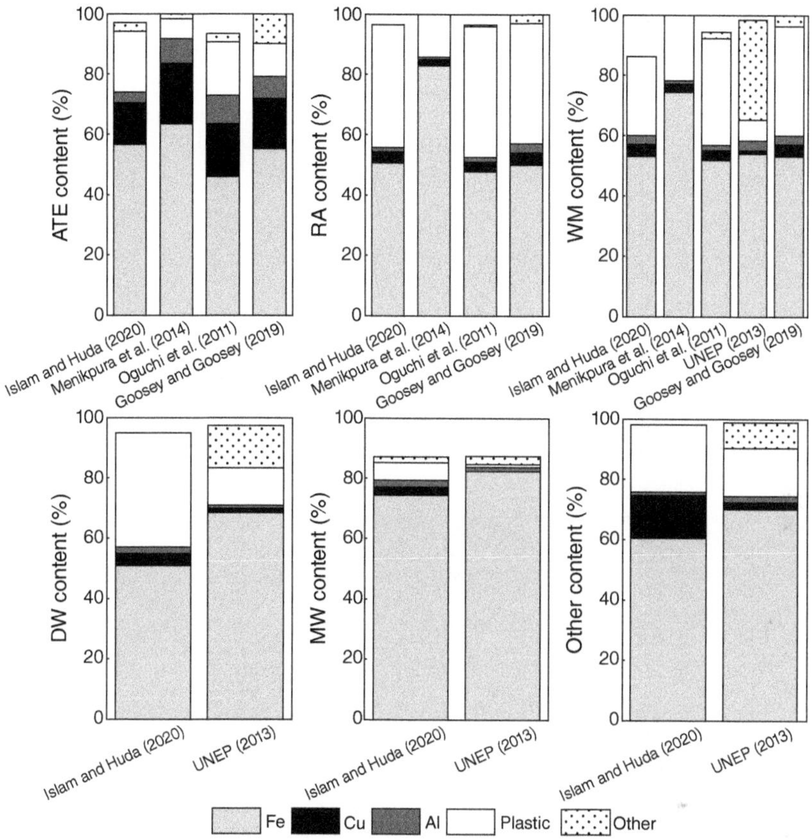

Figure 5.2 Most common large household appliances composition (%). ATE, air treatment equipment; DW, dishwasher; MW, microwave oven; RA, refrigerator appliance; WM, washing machine; Other, dryers, food preparation appliance. (Goosey and Goosey 2019; Islam and Huda 2020; Menikpura et al. 2014; Oguchi et al. 2011; Reuter et al. 2013). Source: Adapted from Goosey and Goosey (2019), Islam and Huda (2020), and Oguchi et al. (2011).

collection and treatment. The variety, complexity, and cost of small household appliance (SHA), along with low market demand, frequently makes reuse and recycling unviable, mostly due to the heterogeneous, compact, and difficult-to-dismantle products (Dimitrakakis et al. 2009).

Regarding composition, plastics assume the majority of weight in percentage, followed by ferrous and nonferrous metals as Table 5.3 expresses. Among plastics, acrylonitrile butadiene styrene (ABS) and polypropylene (PP) account for 75 wt% of SHA, whereas polystyrene (PS) and polyvinyl chloride (PVC) sum up about 15 wt%. Others were found in smaller quantities, e.g. polyamide (PA) and polycarbonate (PC) (Dimitrakakis et al. 2009).

Figure 5.3 shows a few examples of SHA and their composition by weight. It is observed that plastics and Fe dominate most of the cases, followed by Al and Cu. Other metals are contained in the PCB, but in small amounts.

Table 5.3 Small household appliances generic material composition in %.

Material composition	%	Material composition	%
Ferrous metals	8.99	Steel and ferrous metals	50.10
Nonferrous metals	8.22	Mix of plastic	31.30
Plastics	40.70	Motors	7.10
Rubber	0.69	PCB	3.00
Cables	7.55	Cables	2.90
PCB	0.52	Aluminum	1.40
Electronic components	18.41	Others	3.40
Others	14.91		

Source: Adapted from Biganzoli et al. (2015) and Dimitrakakis et al. (2009).

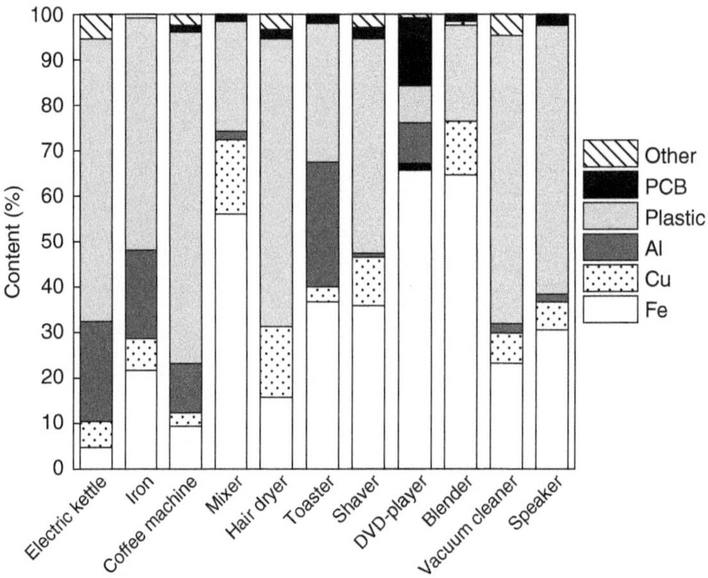

Figure 5.3 Small household appliances examples composition (%). PCB, printed circuit board. Source: Adapted from Parajuly and Wenzel (2017).

5.4 IT and Telecommunications Equipment

5.4.1 Computers and Notebooks

In general, computer equipment is assembled from more than 1000 materials, citing chlorinated and brominated substances, metals, biologically active materials, acids, and polymer additives. The production of semiconductors, PCB, hard disk drives, and monitors uses particular toxic chemicals like dioxins, polychlorinated

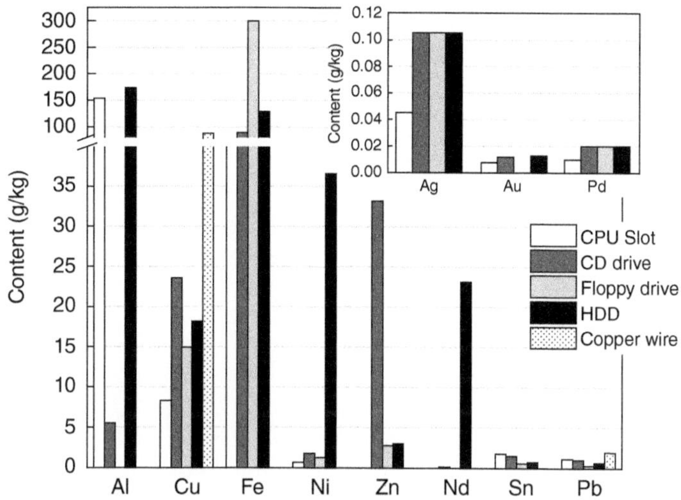

Figure 5.4 Precious and base computer parts metal content (g/kg). Source: Adapted from Habib Al Razi (2016).

biphenyls, cadmium (Cd), chromium (Cr), radioactive isotopes, and mercury (Hg).

In terms of mounting parts, a personal computer consists of three major units: main machine (i.e. CPU, power supplier, fan, PCB, DVD-CD drive, hard disk, casing, etc.), monitor, and keyboard. A mixture of metals can be found in waste computers, such as Cu, Fe, Al, lead (Pb), zinc (Zn), nickel (Ni), tin (Sn), and even precious metals and rare-earth metals, as shown in Figure 5.4. Despite being small in absolute weight, Au, Ag, and Pd can be concentrated in an expressive amount when considering tons of this waste. In comparison, mineral ores that contain these metals have a concentration, in percentage, 0.0001 of Au, 0.01 of Ag, and 0.001 of Pd (Chagnes et al. 2016). In addition, a mixture of polymers, such as polyethylene (PE), PP, polyurethane, and others, are also present. Computers also contain ceramic materials, such as glass, and other inorganic and organic materials. In terms of other valuable metals, Cu is concentrated in significant amount for recycling purposes.

Table 5.4 shows the difference between desktop and notebook regarding the main composition materials. While plastic is nearly absent in desktop computer, in notebook it represents more than a quarter of its weight. The opposite situation is observed in relation to ferrous material given it accounts for approximately half its weight in desktop computers.

5.4.2 Monitors and Screens

Televisions, PC monitors, laptops, tablets, surveillance systems, and other applications have monitors or screens constituted mainly of light-emitting diode (LED), liquid-crystal display (LCD), or cathode-ray tube (CRT) technologies, which contain

Table 5.4 Notebook and desktop computer material content (%).

Materials	Content (%)	
	Notebook	**Desktop**
Plastic	25.8	2.8
Ferrous material	19.5	47.2
Aluminum material	2.4	—
Copper cable and material	1.0	0.9

Source: Adapted from Oguchi et al. (2011).

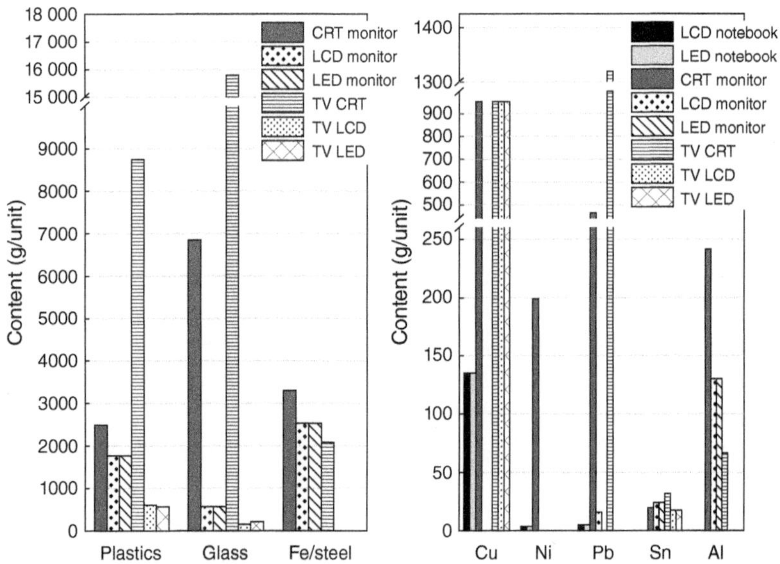

Figure 5.5 Content of most common materials/metals in screens technology and appliances. Source: Adapted from Cucchiella et al. (2015).

specific chemical elements to generate images. Although the evolution in technology has originated successive replacements from CRT to LCD and from LCD to LED in monitors and TV, all three are present in waste flows.

Each of these technologies has been characterized, and variations in chemical compositions have been observed in Figure 5.5. The CRT TV and monitor units are mainly composed of different kinds of glass: panel glass (strontium – Sr/barium – Ba oxides), funnel glass (leaded that covers the CRT unit), and neck glass (highly leaded that covers the electron gun) (Habib Al Razi 2016). Figure 5.6 expresses each of these three glasses' composition. Aside from the glass, the CRT units contain Fe/steel and an expressive amount of plastic, mostly due to the tube-shaped set, which sums up to approximately 17.5 and 16.5 wt% in TV and monitors, respectively. The variety of polymers is reported to sum up to about 10 different types in CRT monitors

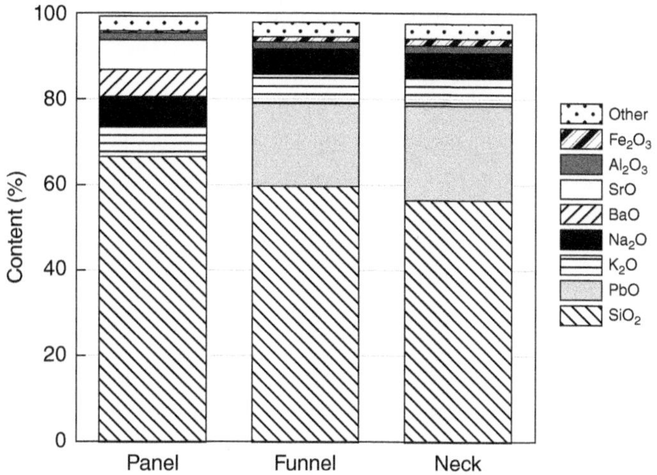

Figure 5.6 CRT monitor oxide content (%). Source: Adapted from Dias et al. (2018).

and televisions, mostly ABS, polycarbonate blends (PC/ABS), and PS (Wang and Xu 2014). The PCB and other components are reported to account for 8 and 13 wt%, respectively, for CRT monitors (Dias et al. 2018). It is also noticed that as the technology evolved (when flat screens were introduced), lower amount of plastics as well as less glass were used.

Indium (In) is the most important material present in LCD screens regarding recycling value. In addition, LCD screens contain other valuable materials (e.g. Cu, Au, and Ag) located in the PCB. LED screens also add gallium (Ga), germanium (Ge), and other rare metals to materials embedded into LCD, present in high concentrations in small components (Cucchiella et al. 2015). The differences in the chemical element content from one technology to the other are mainly concentrated in critical and hazardous metals. For example, from LCD to LED TV, the amount of cerium (Ce) has decreased from 0.005 to $<$0.001 g/unit; the same tendency was observed for europium (Eu), lanthanum (La), praseodymium (Pr), terbium (Tb), and yttrium (Y). While CRT monitors and TV contain on average 464 and 1319 g/unit of Pb, respectively, LCD and LED monitors reported 16 g/unit and zero, respectively (Cucchiella et al. 2015). Cu content remains constant throughout technologies, which is interesting from the point of view of recycling.

5.4.3 Mobile Phones (MP)

Subjected to continuous technological evolution and better services in even more sophisticated devices, mobile phones (MPs) are a part of the lifestyle of increasing numbers of people everywhere, leading to obsolescence and new acquisitions of smaller, lighter, or more modern ones. A 2016 report estimated that 80% of individuals own a mobile phone in the developing world on average, and that this figure can rise up to 98% in high-income countries (World Bank Group and Flagship 2016).

Materials contained in MP are variable based on the model and its manufacturer, with no fixed design or list of contents applicable. For this reason, the sampling characterizations may also be misleading since different/several substances might be used as additives in very low quantities or traces by different manufacturers in the production of microelectronic components, for example (Fontana et al. 2019). However, a typical MP comprises PCB, battery, front and back cases, and a display unit. Studies to date have focused on the characterization of metals within PCB, since they have the highest metal concentrations and are relatively easy to disassemble compared to other components found in MP (Liu et al. 2019). Table 5.5 shows the weight average percentage of each of these mentioned parts of a mobile phone, in which plastics is the dominant material.

Figures 5.7 and 5.8 show metal contents reported by different authors. Both upright graphs are the continuity of the main graphs but with smaller content scale.

Table 5.5 Average material parts (in wt%) of mobile phones.

Mobile phone parts	Content (wt%)
Plastic materials	33
Batteries	24
Electronic parts	23
Metal parts	11
Displays	9

Source: Based on Fontana et al. (2019).

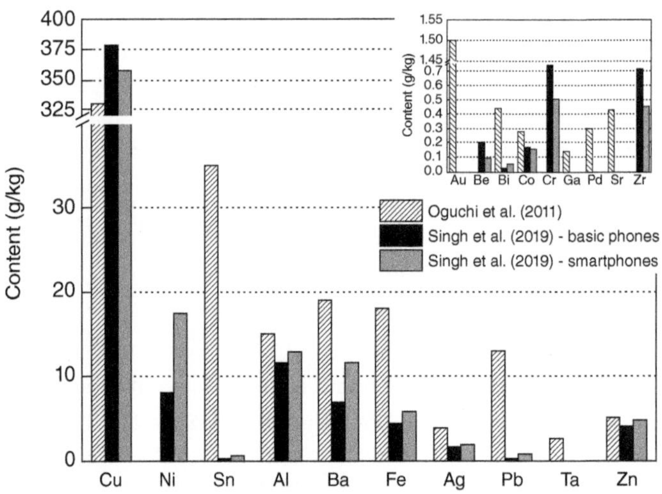

Figure 5.7 Mobile phones content (g/kg). Source: Adapted from Oguchi et al. (2011) and Singh et al. (2019).

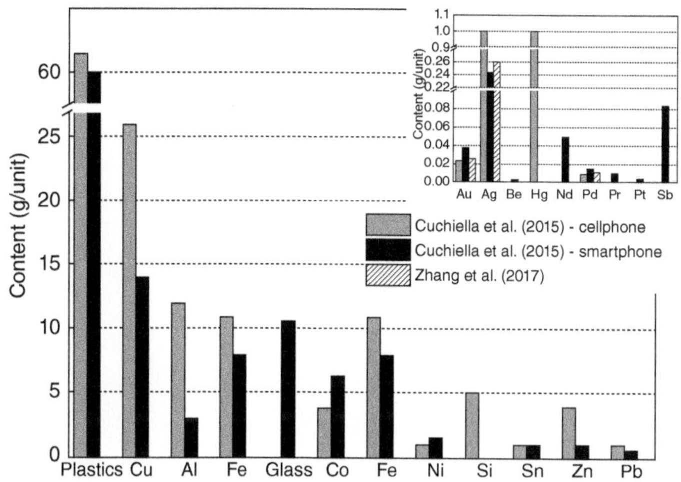

Figure 5.8 Mobile phone content (g/unit). Source: Adapted from Cucchiella et al. (2015) and Zhang et al. (2017).

5.4.4 Printed Circuit Boards (PCB)

In spite of PCB not being an official category in the EU directive nor belonging exclusively to IT equipment, they were given their own subchapter because of their economic relevance, material diversity, and close relation with IT equipment.

Projected for mechanical support and electrical connection to the various electronic components, PCB are essential part of electronic equipment. Thermoplastic material and epoxy resin with flame retardants work as substrate, in which laminated Cu-sheets are etched. The complex mixture of polymers (plastics), silicates (Si-based), nonferrous and ferrous metals, with variable concentrations, is a challenge for recycling. The PCB weight fraction in relation to the total device varies from 2% for large electronic devices to 11% for laptop computers and up to 22% for mobile phones (Işildar et al. 2017).

Metal concentrations in PCB depend on the source, type of the board, manufacturer, and period of production, since manufacturing techniques, device-specific board designs, and soldering technologies are the main factors of diversity in PCB. Nevertheless, Cu is the predominant metal, varying in content from 15 to 35 wt% (Işildar et al. 2017). As an estimate, PCB are generally made up of 40% metals (mainly Cu, Pb, and Ni), 30% organic materials, and various inorganic oxides (ceramics) as a balance (Pietrelli et al. 2019). Other report summarizes PCB average content as Table 5.6.

The electronic components contain different metals of interest (arsenic – As, In, Ga, Ge, antimony – Sb, selenium – Se, Si, tantalum – Ta, tellurium – Te, titanium – Ti) and are attached to the conductive substrate through leads typically made of Pb/Sn alloy (up to 4–6% of the total weight) or occasionally of Au or Pd. Precious metals are

Table 5.6 PCB typical average content (%).

Materials	Content (%)
Fe, Ni, Si, and Al	29
Cu	22–27
Ag	0.2
Au	0.008–0.035
Fiber glass	30
Epoxy resin	15

Source: Based on Adie et al. (2017).

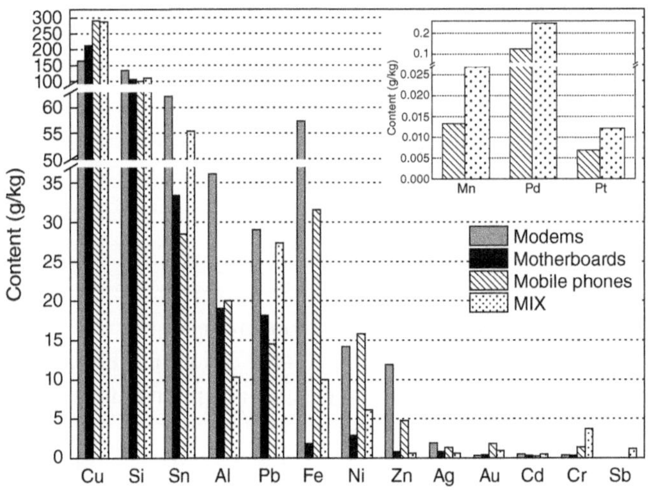

Figure 5.9 PCB content in g/ton of different appliances. Source: Adapted from Pietrelli et al. (2019).

used as contact material in lead relays, switchers, or sensors. Ni is used in its pure form and increasingly used in contacts as additive. Toxic metals, such as Pb, Cd, and Hg, are known to be present in the PCB of PCs (Pietrelli et al. 2019).

Figure 5.9 shows a PCB content, in g/kg, of internet modems, computer motherboards, and mobile phones. The "MIX" content is originated from unknown sources, according to Pietrelli et al. (2019). The amount of Cu is expressive, as expected, reaching nearly 300 g/kg, followed by other already-mentioned metals. Pb content is significantly high in the "MIX" fraction regarding its hazardous potential. Ag and Au have also occurred in significant amounts, nearly 1.7 g/kg of PCB. Figure 5.10 shows parts of a desktop computer PCB content, in which memory modules concentrate the majority of Au, while Cu is spread over the different parts of the board.

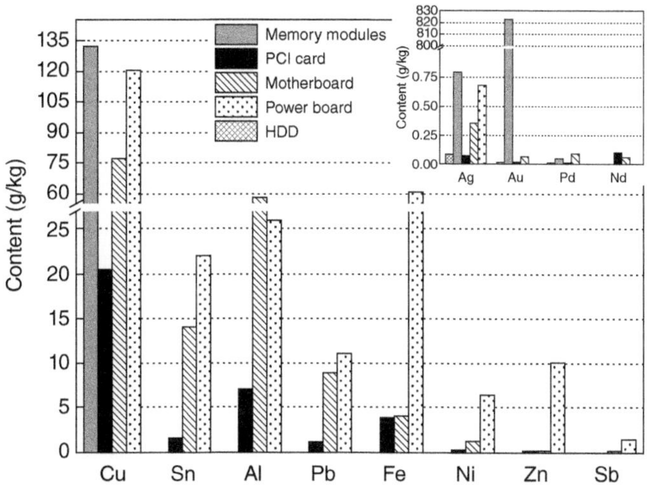

Figure 5.10 Parts of a desktop computer PCB content in g/kg. Source: Adapted from Habib Al Razi (2016).

5.5 Photovoltaic (PV) Panels

These EEE have been recently spotlighted regarding recycling due to the fact that the first significant volumes of photovoltaic (PV) panels installed are now being discarded. Undergoing an evolution in technology, PV panels are nowadays classified in three generations: the first comprehends the monocrystalline or polycrystalline silicon panels, which differ in the purity degree of Si and, consequently, in the efficiency of energy conversion. Second generation is an arrangement of thin films deposited on a substrate (glass, stainless steel, etc.) with one or more layers. Different materials can be used to produce the thin film, including amorphous silicon (a-Si), cadmium telluride (CdTe), or copper indium gallium selenide (CIGS). Finally, the third generation includes emerging technologies such as concentrator photovoltaics (CPVs), dye-sensitized solar cells, organic cells, and hybrid cells (Padoan et al. 2019).

In terms of composition, crystalline silicon PV modules are made from the following materials, in order of mass: glass, Al frame, ethylene vinyl acetate (EVA) copolymer transparent encapsulating layer, photovoltaic cells, installation box, Tedlar® protective foil, and assembly bolts (Dias et al. 2017). Although Cucchiella et al. (2015) had not specified which type of PV panel was analyzed, the reported amount of each part, in grams/unit, is shown in Table 5.7.

The metals found by different authors in PV panels are shown in Figure 5.11. Their reports summed up 14 metals, according to the specific thin films used in layers. The evident discrepancy between the same metal contents reported by different authors can be justified due to the fact that each author considers one or another fraction of the panel, such as Al frame, metal contacts, and junction box (Padoan et al. 2019). Such fractions may or may not be included in the characterized material. In addition, it is known that the photovoltaic market is extremely heterogeneous, and the

Table 5.7 PV panel content in g/unit.

Metals/materials	Content (g/unit)	Metals/materials	Content (g/unit)
Al	1 370	Sn	0.120
Si	226	Ga	0.119
Cu	78	Se	0.119
Cd	0.407	In	0.119
Te	0.406	Glass	6 915
Zn	0.400	Plastics	1 172
Mo	0.295		

Source: Based on Cucchiella et al. (2015).

evolution of the PV technology moves toward decreasing the panel metal content, for example silver (VDMA 2019).

5.6 Lighting Equipment

Since 2010, LED have become affordable enough for general lighting and, as a result, these are expected to replace fluorescent lamps (FLs) and become the dominant lighting technology (Qiu and Suh 2019). However, FLs have still remained as the majority of lighting source in the recycling streams. Table 5.8 shows the main components found in FL lighting equipment and lamps, in which phosphorous powder is the major interest material due to the rare-earth elements contained therein, such as Eu, La, Ce, and Y (Figure 5.12). The replacement for LED lamps, which are more efficient and mercury-free, is not fully completed yet due to higher costs. LED lamps use also REE, making them possible targets for urban mining as well. The content of rare earths in LED is, however, lower when compared to FL, as Figure 5.13 expresses. The content of Y in LED used for background illumination in TV was reported to be 4.9 mg metal/display in contrast to 110 mg metal/display of cold cathode fluorescent lamps (CCFLs). For Eu the content is 0.09 and 8.1 mg metal/display, respectively (Tunsu et al. 2015).

5.7 Toys, Leisure, and Sport

Since the introduction of integrated circuits, toys have undergone a high degree of innovation and addition of technological features. A few examples include puzzles, board games, dolls, or sound and light toys that allow interaction with kids unimagined earlier, as well as educational toys. However, this integration has not been taking into account the reuse and end-of-life perspective. As for the majority of small WEEE, they have been produced as “disposable” items, not having been

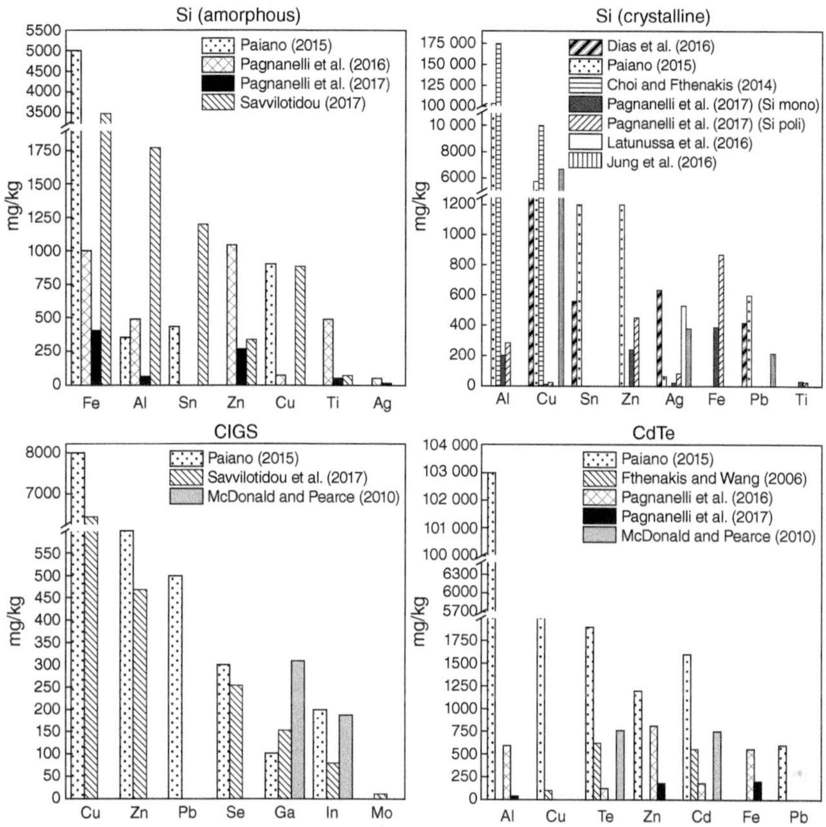

Figure 5.11 Photovoltaic panels composition, in mg/kg. (Choi and Fthenakis 2014; Dias et al. 2016; Fthenakis and Wang 2006; Jung et al. 2016; Latunussa et al. 2016; McDonald and Pearce 2010; Pagnanelli et al. 2016; Pagnanelli et al. 2017; Paiano 2015; Savvilotidou et al. 2017). Source: Adapted from Choi and Fthenakis (2014), Fthenakis and Wang (2006), Jung et al. (2016), and Pagnanelli et al. (2016).

Table 5.8 Components of lighting equipment (wt%).

Components	Content (wt%)
Al	4.0
Glass	80.0
Phosphoric powder	8.0
Ferrous material	5.0
Other	3.0

Source: Based on Biganzoli et al. (2015).

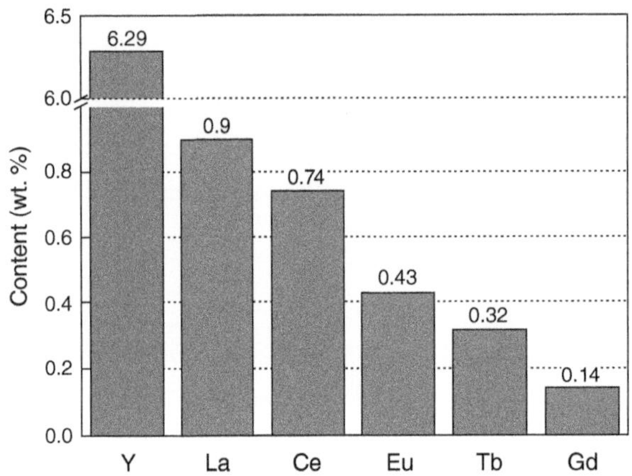

Figure 5.12 Rare-earth content in phosphorous powder of fluorescent lamps (in wt%). Source: Based on Yurramendi et al. (2019).

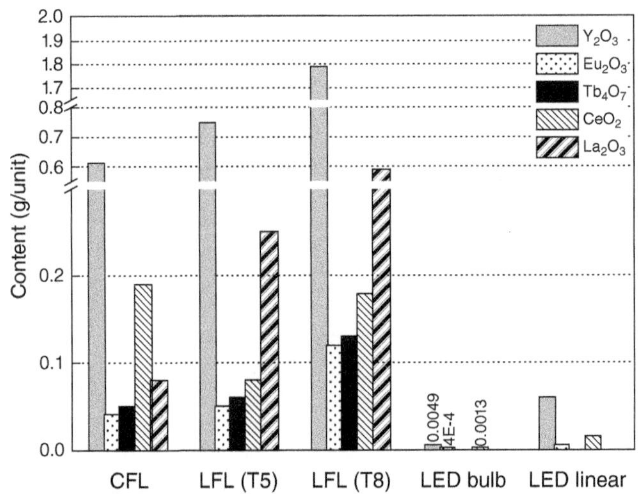

Figure 5.13 Rare-earth oxides content (in g/unit) according to lighting type. CFL, compact fluorescent lamp; LFL (T5), linear fluorescent lamp 5/8 in. diameter; LFL T8, linear fluorescent lamp 8/8 in. diameter; LED, light-emitting diode. Source: Adapted from Qiu and Suh (2019).

manufactured with upgradeability and reuse in mind (Pérez-Belis et al. 2013). This category, however, represented as little as 0.1 wt% of the arising WEEE in the EU in 2005 (Goosey and Goosey 2019).

It is found that electronic components (PCB, wiring, switches, loudspeakers, LED, etc.) respond for 12.07 wt%, whereas plastic, ferrous metals, nonferrous metals, and other materials (rubber, textiles, and others) summed 87.92 wt%. Polymers such as PS, PP, and ABS are commonly found in toys. The weight of each category is shown in Table 5.9. A few other metals, such as Cr, As, Cd, Pb, and Hg, are found in some plastics parts, but quantities are within the limits established in restriction on hazardous substances (RoHSs) Directive (Pérez-Belis et al. 2013), which is

Table 5.9 Toys content in weight %.

Materials	Content (%)	Materials	Content (%)
Ferrous	4.44	*Plastics*	
Nonferrous	0.65	ABS	11.85
Rubber	1.87	PP	16.50
Textile	2.09	PS	47.52
Others	11.27	PE	3.81

Source: Based on Pérez-Belis et al. (2013).

responsible for restricting and setting the maximum amount of substances allowed in EEE. Toys and tools have also ranked among the most frequent brominated equipment, where more than 70% of equipment were found to contain the halogen (Hennebert 2017).

5.8 Future Trends in WEEE – Manufacturing, Design, and Demand

The global production of e-waste will change as economies grow and new technologies are developed; however, the concern with supply and demand of specific chemical elements and compounds, in recent years, has emerged as an issue. For instance, rare-earth metals, platinum group metals, Ga, niobium (Nb), Ta, In are listed as critical hi-tech metals by the USA, EU, Japan, and China, mostly due to an estimate that suggests, for example, that the use of LED in background lighting and in general will increase the demand for Ga 12 times by 2050 (Ueberschaar et al. 2017). The majority of these metals are usually produced as by-products of basic metals and will probably face scarcity of reserves in the near future. In addition, the case of rare earths demands special handling, since their minerals contain also radioactive elements.

The most prominent technology, that is pulling global demand for rare earths, is permanent magnets. Neodymium–iron–boron (NdFeB) magnets are the strongest permanent magnets commercially available today. They are used in a wide variety of applications, such as direct drive wind turbines, electric and hybrid vehicles, and are a key component of the transition toward a low-carbon energy economy (Rollat et al. 2016). Because of their unique physical, chemical, magnetic, luminescent properties, rare-earth elements are responsible for many technological advances, such as greater efficiency (performing at reduced energy consumption), miniaturization, speed, durability, and thermal stability. In recent years, their demand is particularly on rise in energy-efficient gadgets, which are faster, lighter, smaller, and more efficient. These technologies even assist in making analytical instruments smaller and more efficient (Balaram 2019). Currently (2020), few substitutes for rare earth-based materials are commercially available.

In contrast to research on substitutes, research on ways of reducing rare-earth dependence has yielded more promising results. Governments and manufacturers are searching ways of diminishing or avoiding the need for rare earths (Smith Stegen 2015).

A new technology in TV, called QLED, is entering markets and being recognized as key materials for the next generation in displays due to flexible and ultra-high definition characteristics, and for color-controllable and healthy solid-state lighting. The QLED display consists of ITO anodes, lithium-doped aluminum (Li:Al)/Au cathodes, and CTLs with a QD layer, which are sandwiched between the anode and the cathode (Kim et al. 2017).

An evaluation regarding PCB of TV from 1980 to 2005 reports a content decrease in Cu, which varied between 5.1 wt% in 1980 and 12.8 wt% in the mid-1990s and decreased afterwards in PCB of newer model TV. Ag content also shows a decreasing trend, while Au composition seems to fluctuate within years. The main reason is due to thinner metal layers in new products with improved technologies and more compact designs (Adie et al. 2017).

Alongside, the concept of eco-design is getting more attention worldwide, since it supports the innovation of sustainable products. The practice focuses on minimizing material consumption and extending the lifespan of materials. At the same time, there is interest in developing new polymers for EEE that are not derived from petrochemical-based precursors, e.g. biodegradable polymer such as polylactic acid (PLA). This idea of designing with the end of life in mind (eco-design, design for disassembly, and others) has gained thrust especially due to the implementation of producer responsibility legislation such as the WEEE and RoHS Directives, since the manufacturers can reduce their liability at the end of a product lifespan by investing in its beginning (Goosey and Goosey 2019). A bamboo case used by Asus, recycled parts embodied by Dell, and As- and Hg-free in displays by Apple are a few examples of eco-design applied by IT manufacturers.

A report suggests that product eco-design, which uses tools like "Product Data Management (PDM) decisions" and "Design for X," is better in terms of product optimal recovery strategy than recovery process technologies, like X-ray equipment that assesses the material composition and post-shredder Separation Technology. This means that the EU policy to make original equipment manufacturer primarily responsible for recovery, as implemented by the WEEE Directive, is appropriate, but more incentives are needed to reward product eco-design. From a recovery value standpoint, remanufacturing should be the prime recovery strategy, but it requires a high quality of returns. Moreover, the cost of misclassification can be high, so that for remanufacturing a PDM with reliable return quality information is needed (Zuidwijk and Krikke 2008).

Despite the benefits of eco-design, assembly for disassembly and others, EEE manufacturers from outside EU, where nonrestricted or even nonexistent environmental laws are applied, tend not to follow discretions in relation to their manufacturing processes. In most cases, singularities and specific designs are obstacles to the end-of-life WEEE that could be reused, remodeled, or easier to be recycled.

References

Adie, G.U., Sun, L., Zeng, X. et al. (2017). Examining the evolution of metals utilized in printed circuit boards. *Environmental Technology (United Kingdom)* 38 (13–14): 1696–1701. https://doi.org/10.1080/09593330.2016.1237552.

Balaram, V. (2019). Rare earth elements: a review of applications, occurrence, exploration, analysis, recycling, and environmental impact. *Geoscience Frontiers* https://doi.org/10.1016/j.gsf.2018.12.005.

Biganzoli, L., Falbo, A., Forte, F. et al. (2015). Mass balance and life cycle assessment of the waste electrical and electronic equipment management system implemented in Lombardia Region (Italy). *Science of the Total Environment* https://doi.org/10.1016/j.scitotenv.2015.04.041.

Chagnes, A., Cote, G., Ekberg, C. et al. (2016). WEEE recycling: research, development, and policies. In: *WEEE Recycling: Research, Development, and Policies* (eds. A. Chagnes, G. Cote, C. Ekberg, et al.). Amsterdan, Netherlands: Elsevier Inc.

Choi, J.-K. and Fthenakis, V. (2014). Crystalline silicon photovoltaic recycling planning: macro and micro perspectives. *Journal of Cleaner Production* 66: 443–449. https://doi.org/10.1016/j.jclepro.2013.11.022.

Cucchiella, F., D'Adamo, I., Lenny Koh, S.C., and Rosa, P. (2015). Recycling of WEEEs: an economic assessment of present and future e-waste streams. *Renewable and Sustainable Energy Reviews* 51: 263–272. https://doi.org/10.1016/j.rser.2015.06.010.

De Meester, S., Nachtergaele, P., Debaveye, S. et al. (2019). Using material flow analysis and life cycle assessment in decision support: a case study on WEEE valorization in Belgium. *Resources, Conservation and Recycling* https://doi.org/10.1016/j.resconrec.2018.10.015.

Dias, P., De Oliveira, E., and Veit, H. (2018). Lead hazard evaluation for cathode ray tube monitors in Brazil. *Brazilian Journal of Chemical Engineering* 35 (1): 43–49. https://doi.org/10.1590/0104-6632.20180351s20160367.

Dias, P., Javimczik, S., Benevit, M., and Veit, H. (2017). Recycling WEEE: polymer characterization and pyrolysis study for waste of crystalline silicon photovoltaic modules. *Waste Management* 60: 716–722. https://doi.org/10.1016/j.wasman.2016.08.036.

Dias, P., Javimczik, S., Benevit, M. et al. (2016). Recycling WEEE: extraction and concentration of silver from waste crystalline silicon photovoltaic modules. *Waste Management* 57: 220–225.

Dimitrakakis, E., Janz, A., Bilitewski, B., and Gidarakos, E. (2009). Small WEEE: determining recyclables and hazardous substances in plastics. *Journal of Hazardous Materials* https://doi.org/10.1016/j.jhazmat.2008.04.054.

Fan, S.K.S., Fan, C., Yang, J.H., and Liu, K.F.R. (2013). Disassembly and recycling cost analysis of waste notebook and the efficiency improvement by re-design process. *Journal of Cleaner Production* 39: 209–219. https://doi.org/10.1016/j.jclepro.2012.08.014.

Fontana, D., Pietrantonio, M., Pucciarmati, S. et al. (2019). A comprehensive characterization of end-of-life mobile phones for secondary material resources identification. *Waste Management* https://doi.org/10.1016/j.wasman.2019.08.011.

Fthenakis, V.M. and Wang, W. (2006). Extraction and separation of Cd and Te from cadmium telluride photovoltaic manufacturing scrap. *Progress in Photovoltaics: Research and Applications* 14 (4): 363–371. https://doi.org/10.1002/pip.676.

Goosey, E. and Goosey, M. (2019). The materials of waste electrical and electronic equipment. In: *Waste Electrical and Electronic Equipment (WEEE) Handbook* (eds. V. Goodship, A. Stevels and J. Huisman), 231–262. Elsevier https://doi.org/10.1016/B978-0-08-102158-3.00008-2.

Habib Al Razi, K.M. (2016). Resourceful recycling process of waste desktop computers: a review study. *Resources, Conservation and Recycling* 110: 30–47. https://doi.org/10.1016/j.resconrec.2016.03.017.

Hennebert, P. (2017). WEEE plastic sorting for bromine content is essential to enforce EU regulation. *International Waste Management and Landfill Symposium (Sardinia 2017)*, Cagliari, Italy.

Işildar, A., Rene, E.R., van Hullebusch, E.D., and Lens, P.N.L. (2017). Electronic waste as a secondary source of critical metals: management and recovery technologies. *Resources, Conservation and Recycling*, December 2016. doi:https://doi.org/10.1016/j.resconrec.2017.07.031.

Islam, M.T. and Huda, N. (2020). Assessing the recycling potential of "unregulated" e-waste in Australia. *Resources, Conservation and Recycling* https://doi.org/10.1016/j.resconrec.2019.104526.

Jung, B., Park, J., Seo, D., and Park, N. (2016). Sustainable system for raw-metal recovery from crystalline silicon solar panels: from noble-metal extraction to lead removal. *ACS Sustainable Chemistry & Engineering* 4 (8): 4079–4083.

Kim, J., Shim, H.J., Yang, J. et al. (2017). Ultrathin quantum dot display integrated with wearable electronics. *Advanced Materials* 29 (38): 1700217. https://doi.org/10.1002/adma.201700217.

Latunussa, C.E.L., Ardente, F., Blengini, G.A., and Mancini, L. (2016). Life cycle assessment of an innovative recycling process for crystalline silicon photovoltaic panels. *Solar Energy Materials and Solar Cells* 156: 101–111.

Liu, W., Ford, P., Uvegi, H. et al. (2019). Economics of materials in mobile phone preprocessing, focus on non-printed circuit board materials. *Waste Management* https://doi.org/10.1016/j.wasman.2019.01.044.

McDonald, N.C. and Pearce, J.M. (2010). Producer responsibility and recycling solar photovoltaic modules. *Energy Policy* 38 (11): 7041–7047. https://doi.org/10.1016/j.enpol.2010.07.023.

Menikpura, S.N.M., Santo, A., and Hotta, Y. (2014). Assessing the climate co-benefits from waste electrical and electronic equipment (WEEE) recycling in Japan. *Journal of Cleaner Production* https://doi.org/10.1016/j.jclepro.2014.03.040.

Oguchi, M., Murakami, S., Sakanakura, H. et al. (2011). A preliminary categorization of end-of-life electrical and electronic equipment as secondary metal resources. *Waste Management* https://doi.org/10.1016/j.wasman.2011.05.009.

Padoan, F.C.S.M., Altimari, P., and Pagnanelli, F. (2019). Recycling of end of life photovoltaic panels: a chemical prospective on process development. *Solar Energy* https://doi.org/10.1016/j.solener.2018.12.003.

Pagnanelli, F., Moscardini, E., Abo Atia, T., and Toro, L. (2016). Photovoltaic panel recycling: from type-selective processes to flexible apparatus for simultaneous treatment of different types. *Mineral Processing and Extractive Metallurgy* 125 (4): 221–227.

Pagnanelli, F., Moscardini, E., Granata, G. et al. (2017). Physical and chemical treatment of end of life panels: an integrated automatic approach viable for different photovoltaic technologies. *Waste Management* 59: 422–431. https://doi.org/10.1016/j.wasman.2016.11.011.

Paiano, A. (2015). Photovoltaic waste assessment in Italy. *Renewable and Sustainable Energy Reviews* 41: 99–112. https://doi.org/10.1016/j.rser.2014.07.208.

Parajuly, K. and Wenzel, H. (2017). Potential for circular economy in household WEEE management. *Journal of Cleaner Production* https://doi.org/10.1016/j.jclepro.2017.03.045.

Parliament, E. (2012). Directive 2012/19/EU on waste electrical and electronic equipment (WEEE). *Official Journal of the European Union.*

Pérez-Belis, V., Bovea, M.D., and Gómez, A. (2013). Waste electric and electronic toys: management practices and characterisation. *Resources, Conservation and Recycling* https://doi.org/10.1016/j.resconrec.2013.05.002.

Pietrelli, L., Ferro, S., and Vocciante, M. (2019). Eco-friendly and cost-effective strategies for metals recovery from printed circuit boards. *Renewable and Sustainable Energy Reviews* https://doi.org/10.1016/j.rser.2019.05.055.

Qiu, Y. and Suh, S. (2019). Economic feasibility of recycling rare earth oxides from end-of-life lighting technologies. *Resources, Conservation and Recycling* https://doi.org/10.1016/j.resconrec.2019.104432.

Reuter, M.A., Hudson, C., van Schaik, A., et al. (2013). UNEP (2013) metal recycling: opportunities, limits, infrastructure. A report of the working group on the global metal flows to the international resource panel.

Robinson, B.H. (2009). E-waste: an assessment of global production and environmental impacts. *The Science of the Total Environment* 408 (2): 183–191. https://doi.org/10.1016/j.scitotenv.2009.09.044.

Rollat, A., Guyonnet, D., Planchon, M., and Tuduri, J. (2016). Prospective analysis of the flows of certain rare earths in Europe at the 2020 horizon. *Waste Management* 49: 427–436. https://doi.org/10.1016/j.wasman.2016.01.011.

Savvilotidou, V., Antoniou, A., and Gidarakos, E. (2017). Toxicity assessment and feasible recycling process for amorphous silicon and CIS waste photovoltaic panels. *Waste Management* 59: 394–402.

Scace, R.I. (2016). *Electronics.* Encyclopædia Britannica, Inc.

Singh, N., Duan, H., Ogunseitan, O.A. et al. (2019). Toxicity trends in E-waste: a comparative analysis of metals in discarded mobile phones. *Journal of Hazardous Materials* https://doi.org/10.1016/j.jhazmat.2019.120898.

Smith Stegen, K. (2015). Heavy rare earths, permanent magnets, and renewable energies: an imminent crisis. *Energy Policy* 79: 1–8. https://doi.org/10.1016/j.enpol.2014.12.015.

StEP (2014). Solving the E-waste problem (step) white paper: one global definition of E-waste. In *Step Initiative*. http://collections.unu.edu/view/UNU:6120#viewAttachments (accessed 18 March 2021).

Tunsu, C., Petranikova, M., Gergorić, M. et al. (2015). Reclaiming rare earth elements from end-of-life products: a review of the perspectives for urban mining using hydrometallurgical unit operations. *Hydrometallurgy* 156: 239–258. https://doi.org/10.1016/j.hydromet.2015.06.007.

Ueberschaar, M., Otto, S.J., and Rotter, V.S. (2017). Challenges for critical raw material recovery from WEEE – the case study of gallium. *Waste Management* https://doi.org/10.1016/j.wasman.2016.12.035.

VDMA (2019). *International Technology Roadmap for Photovoltaic (ITRPV) – Results 2018*. Germany: VDMA Photovoltaic Equipment.

Wang, R. and Xu, Z. (2014). Recycling of non-metallic fractions from waste electrical and electronic equipment (WEEE): a review. *Waste Management* 34 (8): 1455–1469. https://doi.org/10.1016/j.wasman.2014.03.004.

Widmer, R., Oswald-Krapf, H., Sinha-Khetriwal, D. et al. (2005). Global perspectives on e-waste. *Environmental Impact Assessment Review* 25 (5): 436–458. https://doi.org/10.1016/j.eiar.2005.04.001.

World Bank Group and Flagship, G. (2016). Digital dividends – overview. *World Development Report* 2016: 1–359. https://doi.org/10.1596/978-1-4648-0671-1.

Yurramendi, L., Gijsemans, L., Forte, F. et al. (2019). Enhancing rare-earth recovery from lamp phosphor waste. *Hydrometallurgy* https://doi.org/10.1016/j.hydromet.2019.04.030.

Zhang, S., Ding, Y., Liu, B., and Chang, C. (2017). Supply and demand of some critical metals and present status of their recycling in WEEE. *Waste Management* 65: 113–127. https://doi.org/10.1016/j.wasman.2017.04.003.

Zuidwijk, R. and Krikke, H. (2008). Strategic response to EEE returns: product eco-design or new recovery processes? *European Journal of Operational Research* https://doi.org/10.1016/j.ejor.2007.08.004.

6

Recycling Technologies – Physical Separation

Amit Kumar[1], Maria E. Holuszko[1], and Shulei Song[2]

[1] *University of British Columbia, NBK Institute of Mining Engineering, 6350 Stores Road, Vancouver, British Columbia V6T 1Z4, Canada*
[2] *China University of Mining and Technology, School of Chemical Engineering, Ding No.11 Xueyuan Road, Haidian District, Beijing 100083, People's Republic of China*

6.1 Introduction

Electronic waste has been a growing concern around the world. The amount of e-waste generated has been growing at a rate of 3–4% and is expected to reach 120 million tonnes by 2050. With the development of new technologies, especially in laptops, cellphones, and tablets, older technologies are getting obsolete, reducing the lifespan of electrical and electronics products and thus contributing to a higher rate of waste generation. These staggering facts should be considered as an essential incentive for e-waste recycling. If this waste is recycled correctly, it could offer an opportunity for urban mining for the recovery of copper, gold, silver, palladium, and other metals with an estimated value of 55 billion euros. The concentration of metals in e-waste is also significantly higher than the natural ores.

Various routes are currently being implemented to recover metals from the e-waste stream. The recycling methods for e-waste could be broadly categorized into two sections, preprocessing and end processing (Namias 2013). The preprocessing stage, also known as mechanical recycling, has gained popularity due to its relatively low operating cost. It also recovers the metallic fraction without destroying the nonmetal fraction, which, in turn, helps to promote the recycling of nonmetal fraction and the circular economy (Ning et al. 2017). The physical methods recover the metallic fraction and thus reduce the overall mass to be transported or processed subsequently and are less energy-intensive usually. On the other hand, it requires proper dust control units if processed dry or proper water recycling/treatment systems if processed wet.

The preprocessing techniques deal with physical separation methods focused on separating metals from the nonmetals and thus reduce the amount to be processed in subsequent energy-intensive and costly stages. The end processing uses the metal concentrate obtained from the preprocessing stage and methods such as pyrometallurgy, hydrometallurgy, and biometallurgy to produce high-purity metals.

Electronic Waste: Recycling and Reprocessing for a Sustainable Future, First Edition.
Edited by Maria E. Holuszko, Amit Kumar, and Denise C. R. Espinosa.

Mechanical recycling of waste printed circuit boards (PCBs) is usually based on density separation, magnetic separation, or electrostatic separation. The liberation of metal and nonmetals is achieved by using a combination of shredders (coarse crushing), hammer mills, and knife/cutting mills (fine crushing) (Bizzo et al. 2014; Ghosh et al. 2015; Mou et al. 2004; Ogunniyi et al. 2009; Otsuki et al. 2019; Veit et al. 2006; Wen et al. 2005a; Zhang and Forssberg 1999). After the size reduction, the liberated waste is processed using physical separation methods.

This chapter provides the details of the preprocessing processes used in e-waste recycling. The preprocessing step majorly deals with the disassembly, size reduction, and physical separation of various components such as metals, glass, and plastics from the e-waste.

6.2 Dismantling

The dismantling process is mostly a manual disassembly process to remove the hazardous materials out of the e-waste before recycling. The substances removed during dismantling include (Kaya 2018),

- Inks and toners from printers
- Cathode-ray front panels from cathode tube ray (CRT) screens
- Batteries from laptops
- Degassing of chlorofluorocarbons (CFC), hydrofluorocarbons (HFC), and hydrochlorofluorocarbons (HCFC)
- Coolants from refrigerators
- Mercury-containing switchers and PCB capacitors
- Removal of compressor and deoiling

This stage is also used to reduce cross-contamination and to segregate different streams such as plastics, steel, aluminum, copper, and printed circuit boards (Abdelbasir et al. 2018). Automatic and semiautomatic dismantling systems use has increased in recent years; however, manual dismantling is still the most widely used (Hsu et al. 2019). Robotic systems such as Liam and Daisy have been developed by Apple Inc. for dismantling iPhones that can process several cellphones at once. These automatic systems reduce the risk of worker safety and increase the capacity and processing speed for the processing plant. The dismantled streams are sent to the individual processor for the subsequent stage involving size reduction and metal recovery.

Dismantling helps to improve the quality of streams/products for subsequent steps, has low investment, and reduces/eliminates the dust issue. However, it is harder to dismantle new sophisticated technologies, is time consuming, and increases the risk for worker safety. It also is a significant cost concern, especially in high-labor-cost economies.

6.3 Comminution/Size Reduction

Similar to the mining industry, the comminution step involves reducing the particle size to increase liberation and provide proper particle size for subsequent processing/enrichment. The size reduction in e-waste recycling is achieved by using shredders and hammer mills. The particle breakage is achieved by impact and shear forces (Schubert and Hoberg 1997). Recently, high-voltage pulse fragmentation has also gained attention for size reduction as it reduces the dust issue and provides liberation at a coarser size (Duan et al. 2015).

The comminution unit increases the overall plant throughput and reduces the volume for materials to be transported for subsequent process. The significant drawbacks of the comminution process are the generation of dust and loss of valuable metals with the dust and decreased grade/cross-contamination of streams if performed without prior dismantling.

6.3.1 Shredders

Shredders are the first stage of the size reduction process in e-waste recycling. A shredder utilizes sharp cutting blades mounted on a rotating shaft powered by high-power electric motor for cutting materials into smaller pieces. It relies on shear cutting and tearing forces provided in the form of low-speed and high-torque operation. The two main parts of shredders are the rotating shaft and the cutting knives. A shredder can be mounted with single, double, or four heavy-duty rotating shafts depending on the requirement of the operation. The additional shafts provide second-stage size reduction and can further reduce the particle size by increasing the cutting action at a given time (Ekman 2018). The shafts rotate counterclockwise, and the knives mounted on the shaft provide the required shear forces on the material trapped between the knives (Fitzgerald 2009). The knife design depends on the application. Heavy-duty shredders used for cutting cars and metals require thicker and smaller knives. The number of knives dictates the operation of the shredder. More knives not only increase the probability of shredding actions, but also increase the chances of materials to keep bouncing on the surface of the shredder and sometimes clogging the equipment (Ekman 2018).

The rotational speed of industrial shredders ranges from 10 to 50 RPM and can process up to 150 t of materials per hour; however, the throughput capacity changes widely depending on the application (Fitzgerald 2009). The energy consumption ranges from 3 to 11 kWh/t depending on the feed rate and feed properties. A low-power shredder from UNTHA shredding technology has a single shaft (LR520) powered by an 11 kW motor rotating at a speed of 85 RPM, whereas a two-shaft shredder (S120) from the same manufacturer (high torque shredder) has 88 KW drive power and a rotational speed of 10–25 RPM (UNTHA Shredding Technology 2018).

Figure 6.1 Cross-section of a typical hammer mill. Source: Wills and Finch (2016a).

6.3.2 Hammer Mills

A hammer mill is usually installed after the shredder to reduce the particle size of the shredded product for subsequent processes. It uses a high-speed hammer head mounted on a rotating shaft to provide high-impact forces to feed particles to produce smaller particles (Fitzgerald 2009). The particles are impacted by the rotating hammer heads and subsequently by the grid plates for size reduction. Due to high rotational speed, the hammer head also gets severely damaged in the presence of steel and iron. A cross-belt electromagnet is preferably used before feeding the hammer mills. The hammer mill is also lined with a screen at the product end; hence, particles larger than the screen size are retained and undergo multiple impacts. The hammer head is usually manufactured using manganese steel or cast iron with chromium carbide. Figure 6.1 shows the cross-section of a typical hammer mill.

The size reduction in the hammer mill allows particles to achieve proper liberation for subsequent separation processes. These mills have low cost and high production capacity. The heavy-duty WA-36-H-E-Cycle hammer mill from Schutte is equipped with 100 hp. motor and could process at a unit throughput of 455–1360 kg/h depending on the application (SCHUTTE Hammer Mill 2019).

6.3.3 High-Voltage Fragmentation

High-voltage pulse fragmentation is a relatively new technology that has been developed to utilize electrical energy to break the rocks/ore at the grain boundary. The primary benefit of this technology is its higher selectivity. The electrical fragmentation uses fast pulse rise time ($<1\,\mu s$) and higher voltages ($>90\,kV$) to

Figure 6.2 SELFRAG lab. Source: Courtesy: SELFRAG (2012).

discharge into a solid material, causing an "explosion" within. The higher selectivity of this technology comes from different electrical properties of different materials. Field distortions caused by adjacent minerals with different dielectric constants provide the fastest route to ground for the discharge (Wielen et al. Van Der Wielen et al. 2014).

A comparison of ores crushed using SELFRAG (a high-voltage pulse fragmentation technology, shown in Figure 6.2) and ground using conventional methods at same specific energy showed that the pulse technology generates less fines and a significantly higher proportion of >95% liberated minerals in coarser size fraction (Wielen et al. Van Der Wielen et al. 2014). Shi et al. (2014) showed that an energy reduction of 5 kWh/t was achieved while simulating a grinding circuit with this technology.

SELFRAG (2012) showed that metals from e-waste could be separated without fine grinding as proper liberation is achieved during the fragmentation process. Since it utilizes a water as a process medium, the loss-to-fines is depressed, and a coarser liberation size also improves the recovery of the subsequent processes. Duan et al. (2015) showed that ~98% of the copper gets liberated and concentrated in the −2 mm size when a printed circuit board was processed through the SELFRAG lab at 400 pulses.

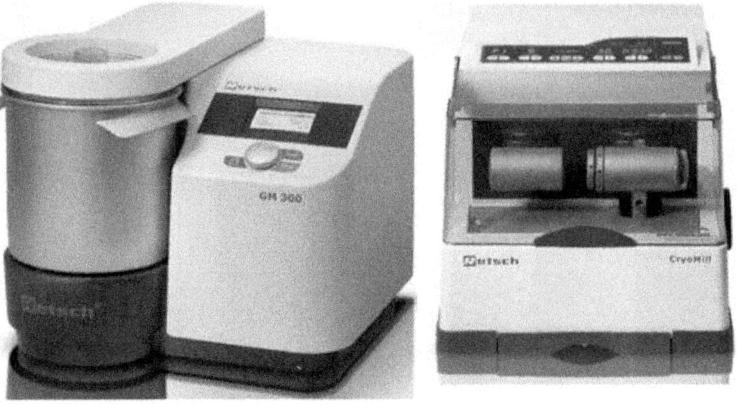

(a) (b)

Figure 6.3 (a) GRINDOMIX GM 300 knife mill (b) CryoMill. Source: Courtesy: Retsch.

6.3.4 Knife Mills

Knife mill is a low-volume cutting mill mostly used in the laboratory environment to create a fine powder. The material is fed to the cutting chamber fitted with sharp blades rotated by a high-speed motor. It is mostly used for grinding plastic and fibrous materials. Figure 6.3a shows the Retsch GRINDOMIX GM 300 knife mill. The motor capacity for this unit peaks at 3 kW, and it is capable of grinding materials from 130 mm to <300 μm. The grinding chamber has a capacity of 4.5 l and can be used dry or wet depending on the requirement.

6.3.5 Cryogrinding

Cryogrinding technology is another process for size reduction, especially suitable for a laboratory environment. The material is frozen at low temperatures using liquid nitrogen or liquid carbon dioxide, which induces the brittle nature in the sample. The sample is then placed in the grinding chamber, which is cooled by circulating liquid nitrogen continuously. The size reduction is mostly achieved by impact forces. The low temperature provided by liquid nitrogen suppresses the recrystallization of particles and thus leads to a finer particle size while reducing the contamination issues. Figure 6.3b shows the Retsch cryomill that can handle 20 ml materials and can reduce particle size from 8 mm to 5 μm. In a study, Zhou et al. (2016) showed that the cooling of the printed circuit board could help to reduce the energy expenditure during the comminution process, and the required liberation could quickly be achieved if the circuit boards are cooled at −30 °C for five minutes. This step makes the boards stiff and brittle, rendering them easier to crush.

6.4 Particle Size Analysis

The particle size plays an essential role in the processing of e-waste or ore. All equipment has an optimum particle size for its efficient functionality. Guo et al. (2011) and Ventura et al. (2018) have shown that metals usually get enriched in

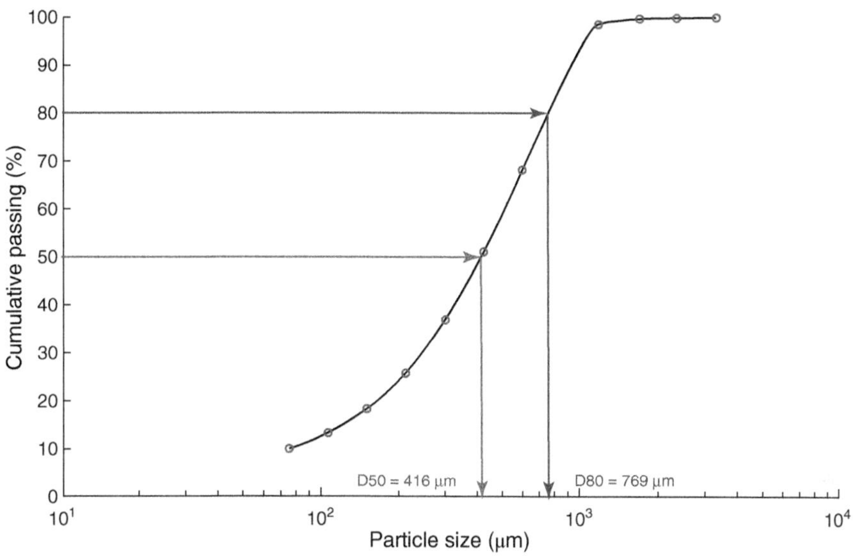

Figure 6.4 Particle size distribution plot.

sizes below 1.25 mm and 1.7 mm, respectively, in crushed waste circuit boards with the highest concentration in the finest fraction, whereas Duan et al. (2015) have suggested that all metals could achieve nearly complete liberation in −1 mm size fraction. Many researchers have suggested a low separation efficiency of the mechanical process for −75 μm particles (Zhao et al. 2012).

Particle size analysis could be conducted using mechanical screens for coarse particles (up to 38 μm) or laser-based sizing for fine particles (below 500 μm). Depending on the particle size, screening can be performed either dry or wet. The efficiency of dry screening decreases with particle size, and wet screening is preferred for fine particles. The particle size of a sample is most widely defined by its 80% or 50% cumulative passing sizes (D_{80} or D_{50}), the screen size at which 80% or 50% of the total mass would pass through the screen. The D_{80} or D_{50} for a sample is obtained using sieve analysis where a given sample mass is placed on a stack of consecutive screens and shaken for 10–20 minutes to allow the particle to segregate based on sizes. The sample retained on each screen is then weighed to determine cumulative passing at each screen size and can be plotted, as shown in Figure 6.4. The D_{80} or D_{50} is obtained by interpolating the screen size and cumulative passing data.

The particle size distributions can also be represented using cumulative distribution functions such as Gaudin–Schuhmann equation for fine particles or Rosin–Rammler distribution for coarse particles. Rosin–Rammler distribution has shown a good fit for waste printed circuit boards, shredded electronics products, and solid waste (Quan et al. 2012; Sun et al. 2015; Vesilind 1980).

$$\text{Gaudin–Schuhmann equation: } F(X) = \left(\frac{X}{X'}\right)^n$$

$$\text{Rosin–Rammler equation: } F(X) = 1 - e^{-\left(\frac{X}{X'}\right)^n}$$

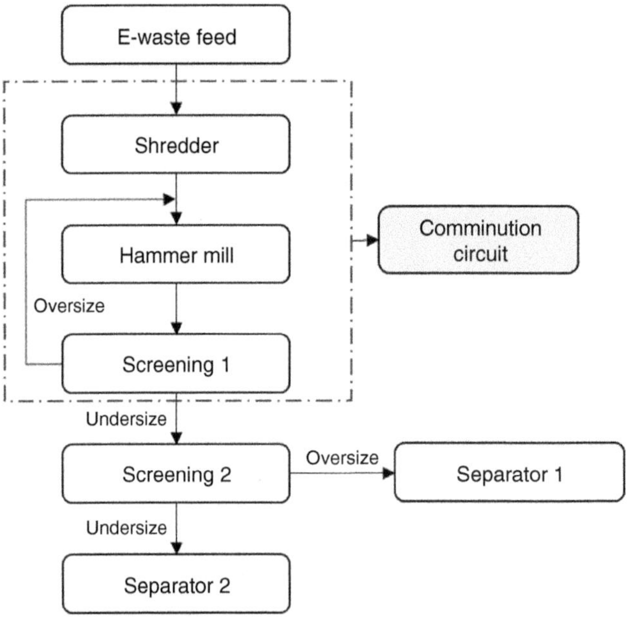

Figure 6.5 An example of a closed-circuit comminution circuit.

Where $F(x)$ is cumulative percent passing at size X, X is the particle size, X' is size modulus defining the theoretical maximum particle size, and n is distribution modulus defining the spread of the distribution (small n represents wide distribution).

6.5 Size Separation/Classification

Size separation is a crucial stage in processing and is used for two main reasons,

- Provide the optimum size range particles to specific downstream equipment
- Maintain a specific top size from the comminution circuit

A typical flowsheet for a comminution circuit is shown in Figure 6.5. It shows that the hammer mill is in the closed circuit with the first screen to provide a fixed top size to the subsequent process. Screen 2 is used to provide an optimum particle size range to individual separators. The two main sizing processes, namely screening and classification (hydrocyclones or air cyclones), could be used for size separation (Menad 2016).

6.5.1 Screening

Screening is the most commonly used process to separate particles by size to provide an optimum size range for the subsequent process to increase their process efficiency. In the case of close circuit comminution, it is used to prevent oversized materials from passing to the next stage. This process is mostly used for coarse applications (up to 1 mm), and the screening efficiency reduces with the reduction in particle size.

The most commonly used screens are the vibratory screens with perforated decks. These screens can be used as single, double, or triple deck depending on the process requirement. The decks have opening holes, and the diameter of these holes depends on the particle cut size. The particle larger than the opening size reports to the oversize, which could be recirculated back to the hammer mill. The particle smaller than the opening size would pass through the screen and report to the undersize, which is sent to the subsequent process. The perforated screen deck could be made of woven wire, polyurethane, rubber, or stainless steel depending on the properties of the materials being processed.

The efficiency of screens is estimated using the recovery of a particle at a given size, and amount of misplaced particles (fine particles reporting to the oversize) and can be calculated using the following equation,

$$R_u = \frac{(f_x - O_x)}{f_x(1 - O_x)} \times 100$$

Where R_u is the efficiency of undersize recovery and f_x and O_x are fractions of material smaller than the cut size in feed and oversize.

The efficiency of screening depends on several machine and operating parameters as well as feed material properties. Some of these factors are:

- Feed rate: inversely proportional
- Screen angle: inversely proportional
- Particle shape: inversely proportional to extreme particle shapes
- Opening area: inversely proportional
- Vibration: excessive vibration would reduce screening efficiency
- Moisture: inversely proportional

The size (length and width) of the screen required could be estimated using Vibrating Screen Manufacturers Association (VSMA) equation for screening surface area calculation or Metso equation and is shown in the equation below (Olsen and Carnes 2016):

$$\text{Screen area (SA)} = \frac{\text{U.S}}{\text{A} \times \text{B} \times \text{C} \times \text{D} \times \text{E} \times \text{F} \times \text{G} \times \text{H} \times \text{J}}$$

Where U is undersize tonnes per hour, S is safety factor, A is basic capacity factor, B is oversize percentage factor, C is half-size factor, D is deck location factor, E is wet screening factor, F is bulk density factor, G is screen surface open area factor, H is opening shape factor, and J is screen efficiency factor.

Once the required area is estimated, the screen would be selected based on the screen information available from different vendors.

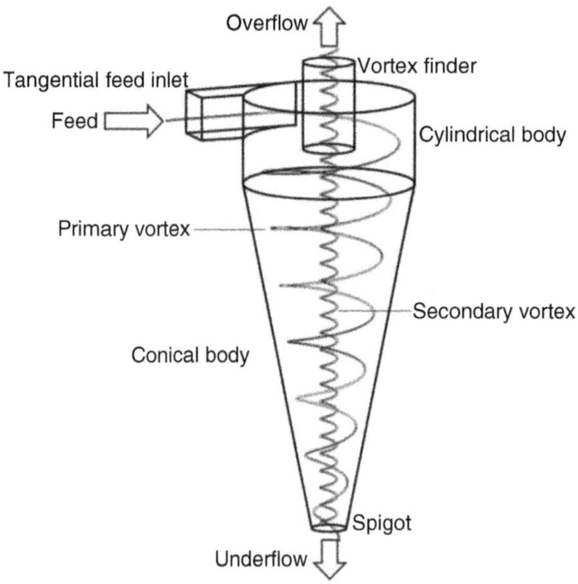

Figure 6.6 Schematic of a hydrocyclone. Source: Vega-Garcia et al. (2018). Licensed under CC BY-4.0.

6.5.2 Classification

Classification is a size separation process based on the particle behavior in a fluid such as air or water. The process is suitable for fine particles (<1 mm) and mostly used in the final stage of comminution as a size control device before beneficiation.

The classification process depends on the forces acting on a particle in a fluid. The net resultant force on the particle dictates the movement of the particle. To achieve proper size classification, the direction of the net resulting force should be different for coarse and fine particles. The forces acting on the particle in the fluid are gravitational force, drag force, and centrifugal force. Depending on the solid density (amount of solids in the slurry by weight), the particle-settling behavior in the fluid can be defined as free settling or hindered settling. If particles of the same density are placed in a water medium, the coarser particle moves faster than finer particles, thus achieving separation. The classification units can be classified as gravitational or centrifugal depending on the forces acting on the particles.

6.5.2.1 Centrifugal Classifier

The most widely used centrifugal classifier in the processing industry is hydrocyclone. These are robust and relatively efficient size separators. The schematic of a hydrocyclone is shown in Figure 6.6. A hydrocyclone consists of a feed inlet, an apex/spigot (underflow discharge), a vortex finder (overflow discharge), and a cylindroconical body. The slurry is tangentially fed through the feed inlet under pressure and high velocity. The tangential feed inlet provides the rotational motion to the slurry that helps to create opposing forces, centrifugal force, and drag force. The high centrifugal force on the coarse particle moves it toward the outside wall, where it spirals downwards and reports to the underflow through apex/spigot. The finer particles remain in the center and move upward with the secondary vortex and then exit as overflow through the vortex finder. The cyclone diameter could vary from 13 mm to 900 mm depending on the requirement and could be used for classification down to 20 μm or sometimes lower.

The cut point of the hydrocyclone is the size that has an equal probability of reporting in underflow or overflow stream. The cut size for a hydrocyclone depends on the cyclone diameter and flow rate. The cut size increases with the cyclone diameter but decreases with flow rate. The efficiency of the cyclone is quantified using partition curves, which defines the percentage of a size fraction in the feed reporting to underflow. Ideally, the particles larger than cut size should report to underflow, whereas particles smaller than the cut size should report to overflow. However, in real operations, some particles are always misplaced. Some coarse particles that did not have sufficient opportunity to move to the underflow would report to overflow as short circuit, whereas some fine particles could become trapped among coarse particles and would report to the underflow as entrainment. The efficiency of separation is estimated using the following equation,

$$\text{Imperfection } (I) = \frac{d_{75} - d_{25}}{2 \times d_{50}}$$

Where d_{75}, d_{50}, and d_{25} are particle sizes at the partition value of 0.75, 0.50, and 0.25. For the ideal scenario, imperfection would be 0.

Another criterion for cyclone performance is the amount of water in the underflow. In the ideal scenario, no water should report to underflow; however, the water content of $<30\%$ in underflow is considered to be good.

6.5.2.2 Gravitational Classifiers

Gravitational classifiers are simple and low-energy consuming classifiers that work on particle settling by gravity. The rake and spiral classifiers are the most common examples of gravitational classifiers. It comprises a tank with an inclined surface and a system to move settled material in an upwards motion. The coarse particles are settled due to their weight and get continuously extracted with the help of a mechanical device (Dunne 2019; Zhou et al. 2012).

In the case of spiral classifiers, one or two screws move the settled materials upward, whereas, in rake classifiers, an array of tines is used to separate the coarser material from the finer ones, as shown in Figure 6.7. The fine particles stay suspended in water and get discharged by overfilling the tank.

Some other gravitational classifiers are hydraulic classifiers such as Linatex, Reflux, Crossflow, Allux, and Floatex classifiers. These devices use gravity for separating coarse particles from fine particles instead of a mechanical system. The coarse particles form a bed at the bottom of the classifiers, and the fine particles are moved upward with water injection through a perforated bottom plate that moves fine particles to the overflow launder. The coarse particles are discharged through the lower part of the tank as thick pulp (Luttrell et al. 2006; Mankosa et al. 2019).

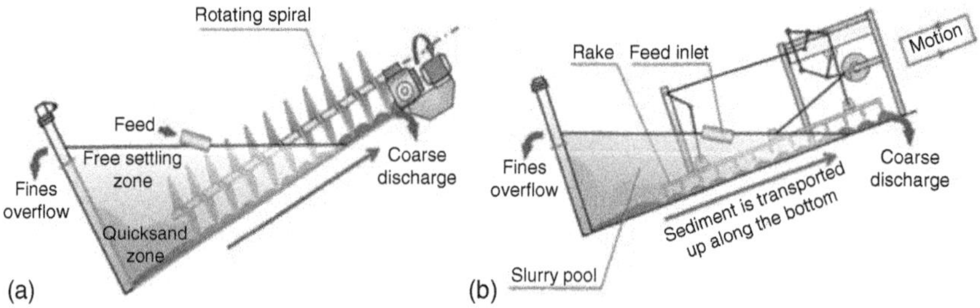

Figure 6.7 Mechanical size classification equipment: (a) spiral; and (b) rake. Source: Wills and Finch (2016b).

6.6 Magnetic Separation

Magnetic separation has been utilized to remove the ferromagnetic elements, iron and nickel, and other iron scraps. It uses the difference in particle behavior in a magnetic field for separation. All particles can be classified as diamagnetic, paramagnetic, and ferromagnetic, depending on their magnetic properties.

Diamagnetic particles such as copper, gold silver, plastic, fibers are repelled from the magnetic field lines and would report to the nonmagnetic tailings bin. The paramagnetic particles such as platinum, manganese, tungsten are attracted to the magnetic field lines and would report to the magnetic concentrate bin during separation. The ferromagnetic particles such as iron, cobalt, and nickel are a special case of paramagnetic behavior where the particle retains magnetic property even after the removal of magnetic field lines. These particles would also report to the magnetic concentrate bin during separation.

A magnetic separator consists of a permanent magnet or electromagnet to generate the magnetic field and a conveyor to move particle over the field lines. The types of magnetic separators can broadly be classified into low- and high-intensity magnetic separators depending on their magnetic field intensity.

6.6.1 Low-Intensity Magnetic Separators

Low-intensity magnetic separators are used to separate ferromagnetic materials such as iron and nickel from the crushed e-waste. The magnetic field intensity used is $<0.3\,T$, which is sufficient to take advantage of the high magnetic susceptibility of iron particles. The primary utilization of the low-intensity separator is in the form of cross-belt separators located over the conveyor. It is widely used to remove the coarse magnetic iron particles before the materials are fed to the hammer mills.

Another major type to the low-intensity separator is the rare-earth roll separator used in dry applications. The feed is moved using a conveyor, and a rare-earth magnetic roll is installed at one discharge end of the conveyor. The magnetic particles are attracted to the roll, move with the belt, and get discharged into the magnetic bins, whereas the nonmagnetic particles fly away from the belt due to the centrifugal forces from the rotating rolls and fall in the nonmagnetic bin as shown in Figure 6.8 (Wills and Finch 2016c). The roll speed, feed rate, field strength, and particle size

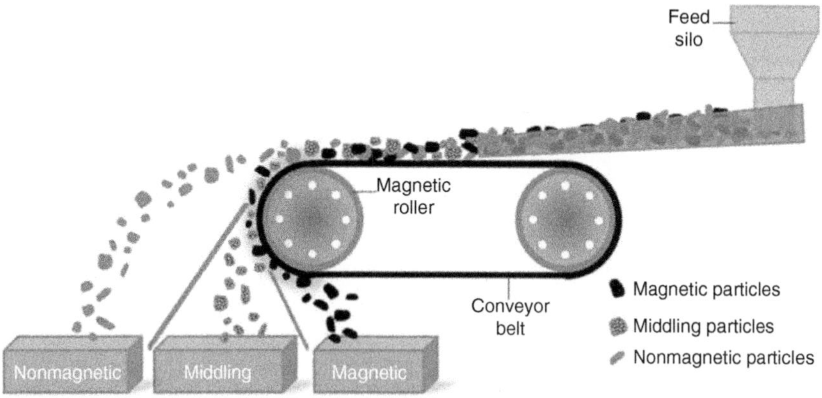

Figure 6.8 Rare-earth roll dry low-intensity magnetic separator. Source: Miceli et al. (2017).

are the major factors affecting the separation. In the case of wet applications, drum separators are the most widely used. It consists of a nonmagnetic drum fitted with magnets that rotate in a pool of slurry. The magnetic particles get pinned with the drum and lifted to the concentrate bin, whereas the nonmagnetic particles move with the slurry and get discharged into the tailings (Haldar 2018).

6.6.2 High-Intensity Magnetic Separators

These types of magnetic separators are useful for the separation of weakly magnetic particles (paramagnetic materials). The magnetic field could be generated using rare-earth alloy magnets or electromagnetic coils and is more useful for fine particles and low iron-containing materials.

The induced roll separator is one of the most common dry high-intensity magnetic separators where the material is fed directly on the rotating roll. The magnetic materials get pinned with the roll, whereas the nonmagnetic materials fly off the roll due to the centrifugal forces induced due to the roll rotation (Figure 6.9). Magnetic field strength up to 2 T could be achieved with this separator. Particle size and feed rate are the major factors for particle separation.

Wet high-intensity magnetic separators have successfully been used for fine hematite ores to produce iron ore concentrates and can be used for ultra-fine particles too. Other magnetic separators such as high-gradient separators and superconducting separators are used for particles of very low magnetic susceptibility.

Yoo et al. (2009) used magnetic separation to recover 83% of iron and nickel as magnetic concentrate from shredder printed circuit boards. Veit et al. (2005) showed that a high iron and nickel concentration in the magnetic concentrate for printed circuit boards and electronic component scrap could be achieved at a low magnetic field strength of 0.65 T.

The iron concentration in the magnetic materials was increased to 41% compared to 27% in the feed with 89% iron recovery for fine (−0.6 mm) e-waste powder in a

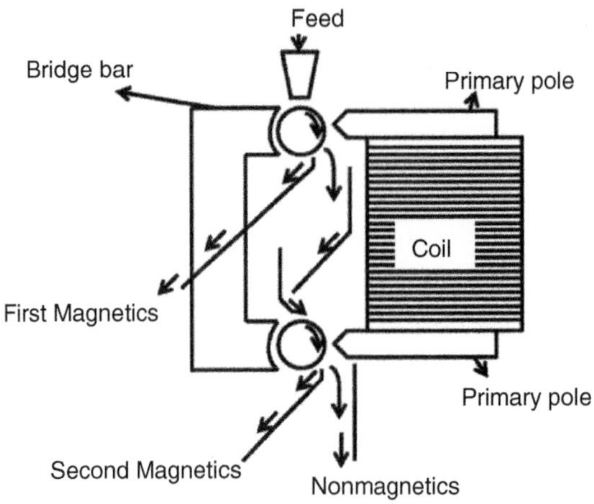

Figure 6.9 Induced roll magnetic separator. Source: Tripathy et al. (2014).

study conducted by Lee and Mishra (2018). In another study, up to 89% nickel and 98% iron were recovered for relatively coarse (+0.5 mm) circuit board particles at 0.1 and 0.8 T field strength (Fujita et al. 2014). Recently, Zhu et al. (2020) used a wet low-intensity magnetic separator for fine waste printed circuit board particles (−0.5 mm) to recover iron particles and suggested that magnetic separation is an effective method to recover high-grade magnetic particles.

6.7 Electrical Separation

Electrical separation is one of the most used techniques to recycle metals such as copper from nonmetals such as plastic and fibers. It uses the differences in electrical conductivity among various materials present in e-waste. One of the significant challenges with the process is its limitation to dry feed. For most efficient operation, the feed should be in monolayer, which also restricts its throughput. The separator needs to charge the particles to use the conductivity differences selectively. There are three main mechanisms of particle charging, ion bombardment (corona electrostatic), frictional charging (triboelectric), and eddy current. Corona electrostatic separation has been proven useful to separate metals from nonmetal streams in dry environment, whereas eddy current separation has been successfully utilized to extract aluminum from the nonferrous metal stream (Abdelbasir et al. 2018; Bizzo et al. 2014; Burat and Özer 2018; Das et al. 2009; Duan et al. 2009; Hsu et al. 2019; Kasper et al. 2015; Kaya 2016, 2018; Menad 2016; Veit et al. 2014).

6.7.1 Corona Electrostatic Separation

Corona separator uses ion bombardment as the method for particle charging. The charging process uses high voltage between two electrodes that ionize the gases near the electrodes forming a continuous flow of gaseous ions called the corona. When the feed particles pass through this corona, all particles receive a surface charge. When the particles move out of the corona, the nonconductive particles maintain the surface charge and get pinned to the rotor surface. Particles of high conductivity lose their surface charge as they quickly dissipate to the earthed rotor, and the centrifugal force induced due to the rotor rotation throws the particles away from the rotor/roll (Han et al. 2014).

Guo et al. (2011) obtained over 90% recovery for copper using electrostatic separation with a copper grade of 71%, whereas Veit et al. (2005) used electrostatic separation to enrich metal concentration to 50% Cu, 25% Sn, and 7% lead in printed circuit boards. Li et al. (2007) showed that the optimum particle size for electrostatic separation is 0.6–1.2 mm in circuit boards.

6.7.2 Triboelectric Separation

The triboelectric separator uses particle contact/friction as the method for particle charging. When two particles with different electrical properties come in contact and rub against each other, there is an exchange of electrons resulting in positively and negatively charged particles. If these charged particles are subjected to an electric field, the particles move toward oppositely polarized electrodes and hence provide the needed separation. The rubbing/friction between the particles is achieved by using a fluidized bed, and then the charged particles are allowed to free fall between two oppositely charged electrode plates. This process is more useful for plastics and other nonmetallic materials.

Saeki (2008) used triboelectric separation to recover 95% polyvinyl chloride (PVC) with 99% purity from a mixture of PVC, polyethylene terephthalate (PET), and polyethylene (PE), whereas 90% recovery of acrylonitrile butadiene styrene (ABS) with 99% purity was achieved from a mixture of ABS, polypropylene (PP), and polystyrene (PS). Dodbiba et al. (2002b) listed that the triboelectric charging sequence for various plastics is,

(positive end) ABS–PP–PC–PET–PS–PE–PVC–PTFE (negative end)

PC is polycarbonate and PTFE is polytetrafluoroethylene.

The sequence suggests that if two plastics (i.e. PE and PS) are rubbed against each other, the one on the left of the sequence (PS) gets a positive charge and one on the right of the sequence (PE) gets a negative charge. In a two-component mix of PET and PE, triboelectric separation could achieve over 95% grade of individual plastic. Other researchers have also concluded high grades and high recoveries for different plastics using triboelectric separation (Lee and Shin 2002; Li et al. 2015; Xiao et al. 2003; Zelmat et al. 2013).

Figure 6.10 shows a simplified schematic for a triboelectric separation system where particle charging is achieved via tribo-cyclone. The charged particles are gravity fed to the separation system and products are collected in the collecting bins.

In the field of nonmetal fraction from printed circuit boards recycling, Zhang et al. (2017) and Yang et al. (2019) used triboelectric separation to separate inorganics

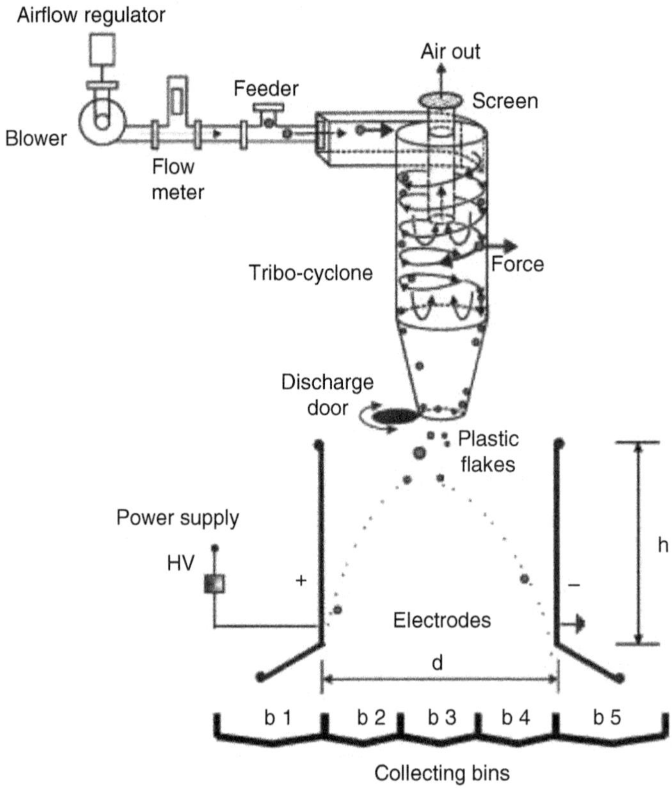

Figure 6.10 Design of a triboelectric separation system. Source: Dodbiba et al. (2005).

from the nonmetal fraction to improve the quality of the product. Overall, a concentrate yield of 47% was achieved with an organic content of 69%, recovery of 59%, and an upgradation ratio of 1.27 by Zhang et al. (2017). In the case of Yang et al. (2019), an upgradation ratio of 1.17 was achieved.

6.7.3 Eddy Current Separation

Eddy current separation is widely used in the recycling industry to recover nonferrous metals such as aluminum and copper from electronic waste (Jujun et al. 2014; Ruan et al. 2017). The process is dictated from the generation of eddy current in particles when passing over a rotating magnetic field. When a nonferrous metal is placed in a high-speed rotating magnetic field, eddy currents generated in the nonferrous materials create a secondary magnetic field around itself. These particles are then repelled due to the same polarity of the two magnetic fields and hence separating the nonferrous metals from the mixture, as shown in Figure 6.11. It can be used to separate particles in size range of 3 mm–350 mm (Smith et al. 2019).

Eddy current separators were used to recovery over 90% aluminum from electronic scrap by Zhang et al. (1998). Bi et al. (2019) used eddy current separation to

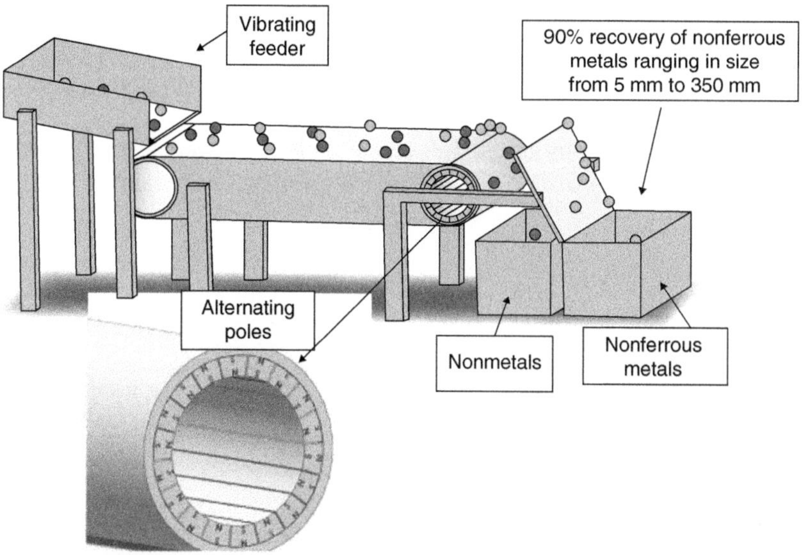

Figure 6.11 Schematic of eddy current separator. Source: Smith et al. (2019).

recover copper and aluminum from the crushed spent lithium iron phosphate batteries in size range of 2–20 mm. Li et al. (2017) used eddy current separator to recover printed circuit boards from crushed cell phones with over 95% separation efficiency. Ruan and Xu (2011) showed that this process is also an efficient method to separate aluminum from plastic in crushed waste toner cartridges.

6.8 Gravity Separation

Gravity separation is one of the most common low-cost physical processing methods and has already being used for the separation of metals from nonmetals (Eswaraiah et al. 2008; Forssberg et al. 1998; Hanafi et al. 2012; He et al. 2015; Meng et al. 2018; Nekouei et al. 2018; Sarvar et al. 2015; Veit et al. 2014; Ventura et al. 2018; Wen et al. 2005b; Zhou et al. 2016). It has been successfully used in the separation of different types of plastic wastes too. Processes such as air classifiers, jigs, hydrocyclones, and float–sink have been widely studied in the separation of different types of plastics (Bauer et al. 2018; Ruj et al. 2015; Serranti and Bonifazi 2019).

A gravity separator utilizes the differences in the density of various metals and nonmetals as the basis of separation. If particles of same size are placed in a fluid, the high-density particles sink faster than the light-density particles. Since the density of most of the metals is higher compared to the nonmetals present in the e-waste, gravity separation is the most successful technique for the physical recycling of e-waste. The ease of gravity separation can be expressed by the concentration criterion and can be estimated using the equation below,

$$\Delta\rho = \frac{\rho_h - \rho_f}{\rho_l - \rho_f}$$

Where ρ_h, ρ_l, and ρ_f are the density of heavy material, light material, and the fluid medium.

Generally, if the concentration criterion is greater than 2.5, the gravity separation is relatively easy, and the efficiency of separation decreases with decreasing value of concentration criterion. Various gravity separation units are available at industrial scale for metal recycling from e-waste. Some of the most common separators are discussed in this section.

6.8.1 Jigs

Jigs are one of the oldest gravity separation technologies used in the mineral processing industry. The jigging process is a stratification of a particle bed by fluidization achieved by a pulsating flow of medium (Ambrós et al. 2019). Traditional jigs operate using pulsating water; however, air jigs have also been employed for dry application where constant airflow is used for stratification. Each pulsation cycle causes the particle bed to lift and then cause the heavy particle to settle first, followed by light particles causing the formation of a stratified bed. Ambrós et al. (2019) showed that particle-size distribution has a significant effect on the separation efficiency of a jig due to the packing effects. Particle shape, pulsation frequency, water flow rate, and feed density also affect efficiency (Sarvar et al. 2015).

Phengsaart et al. (2018) used a jig to achieve up to 98% copper recovery from shredded small home appliances. Sarvar et al. (2015) showed that the wet jigging could produce a metal concentrate with 94% recovery and 41% metal grade. The jigging process has also shown success in the separation of mixed plastics (Pita and Castilho 2016). Hori et al. (2009) used jigs with water pulsation to separate polyvinyl chloride (PVC), polyethylene terephthalate (PET), and polyethylene (PE) with a specific gravity difference as low as 0.11. Ito et al. (2010) showed that a bench-scale continuous jig could be used to separate PP and high-density polyethylene (HDPE) plastics with 99% product recovery.

6.8.2 Spirals

Spirals are simple, robust, and low-cost gravity separators where the slurry is fed at a solid density of ~20–25% at the top of the spiral. As the slurry flows down the curved channel, lighter particles due to action of drag forces combined with the water flow report to the outer area of the spiral as tailings, while the heavier particles are pulled inward to the inner area of the spiral as a concentrate, thus effecting the separation. Wash water is continuously added at the inner area to flush any entrained light particles. Modern spirals are constructed from fiberglass and plastic and can treat particles down to 75 μm. The operating parameters are feed rate, pulp density, feed grade, splitter openings, the position of distributors, and the diameter of spirals.

No researchers have yet used spirals in the field of e-waste recycling; however, spiral concentrators have shown success in iron, coal, and rare-earth industry.

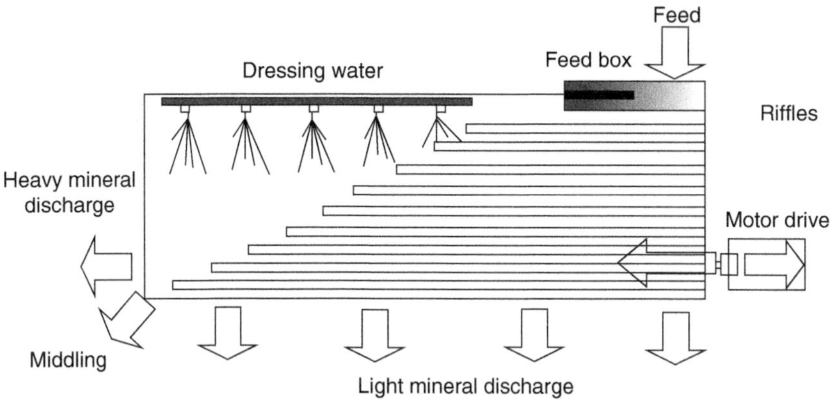

Figure 6.12 Schematic of a wet shaking table. Source: Shariati et al. (2015). Licensed under CC BY-4.0.

6.8.3 Shaking Tables

The shaking table is the most widely used gravity separator in the recycling industry. The table consists of a riffled deck enabled to move in a reciprocating motion in a direction parallel to the riffler (Figure 6.12). The slope of the table could be adjusted depending on the requirement. When the table is given the reciprocating motion at a right angle to the flow of water, heavier minerals settle down in riffles and are carried along the diagonal line of the table as a concentrate. The lighter minerals do not settle in riffles and are washed along with the water as tailings. The riffles on the table form small jigging motion to stratify particles based on their density (Manser et al. 1991). The operating parameters are water flow rate, deck slope, feed throughput, stroke length, and pulp density. The solid density is generally maintained at is 25–30% (wt/wt).

Air tables have also been incorporated in the industry for dry gravity separation. Compressed air is pumped below the vibrating table with the surface fitted with a perforated cloth. The feed is introduced near the top of the inclined table. Lighter particles are lifted by the compressed air and flow downwards as tailings. The oscillating motion of the table causes the heavy minerals in contact with the table surface to move upward and are collected as a concentrate (Dodbiba and Fujita 2015).

Veit et al. (2014) used Mozley concentrator, a variance of shaking table, to recover 85% Cu, 95% Sn, 96% Ni, and 98% Ag from printed circuit boards at $-1+0.25$ mm size. The shaking table could also be used as a preconcentration stage to recover high-grade concentrate from subsequent processes (Das et al. 2009; Ventura et al. 2018; Xia et al. 2017). Liu et al. (2014) showed that the optimum size range for the shaking table is $-1+0.074$ mm, and recoveries over 90% could be achieved for Cu, Sn, and Al. A recovery of ~80% of Cu and other metals was achieved for particles smaller than 1 mm using a wet shaking table by Wu et al. (2012) and Zhu et al. (2020). Pita and Castilho (2018) showed that wet shaking tables could also be used to separate copper from electric cable waste with high recoveries and high grades.

Researchers have also shown the applicability of air tables for metal recoveries from crushed printed circuit boards. Zhou et al. (2016) achieved recovery of nearly 95% with 60% metal grade in the air table concentrate. Air tables have also shown their applicability for the separation of low-density PP from high-density PET and PVC mixture with high recoveries and high grades (Dodbiba et al. 2005).

6.8.4 Zig-Zag Classifiers

Zig-zag classifiers are a dry classification system that uses an air stream to separate light particles from heavy particles. The classifier consists of several sections with rectangular cross-sections connected to form a zig-zag shaped channel. A current of upflowing air carries light particles upward while the heavy particles move downwards, thus achieving required separation (Senden 1980). The efficiency of the classifier is highly dependent on the particle size, airflow rate, feed rate, and the angle of the sections.

Yoo et al. (2009) used a zig-zag classifier on various size fractions at an optimized airflow rate to obtain light and heavy streams and achieved high recoveries up to 95% for particles larger than 0.6 mm. Recently, Da Silva et al. (2019) showed that zig-zag classifier is an economical and energy-efficient alternative for metal recovery from waste printed circuit boards.

6.8.5 Centrifugal Concentrators

Centrifugal concentrates such as Knelson concentrator and Falcon concentrator are highly efficient processes used to enhance the gravitational forces experienced by particle by adding high rotational speed, especially for fine particles. The feed slurry is pumped to the top of the feed inlet of the concentrator. In the centrifugal concentrator, the particles are accelerated up to 60–300 times the regular g-forces from the rotating bowl trapping the heavier particles in the grooves located in the bowl, whereas the lighter particles are flushed with the water and report to the tailing launder as shown in Figure 6.13. The centrifugal concentrator in most of the gold processing plants is batch operated with an optimum concentrate cycle. After the completion of the cycle, the heavy concentrate is flushed from the bowl into the concentrate launder (Chen et al. 2020; Oruç et al. 2010).

Zhang et al. (2018) showed that the high density of lithium cathode oxide could help to achieve a separation of lithium cathode oxide from spent lithium-ion batteries with over 80% grade and recovery using Falcon concentrator. The optimum particle size was determined to be $-90+45\,\mu m$. Duane et al. (Duan et al. 2009), Wen et al. (2005b), and Yue-min (2006) showed that the Falcon concentrator could be used for metal recovery from crushed printed circuit boards with high recovery and high grades. In a study conducted by Kwon et al. (2016), Kelson concentrator was used to recover tantalum from the electronic components of printed circuit boards obtained from a laptop with ~80% grade and recovery.

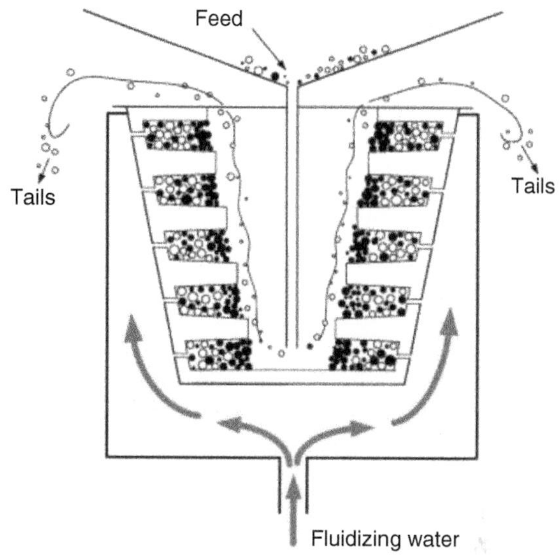

Figure 6.13 Schematic of a Knelson centrifugal concentrator. Source: Chen et al. (2020).

6.8.6 Dense Medium Separation (DM Bath/Cyclone)

Dense media separation is the most common approach used in the coal industry to remove heavy inorganic materials from the light cleaner coal particles. The process uses a fluid medium at a given separation density to achieve separation. When a particle is placed in a fluid, it would either float or sink depending on fluid density. If the particle density is less than the fluid density, it floats to the surface, or else it would sink to the bottom. The fluid used for this process is mostly organic liquid of different specific gravities at lab scale; however, at the industrial scale, the required fluid density is achieved by mixing fine heavy media such as magnetite or ferrosilicon into water. This media could then be recycled using simple drum magnetic separators and reused.

The coarse material is mostly processed in a static bath (dense media bath) where the heavy particles are settled at the bottom and then carried out to the heavy launder by a removal mechanism such as a conveyor belt, whereas the light particles overflow or are scraped from the surface. For finer particle size, a dense medium cyclone is used. The working principle is similar to that of a hydrocyclone. The heavy particle moves toward the cyclone wall due to centrifugal forces and move to the bottom of the cyclone and get discharged through the spigot. The light particles are moved to the center and get discharged through the vortex finder. The separation density and theoretical grade and recovery for a gravity separator are established from a float–sink test.

A float–sink test is a simple lab-scale technique used to determine the gravity separation behavior for a given sample where the sample is subjected to increasing density liquids, and the floated and sunk fractions are analyzed from their properties. Previous researchers have utilized heavy liquids of higher densities (~2.89 g/cm^3) to separate the metallic fraction from the nonmetallic fraction (Bizzo et al. 2014). A lab-scale float–sink test was able to recover 75% of metallic fraction using organic

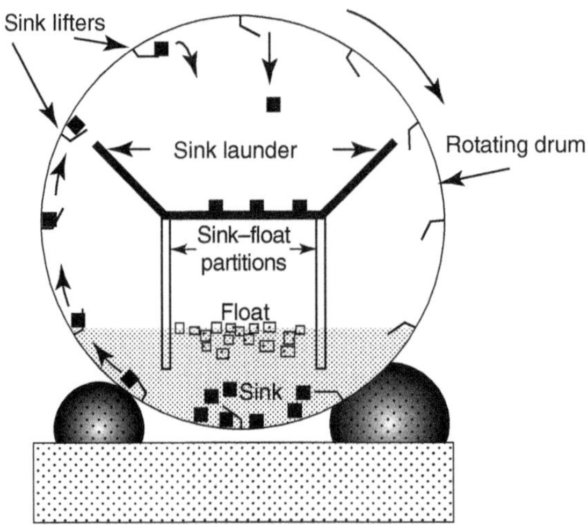

Figure 6.14 End view of the drum separator. Source: Dodbiba et al. (2002a).

liquids (Nekouei et al. 2018). Cui and Forssberg (2007) showed that over 88% of copper in the shredded television scrap could be recovered at a separation density of 1.4 g/cm^3 with 42% copper grade.

Float–sink test based on heavy liquids has also been used to separate waste plastics into <1.0, 1.0–1.1, and 1.1–1.2 specific gravity ranges by Lee (2007). Pongstabodee et al. (2008) used a three-stage float-and-sink method to classify six different plastics into low (specific gravity: 0.90–1.00), medium (specific gravity: 1.00–1.20), and high (specific gravity: 1.20–1.40) density ranges. Choi (2004) used gravity separation to recover PVC (specific gravity 1.32–1.42) from a mix of polyolefin plastic waste (specific gravity: 0.90–1.06). Dodbiba et al. (2002a) used a drum separator at a speed of 1 RPM at a specific gravity of 1.11 to separate PET from polypropylene (PP). The schematic of the drum separator used by Dodbiba et al. (2002a) is shown in Figure 6.14. The light product overflows to the discharge launder, whereas the heavy sink product is carried up by the sink lifters and then discharged to the sink launder.

6.9 Froth Flotation

Froth flotation is a physicochemical separation process that utilizes the differences in the surface properties of various materials for separation in the water medium. This process depends on the selective attachment of air bubbles to one specific material/mineral in the sample and brings it to the surface where it is skimmed off to the concentrate launder.

The tendency of the air bubble attachment to a particle can be expressed using the contact angle. The contact angle depends on the work of adhesion, cohesion, and surface tension at the bubble, water, and solid interface. A contact angle of 0° represents complete wetting rendering the surface hydrophilic (water attracting), whereas a contact angle of 180° represents complete dewetting, rendering the surface hydrophobic (water repelling). The bubble selectively attaches to the hydrophobic particle leaving the hydrophilic particles in the slurry as tailings.

However, the contact angle measurement is possible for smooth and coarse particles. The surface energy or the bubble particle interaction could also be measured using inverse gas chromatography (IGC). The surface tension at three interfaces is highly affected by the surface properties of the solids and surface energy is one of the most important surface properties (Ali et al. 2013). One of the benefits of using an IGC for estimating the hydrophilicity/hydrophobicity of particles that it can be used on fine particles, powders, and rough/uneven surfaces. Mohammadi-Jam and Waters (2016) and Mohammadi-Jam et al. (2014) showed that decreasing the surface energy of particles decreased the work of adhesion with water and increased the floatability of particles and thus increasing hydrophobicity for different minerals.

Some particles such as plastics and nonmetals in e-waste are naturally hydrophobic, whereas the metals are naturally hydrophilic, making flotation a useful process for metal recycling. Inverse/reverse flotation is most studied for flotation for e-waste recycling where the unwanted particles are selectively floated, and the valuable materials are left behind in the slurry. Various reagents are added to the flotation process to aid with the flotation. The reagents can be classified as,

1) Collectors: These are the chemicals that selectively adsorb on the surface of a particle, making it hydrophobic, and facilitates the bubble particle attachment. Examples: Xanthates, amines, oils, sulfonates, etc.
2) Frother: This chemical helps with the formation of stable bubbles by interacting with the air–water interface and decreasing water/air surface tension. Examples: pine oil, methyl isobutyl carbinol, dow froth, etc.
3) Activators: These are the chemicals that help with altering the chemical structure of the particle surface, thus helping collectors to adsorb on the given particle. Examples: metal ions, amines, etc.
4) pH regulators: This group of chemicals is added to the slurry to alter the pH of the slurry and thus changing the surface charge and help with the adsorption process. Examples: lime, sodium hydroxide, etc.
5) Dispersants: It is used to prevent fine particles to agglomerate by controlling the surface charge. Examples: polyphosphates, sodium silicate, starch, guar gum, etc.
6) Depressants: It is used to prevent the adsorption of collectors on the particles rendering them hydrophilic. Examples: humic acids, sodium bisulfide, cyanide, etc.

A combination of these chemicals is used to help with the froth flotation process. However, researchers have shown high recovery without needing any additional chemical due to the nature of particles in the e-waste system. The nonmetal fractions such as fiberglass and plastics are naturally hydrophobic, whereas the metallic fraction is commonly hydrophilic. It should be noted that the flotation is highly dependent on particle size, reagent type, and dosage; hence, very carefully designed experiments at the lab scale are needed to optimize the process.

The flotation is mostly carried out in mechanical flotation cells in the industry as shown in Figure 6.15. It consists of a tank with a rotating impeller. The impeller design changes widely with the vendors; however, it delivers the same purpose. It helps with the agitation of the slurry and create air bubbles by breaking a stream of air. The air bubble rises in the slurry carrying the hydrophobic particles with it, which is skimmed from the surface. The largest flotation cell from Outotec has a capacity of 630 m^3. Usually, flotation cells are installed in banks consisting of several (5–8) individual cells.

Column flotation cells are another type of flotation device commonly used at cleaner and recleaner stage to improve the metal grade. The bubble in the columns is generated directly through air spargers. Normally, the feed to the column enters one-third way down from the top and moves downwards where it encounters upward moving air bubbles. The hydrophobic particles are carried to the surface with the bubbles. Wash water is added at the surface using an array of perforated pipes to remove any entrained hydrophilic particles, thus increasing the grades.

Flotation without any reagent has shown a good metal recovery from waste printed circuit boards (Mäkinen et al. 2015; Ogunniyi and Vermaak 2009). Flotation using kerosene and terpenic oil showed a metallic recovery of over 84% at a particle size below 75 μm (He and Duan 2017), whereas soapy collector was used in the

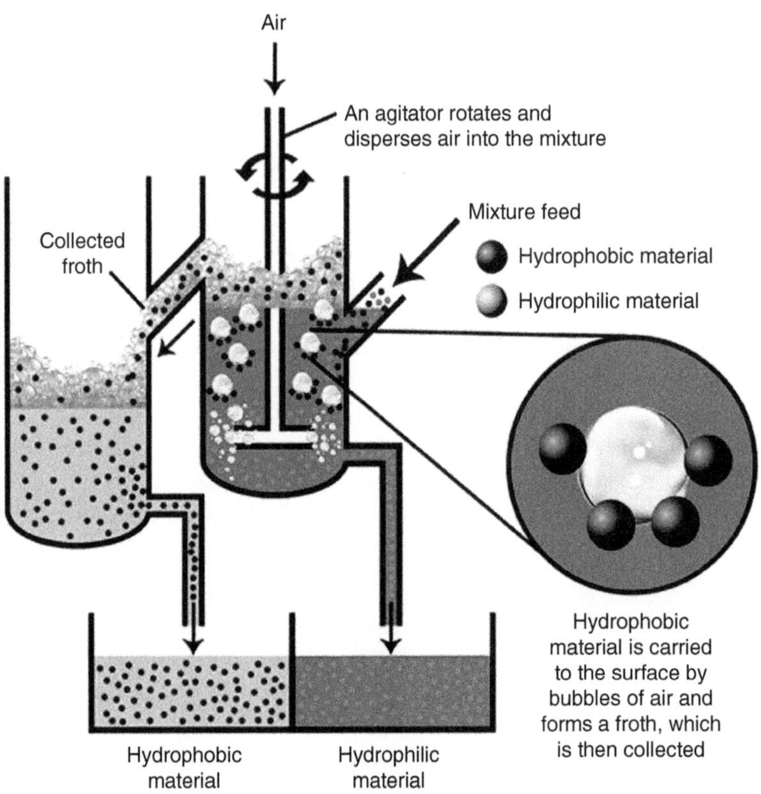

Figure 6.15 The process of froth flotation. Source: Crawford and Quinn (2017).

flotation experiments conducted by Zhu et al. (2019). He and Duan (2017) reported a copper recovery of 79%, whereas the recovery for Zhu et al. (2019) was only 47%, most likely due to larger particle size (−1 + 0.5 mm). Vidyadhar and Das (2013, 2012) showed that over 90% of the total metal in printed circuit boards could be recovered with a single-stage flotation process at 1 mm grind size. Frother flotation with fuel oil as collector and pine oil as frother showed over 95% metal recovery (Sarvar et al. 2015). Gallegos-Acevedo et al. (2014) showed that a metal-rich and a nonmetal rich stream could be obtained with the addition of 5 ppm frother with over 90% metal recovery.

Separation of different plastics using froth flotation is another avenue for extensive research. Researchers have shown that the surface properties of plastics can be modified to increase or decrease the hydrophobicity of plastic particles, thus aiding with the flotation process. Table 6.1 lists some of the research conducted in the field of plastics flotation.

6.10 Sensor-Based Sorting

Sensor-based sorting is an automated material detection and removal system that uses high-precision sorter to identify the materials based on their physical and chemical signature and then remove it using an ejection system. It is an online, contactless real-time system that analyzes the particle properties along with its location. The sensor-based system has four major components (Wills and Finch 2016d),

i) Particle presentation: Particle is fed to the sensor using a conveyor belt in a monolayer distribution.
ii) Sensing (particle examination/detection): Various sensor systems (listed in Table 6.2) have been developed to analyze the particle and capture the particle properties.
iii) Electronic processing (data analysis): The particle properties are processed using a lab-developed algorithm based on sensor type and ore properties.
iv) Separation: The particle rejection is achieved by a stream of air jets. Mechanical ejectors have also been employed for some sensing technologies.

6.11 Example Flowsheets

The European Recycling Platform have published the recycling process for some of the common e-waste items as shown in Figure 6.16. It shows that the recycling process starts with manual dismantling and removal of hazardous chemicals followed by the shredding and size reduction process. The separation of various components is achieved by a combination of physical separation processes and the recovered items are then sent for subsequent processes.

Table 6.1 Research in the field of flotation separation of various plastics present in e-waste.

Authors	Materials[a)]	Reagents	Frother	Particle size
Abbasi et al. (2010)	PET and PVC	Tannic acid	MIBC	−6.23 + 2.78 mm
Barbakadze et al. (2018)	PVC, nylon, and HDPE	Ca lignosulfonate, tannic acid, methylcellulose, Triton X-100	MIBC	0.5–0.7 cm
Buchan and Yarar (1995)	PET, PVC, PE, and PP	Methanol	MIBC	−1.7 mm
Carvalho et al. (2010)	PE, PP, PS, PET, and PVC	Ca lignosulfonate, hostaphat	MIBC	3–12 mm
Censori et al. (2016)	PS and ABS	Tannic acid	MIBC	2–4 mm
Dodbiba et al. (2002a)	PET and PE or PP	Dodecylamine acetate, Polyvinyl alcohol	MIBC	2–8 mm
Drelich et al. (1998, 1999)	PET and PVC	–	Rhodasurf 91-6	2–10 mm
Guern et al. (Le Guern et al. 2001; Le Guern et al. 2000)	PET and PVC	Na and NH_4^+ lignosulfonate	–	3–10 mm
Güney et al. (2015)	PET and PVC	Diethylene glycol, lignin alkali	MIBC	−3.36 + 2 mm
Marques and Tenório (2000)	PET and PVC	Ca lignosulfonate	Pine oil and MIBC	−4.70 + 1.70 mm
Martin et al. (2011)	PET and PVC	–	Polyethylene glycol	5 × 5 mm
Negari et al. (2018)	PS, PET, and PVC	Polyethylene glycol, polyvinyl alcohol, methylcellulose, tannic acid	MIBC	5 mm diameter
Pascoe (2005)	ABS and HIPS	Acetic acid, methanol carboxymethyl cellulose, Quebracho	Pine oil	5–10 mm
Pita and Castilho (2017)	PS, PMMA, PET, and PVC	Tannic acid	MIBC	−5.6 mm
Pongstabodee et al. (2008)	PET, HDPE, PVC, PP, PS, and ABS	Ca lignosulfonate	MIBC	0.3–0.5 cm

Table 6.1 (Continued)

Authors	Materials[a]	Reagents	Frother	Particle size
Saisinchai (2014)	PET and PVC	Ca lignosulfonate	Pine oil	5 × 5 mm
Saitoh et al. (1976)	PP, PE, PS, and PVC	Na lignosulfonate, tannic acid	–	5 mm
Salerno et al. (2018)	PS and PVC	—	MIBC	2–4 mm
Shen et al. (2001, 2002)	PET, PC, PMMA, PVC, PS, POM, and ABS	Tannic acid, alkyl ethoxylated nonionic surfactant (Tergitol 15-S-7)	MIBC	−5.6 mm −2.38 + 2.00 mm
Shibata et al. (1996)	PVC, PC, POM, and PPE	Na lignosulfonate, tannic acid, polyoxyethylene lauryl ether, aerosol OT, sorbitan monolaurate, polyvinyl alcohol, saponin	Pine oil	−4 + 2 mm
Singh (1998)	PVC and POM	Na lignosulfonate, sorbitan monolaurate	Pine oil	2–4 mm
Takoungsakdakun and Pongstabodee (2007)	PET, POM, and PVC	Ca lignosulfonate polyvinyl alcohol	MIBC	0.3–0.5 cm
Wang et al. (2013)	PE, PVC, PP, PS, PET, ABS, and PC	–	MIBC, Terpineol Diethyl phthalate	−4 mm
Wang et al. (2014)	PP, PE, PTFE, ABS, PS, PVC, PET, and PC	Lignin sulfonate, tannic acid, methylcellulose, Triton X-100	–	−5 mm
Yenial et al. (2013)	PET and PVC	Diethylene glycol dibenzoate, Triton XL-100 N	MIBC	−3.36 + 2 mm
Zhao et al. (2019)	PET, PVC, and PE	Na lignosulfonate, tannic acid	MIBC, glycol	–

a) ABS – Acrylonitrile–butadiene–styrene, HDPE – High-density polyethylene, HIPS – High-impact polystyrene, PC – Polycarbonates, PE – Polyethylene, PET – Polyethylene terephthalate, PMMA – Polymethyl methacrylate, POM – Polyoxymethylene or Polyacetal, PP – Polypropylene, PPE – Polyphenylene ether, PS – Polystyrene, PTFE – Polytetrafluoroethylene, PVC – Polyvinyl chloride.

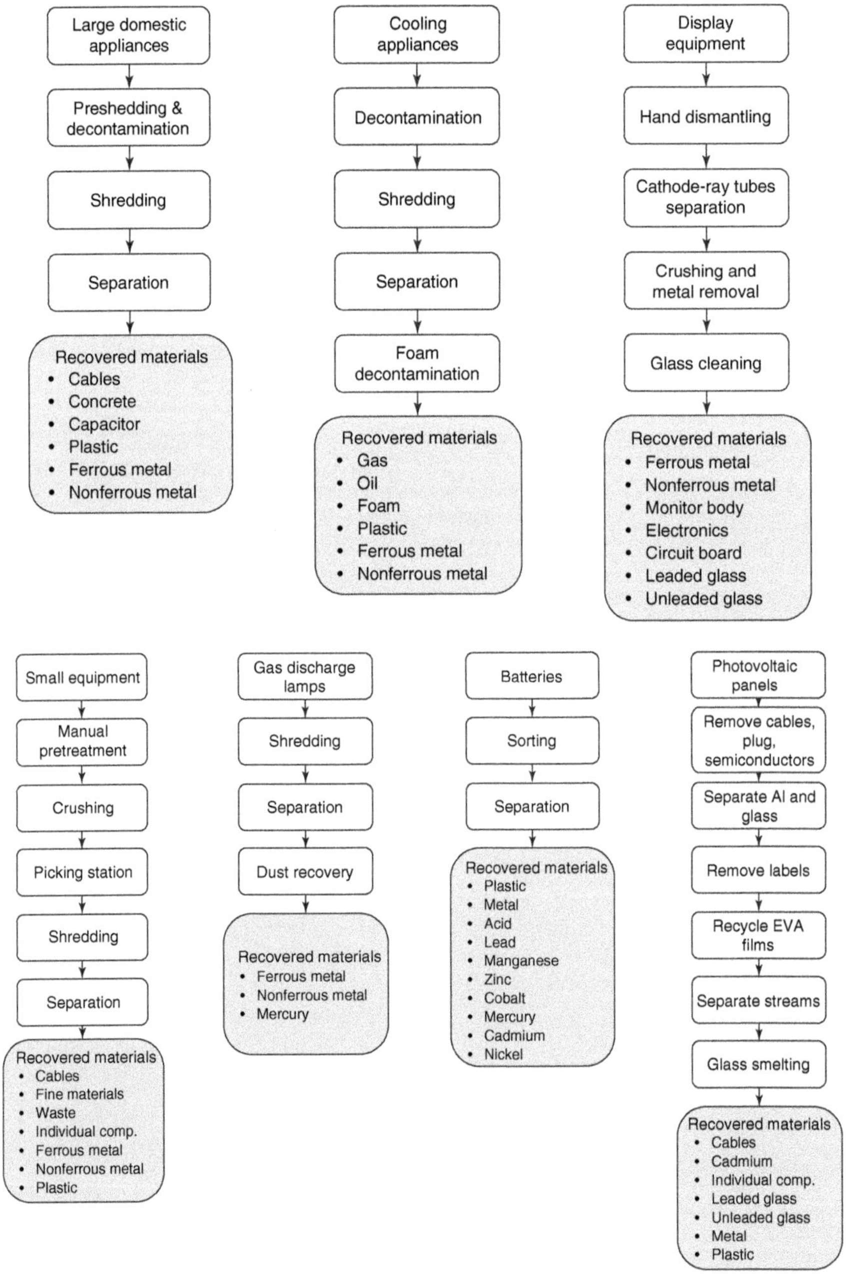

Figure 6.16 Recycling process and recovered materials from various e-waste units. Source: Based on European Recycling Platform (2017).

Table 6.2 Sensor technologies.

Sensors	Physical properties	Application
Prompt gamma neutron activation analysis	Elemental composition	Metals
X-ray transmission	Density	Metal, glass, plastic
X-ray fluorescence	Elemental composition	Metal separation
Photometric	Color	Circuit board, glass, plastic
Near-infrared spectroscopy	Infrared absorption	Plastic, rubber, polymers
Electromagnetic induction	Conductivity	Metal separation
Laser-induced breakdown spectroscopy	Elemental composition	Metal separation, flame retardants

Sources: Based on Hilscher et al. (2017) and Wills and Finch (2016d).

References

Abbasi, M., Salarirad, M.M., and Ghasemi, I. (2010). Selective separation of PVC from PET/PVC mixture using floatation by tannic acid depressant. *Iranian Polymer Journal* 19 (7): 483–489. https://www.researchgate.net/publication/229057122_Selective_Separation_of_PVC_from_PETPVC_Mixture_Using_Floatation_by_Tannic_Acid_Depressant (accessed 18 March 2021).

Abdelbasir, S.M., Hassan, S.S.M., Kamel, A.H., and El-Nasr, R.S. (2018). Status of electronic waste recycling techniques: a review. *Environmental Science and Pollution Research* 25 (17): 16533–16547. https://doi.org/10.1007/s11356-018-2136-6.

Ali, S.S.M., Heng, J.Y.Y., Nikolaev, A.A., and Waters, K.E. (2013). Introducing inverse gas chromatography as a method of determining the surface heterogeneity of minerals for flotation. *Powder Technology* 249: 373–377. https://doi.org/10.1016/J.POWTEC.2013.09.004.

Ambrós, W.M., Sampaio, C.H., Cazacliu, B.G. et al. (2019). Some observations on the influence of particle size and size distribution on stratification in pneumatic jigs. *Powder Technology* 342: 594–606. https://doi.org/10.1016/J.POWTEC.2018.10.029.

Barbakadze, K., Brostow, W., Granowski, G. et al. (2018). Separation of metal and plastic wastes from wire and cable manufacturing for effective recycling. *Resources, Conservation and Recycling* 139: 251–258. https://doi.org/10.1016/J.RESCONREC.2018.06.022.

Bauer, M., Lehner, M., Schwabl, D. et al. (2018). Sink–float density separation of post-consumer plastics for feedstock recycling. *Journal of Material Cycles and Waste Management* 20 (3): 1781–1791. https://doi.org/10.1007/s10163-018-0748-z.

Bi, H., Zhu, H., Zu, L. et al. (2019). A new model of trajectory in eddy current separation for recovering spent lithium iron phosphate batteries. *Waste Management* 100: 1–9. https://doi.org/10.1016/J.WASMAN.2019.08.041.

Bizzo, W., Figueiredo, R., and de Andrade, V. (2014). Characterization of printed circuit boards for metal and energy recovery after milling and mechanical separation. *Materials* 7 (6): 4555–4566. https://doi.org/10.3390/ma7064555.

Buchan, R. and Yarar, B. (1995). Recovering plastics for recycling by mineral processing techniques. *Journal of The Minerals Metals & Materials Society* 47 (2): 52–55. https://doi.org/10.1007/BF03221410.

Burat, F. and Özer, M. (2018). Physical separation route for printed circuit boards. *Physicochemical Problems of Mineral Processing* 54 (2): 554–566. https://doi.org/10.5277/ppmp1858.

Carvalho, T., Durão, F., and Ferreira, C. (2010). Separation of packaging plastics by froth flotation in a continuous pilot plant. *Waste Management* 30 (11): 2209–2215. https://doi.org/10.1016/J.WASMAN.2010.05.023.

Censori, M., La Marca, F., and Carvalho, M.T. (2016). Separation of plastics: the importance of kinetics knowledge in the evaluation of froth flotation. *Waste Management (New York, N.Y.)* 54: 39–43. https://doi.org/10.1016/j.wasman.2016.05.021.

Chen, Q., Yang, H., Tong, L. et al. (2020). Research and application of a Knelson concentrator: a review. *Minerals Engineering* 152: 106339. https://doi.org/10.1016/J.MINENG.2020.106339.

Choi, W.Z. (2004). Removal of PVC from mixed plastic waste by gravity separation process. *Geosystem Engineering* 7 (3): 57–62. https://doi.org/10.1080/12269328.2004.10541221.

Crawford, C.B. and Quinn, B. (2017). Microplastic separation techniques. In: *Microplastic Pollutants* (eds. C.B. Crawford and B.B.T.-M.P. Quinn), 203–218. Elsevier https://doi.org/https://doi.org/10.1016/B978-0-12-809406-8.00009-8.

Cui, J. and Forssberg, E. (2007). Characterization of shredded television scrap and implications for materials recovery. *Waste Management* 27 (3): 415–424. https://doi.org/10.1016/J.WASMAN.2006.02.003.

Da Silva, M.D.F., Dutra, A.J.B., and Mansur, M.B. (2019). Enrichment of copper, lead, and tin by mechanical dry processing of obsolete printed circuit board residues. *Materials Research* 22 (5) https://doi.org/10.1590/1980-5373-MR-2019-0341.

Das, A., Vidyadhar, A., and Mehrotra, S.P. (2009). A novel flowsheet for the recovery of metal values from waste printed circuit boards. *Resources, Conservation and Recycling* 53 (8): 464–469. https://doi.org/10.1016/j.resconrec.2009.03.008.

Dodbiba, G., Haruki, N., Shibayama, A. et al. (2002a). Combination of sink–float separation and flotation technique for purification of shredded PET-bottle from PE or PP flakes. *International Journal of Mineral Processing* 65 (1): 11–29. https://doi.org/10.1016/S0301-7516(01)00056-4.

Dodbiba, G., Sadaki, J., Okaya, K. et al. (2005). The use of air tabling and triboelectric separation for separating a mixture of three plastics. *Minerals Engineering* 18 (15): 1350–1360. https://doi.org/10.1016/J.MINENG.2005.02.015.

Dodbiba, G. and Fujita, T. (2015). Air tabling—a dry gravity solid–solid separation technique. In: *Progress in Filtration and Separation* (ed. S. Tarleton), 527–555. Academic Press https://doi.org/10.1016/B978-0-12-384746-1.00012-4.

Dodbiba, G., Shibayama, A., Miyazaki, T., and Fujita, T. (2002b). Electrostatic separation of the shredded plastic mixtures using a tribo-cyclone. *Magnetic and Electrical Separation* 11 (1–2): 63–92. https://doi.org/10.1080/07313630290002626.

Drelich, J., Kim, J.H., Payne, T. et al. (1999). Purification of polyethylene terephthalate from polyvinyl chloride by froth flotation for the plastics (soft-drink bottle) recycling industry. *Separation and Purification Technology* 15 (1): 9–17. https://doi.org/10.1016/S1383-5866(98)00047-1.

Drelich, J., Payne, T., Kim, J.H. et al. (1998). Selective froth flotation of PVC from PVC/PET mixtures for the plastics recycling industry. *Polymer Engineering & Science* 38 (9): 1378–1386. https://doi.org/10.1002/pen.10308.

Duan, C., Wen, X., Shi, C. et al. (2009). Recovery of metals from waste printed circuit boards by a mechanical method using a water medium. *Journal of Hazardous Materials* 166 (1): 478–482. https://doi.org/10.1016/j.jhazmat.2008.11.060.

Duan, C.L., Diao, Z.J., Zhao, Y.M., and Huang, W. (2015). Liberation of valuable materials in waste printed circuit boards by high-voltage electrical pulses. *Minerals Engineering* 70: 170–177. https://doi.org/10.1016/j.mineng.2014.09.018.

Dunne, R.C. (2019). Mechanical classifiers. In: *SME Mineral Processing & Extractive Metallurgy Handbook* (eds. S.K. Kawatra and C.A. Young), 625–632. Society for Mining, Metallurgy, and Exploration. https://app.knovel.com/web/view/khtml/show.v/rcid:kpSMEMPEM1/cid:kt011IX673/viewerType:khtml//root_slug:sme-mineral-processing/url_slug:mechanical-classifiers?page=last&view=collapsed&zoom=1 (accessed 18 March 2021).

Ekman, R. (2018). Development of a Plastic Shredder. Lund University.

Eswaraiah, C., Kavitha, T., Vidyasagar, S., and Narayanan, S.S. (2008). Classification of metals and plastics from printed circuit boards (PCB) using air classifier. *Chemical Engineering and Processing: Process Intensification* 47 (4): 565–576. https://doi.org/10.1016/J.CEP.2006.11.010.

European Recycling Platform. (2017). *What happens to your waste • ERP UK*. https://erp-recycling.org/uk/learning-centre/what-happens-to-your-waste (accessed 18 March 2021).

Fitzgerald, G. C. (2009). Technical and Economic Analysis of Pre-Shredding Municipal Solid Wastes Prior to Disposal. Columbia University.

Forssberg, E., Zhang, S., & Björkman, B. (1998). Metals recycling from electronic scrap by air table separation: theory and application. EPD Congress, 497–515. http://www.diva-portal.org/smash/record.jsf?pid=diva2%3A1003563&dswid=1889 (accessed 18 March 2021).

Fujita, T., Ono, H., Dodbiba, G., and Yamaguchi, K. (2014). Evaluation of a recycling process for printed circuit board by physical separation and heat treatment. *Waste Management* 34 (7): 1264–1273. https://doi.org/10.1016/J.WASMAN.2014.03.002.

Gallegos-Acevedo, P.M., Espinoza-Cuadra, J., and Olivera-Ponce, J.M. (2014). Conventional flotation techniques to separate metallic and nonmetallic fractions from waste printed circuit boards with particles nonconventional size. *Journal of Mining Science* 50 (5): 974–981. https://doi.org/10.1134/S1062739114050172.

Ghosh, B., Ghosh, M.K., Parhi, P. et al. (2015). Waste printed circuit boards recycling: an extensive assessment of current status. *Journal of Cleaner Production* 94: 5–19. https://doi.org/10.1016/J.JCLEPRO.2015.02.024.

Güney, A., Özdilek, C., Kangal, M.O., and Burat, F. (2015). Flotation characterization of PET and PVC in the presence of different plasticizers. *Separation and Purification Technology* 151: 47–56. https://doi.org/10.1016/J.SEPPUR.2015.07.027.

Guo, C., Wang, H., Liang, W. et al. (2011). Liberation characteristic and physical separation of printed circuit board (PCB). *Waste Management* 31 (9–10): 2161–2166. https://doi.org/10.1016/J.WASMAN.2011.05.011.

Haldar, S.K. (2018). Mineral processing. In: *Mineral Exploration*, 2e, 259–290. Elsevier https://doi.org/10.1016/B978-0-12-814022-2.00013-7.

Han, S., Park, S., Kim, S., and Park, J. (2014). Simulation on the PCB particle trajectories in Corona-discharge electrostatic separator. *Journal of the Korean Institute of Resources Recycling* 23 (6): 30–39. https://doi.org/10.7844/kirr.2014.23.6.30.

Hanafi, J., Jobiliong, E., Christiani, A. et al. (2012). Material recovery and characterization of PCB from electronic waste. *Procedia – Social and Behavioral Sciences* 57: 331–338. https://doi.org/10.1016/J.SBSPRO.2012.09.1194.

He, J. and Duan, C. (2017). Recovery of metallic concentrations from waste printed circuit boards via reverse floatation. *Waste Management* 60: 618–628. https://doi.org/10.1016/J.WASMAN.2016.11.019.

He, J.F., Duan, C.L., He, Y.Q., and Zhang, H.J. (2015). Recovery of valuable metal concentrate from waste printed circuit boards by a physical beneficiation technology. *International journal of Environmental Science and Technology* 12 (8): 2603–2612. https://doi.org/10.1007/s13762-014-0664-2.

Hilscher, B., Nayak, P., Lorio, L., & Yoon, N. (2017). Technical Review and Evaluation of Ore Sorting Technologies on Precious Metals Operations – CEEC (Coalition for Eco Efficient Comminution). *49th Annual Conference of Metallurgists.*

Hori, K., Tsunekawa, M., Ueda, M. et al. (2009). Development of a new gravity separator for plastics – a hybrid jig. *Materials Transactions* 50 (12): 2844–2847. https://doi.org/10.2320/matertrans.M-M2009825.

Hsu, E., Barmak, K., West, A.C., and Park, A.-H.A. (2019). Advancements in the treatment and processing of electronic waste with sustainability: a review of metal extraction and recovery technologies. *Green Chemistry* 21 (5): 919–936. https://doi.org/10.1039/C8GC03688H.

Ito, M., Tsunekawa, M., Ishida, E. et al. (2010). Reverse jig separation of shredded floating plastics – separation of polypropylene and high density polyethylene. *International Journal of Mineral Processing* 97 (1–4): 96–99. https://doi.org/10.1016/J.MINPRO.2010.08.007.

Jujun, R., Yiming, Q., and Zhenming, X. (2014). Environment-friendly technology for recovering nonferrous metals from e-waste: eddy current separation. *Resources, Conservation and Recycling* 87: 109–116. https://doi.org/10.1016/J.RESCONREC.2014.03.017.

Kasper, A.C., de Juchneski, N.C., F., and Veit, H.M. (2015). Mechanical processing. In: *Electronic Waste: Recycling Techniques Recycling Techniques* (eds. H.M. Veit and A.M. Bernardes), 19–38. Springer https://doi.org/10.1007/978-3-319-15714-6_4.

Kaya, M. (2016). Recovery of metals and nonmetals from electronic waste by physical and chemical recycling processes. *Waste Management* 57: 64–90.

Kaya, M. (2018). Current WEEE recycling solutions. In: *Waste Electrical and Electronic Equipment Recycling* (eds. F. Vegliò and I. Birloaga), 33–93. Woodhead Publishing https://doi.org/10.1016/B978-0-08-102057-9.00003-2.

Kwon, S., Park, S., Kim, S. et al. (2016). Separation of tantalum from electronic components on laptop printed circuit board assembly. *Journal of the Korean Institute of Resources Recycling* 25 (1): 24–30. https://doi.org/10.7844/kirr.2016.25.1.24.

Le Guern, C., Baillif, P., Conil, P., and Houot, R. (2001). Adsorption of a lignosulphonate polymer onto PVC and PET surfaces: evaluation by XPS. *Journal of Materials Science* 36 (6): 1547–1554. https://doi.org/10.1023/A:1017565201410.

Le Guern, C., Conil, P., and Houot, R. (2000). Role of calcium ions in the mechanism of action of a lignosulphonate used to modify the wettability of plastics for their separation by flotation. *Minerals Engineering* 13 (1): 53–63. https://doi.org/10.1016/S0892-6875(99)00149-1.

Lee, H. and Mishra, B. (2018). Selective recovery and separation of copper and iron from fine materials of electronic waste processing. *Minerals Engineering* 123: 1–7. https://doi.org/10.1016/J.MINENG.2018.04.021.

Lee, J.-K. and Shin, J.-H. (2002). Triboelectmstatic separation of PVC materials from mixed plastics waste plastic recycling. *Korean Journal of Chemical Engineering* 19 (2): 267–272.

Lee, K.-H. (2007). Pyrolysis of municipal plastic wastes separated by difference of specific gravity. *Journal of Analytical and Applied Pyrolysis* 79 (1–2): 362–367. https://doi.org/10.1016/J.JAAP.2006.12.020.

Li, J., Jiang, Y., and Xu, Z. (2017). Eddy current separation technology for recycling printed circuit boards from crushed cell phones. *Journal of Cleaner Production* 141: 1316–1323. https://doi.org/10.1016/J.JCLEPRO.2016.09.144.

Li, J., Wu, G., and Xu, Z. (2015). Tribo-charging properties of waste plastic granules in process of tribo-electrostatic separation. *Waste Management* 35: 36–41. https://doi.org/10.1016/J.WASMAN.2014.10.001.

Li, J., Xu, Z., and Zhou, Y. (2007). Application of corona discharge and electrostatic force to separate metals and nonmetals from crushed particles of waste printed circuit boards. *Journal of Electrostatics* 65 (4): 233–238. https://doi.org/10.1016/J.ELSTAT.2006.08.004.

Liu, W., Liang, C., Qin, W., and Jiao, F. (2014). A new technology for recovery of metals from waste printed circuit boards. *Applied Mechanics and Materials* 675–677: 698–703. https://doi.org/10.4028/http://www.scientific.net/AMM.675-677.698.

Luttrell, G.H., Honaker, R.Q., Bratton, R.C. et al. (2006). *In-Plant Testing of High-Efficiency Hydraulic Separators*. Virginia Polytechnic Institute and State University https://doi.org/10.2172/907775.

Mäkinen, J., Bachér, J., Kaartinen, T. et al. (2015). The effect of flotation and parameters for bioleaching of printed circuit boards. *Minerals Engineering* 75: 26–31. https://doi.org/10.1016/J.MINENG.2015.01.009.

Mankosa, M.J., Kohmuenc, J.N., and Honaker, R.Q. (2019). Fluidized-bed classifiers. In: *SME Mineral Processing & Extractive Metallurgy Handbook* (eds. S.K. Kawatra and C.A. Young), 611–624. Society for Mining, Metallurgy, and Exploration.

Manser, R.J., Barley, R.W., and Wills, B.A. (1991). The shaking table concentrator – the influence of operating conditions and table parameters on mineral separation – the development of a mathematical model for normal operating conditions. *Minerals Engineering* 4 (3–4): 369–381. https://doi.org/10.1016/0892-6875(91)90142-I.

Marques, G.A. and Tenório, J.A.S. (2000). Use of froth flotation to separate PVC/PET mixtures. *Waste Management* 20 (4): 265–269. https://doi.org/10.1016/S0956-053X(99)00333-5.

Martin, N., Skvarla, J., and Sisol, M. (2011). A possibility of using the flotation process to separate plastics. *International Journal of Engineering*. https://www.researchgate.net/publication/283601296_A_Possibility_of_using_the_flotation_process_to_separate_plastics (accessed 18 March 2021) 9 (3): 275–278.

Menad, N.-E. (2016). Physical separation processes in waste electrical and electronic equipment recycling. In: *WEEE Recycling* (eds. A. Chagnes, G. Cote, C. Ekberg, et al.), 53–74. Elsevier. https://doi.org/10.1016/B978-0-12-803363-0.00003-1.

Meng, L., Zhong, Y., Guo, L. et al. (2018). Recovery of Cu and Zn from waste printed circuit boards using super-gravity separation. *Waste Management* 78: 559–565. https://doi.org/10.1016/J.WASMAN.2018.06.022.

Miceli, H., Rossi, M.G., Neumann, R., and Tavares, L.M. (2017). Contaminant removal from manufactured fine aggregates by dry rare-earth magnetic separation. *Minerals Engineering* 113: 15–22. https://doi.org/10.1016/j.mineng.2017.07.017.

Mohammadi-Jam, S., Burnett, D.J., and Waters, K.E. (2014). Surface energy of minerals – applications to flotation. *Minerals Engineering* 66–68: 112–118. https://doi.org/10.1016/J.MINENG.2014.05.002.

Mohammadi-Jam, S. and Waters, K.E. (2016). Inverse gas chromatography analysis of minerals: pyrite wettability. *Minerals Engineering* 96–97: 130–134. https://doi.org/10.1016/J.MINENG.2016.05.007.

Mou, P., Wa, L., Xiang, D. et al. (2004). A physical process for recycling and reusing waste printed circuit boards. *IEEE International Symposium on Electronics and the Environment*: 237–242. https://doi.org/10.1109/isee.2004.1299722.

Namias, J. (2013). *The Future of Electronic Waste Recycling in the United States: Obstacles and Domestic Solutions*. New York, NY: Columbia University.

Negari, M.S., Ostad Movahed, S., and Ahmadpour, A. (2018). Separation of polyvinylchloride (PVC), polystyrene (PS) and polyethylene terephthalate (PET) granules using various chemical agents by flotation technique. *Separation and Purification Technology* 194: 368–376. https://doi.org/10.1016/J.SEPPUR.2017.11.062.

Nekouei, R.K., Pahlevani, F., Rajarao, R. et al. (2018). Two-step pre-processing enrichment of waste printed circuit boards: mechanical milling and physical separation. *Journal of Cleaner Production* 184: 1113–1124. https://doi.org/10.1016/J.JCLEPRO.2018.02.250.

Ning, C., Lin, C.S.K., Hui, D.C.W., and McKay, G. (2017). Waste printed circuit board (PCB) recycling techniques. *Topics in Current Chemistry* 375 (2): 43.

Ogunniyi, I.O. and Vermaak, M.K.G. (2009). Investigation of froth flotation for beneficiation of printed circuit board comminution fines. *Minerals Engineering* 22 (4): 378–385.

Ogunniyi, I.O., Vermaak, M.K.G., and Groot, D.R. (2009). Chemical composition and liberation characterization of printed circuit board comminution fines for beneficiation investigations. *Waste Management (New York, N.Y.)* 29 (7): 2140–2146. https://doi.org/10.1016/j.wasman.2009.03.004.

Olsen, L., & Carnes, B. (2016). Screen capacity calculation. Technical Paper T-JCI-201. https://www.911metallurgist.com/blog/wp-content/uploads/2016/07/Screen-Capacity.pdf (accessed 14 January 2021).

Oruç, F., Özgen, S., and Sabah, E. (2010). An enhanced-gravity method to recover ultra-fine coal from tailings: Falcon concentrator. *Fuel* 89 (9): 2433–2437. https://doi.org/10.1016/J.FUEL.2010.04.009.

Otsuki, A., Pereira Gonçalves, P., and Leroy, E. (2019). Selective milling and elemental assay of printed circuit board particles for their recycling purpose. *Metals* 9 (8): 899. https://doi.org/10.3390/met9080899.

Pascoe, R.D. (2005). The use of selective depressants for the separation of ABS and HIPS by froth flotation. *Minerals Engineering* 18 (2): 233–237. https://doi.org/10.1016/J.MINENG.2004.07.006.

Phengsaart, T., Ito, M., Hamaya, N. et al. (2018). Improvement of jig efficiency by shape separation, and a novel method to estimate the separation efficiency of metal wires in crushed electronic wastes using bending behavior and entanglement factor. *Minerals Engineering* 129: 54–62. https://doi.org/10.1016/J.MINENG.2018.09.015.

Pita, F. and Castilho, A. (2016). Influence of shape and size of the particles on jigging separation of plastics mixture. *Waste Management* 48: 89–94. https://doi.org/10.1016/J.WASMAN.2015.10.034.

Pita, F. and Castilho, A. (2017). Separation of plastics by froth flotation. The role of size, shape and density of the particles. *Waste Management* 60: 91–99. https://doi.org/10.1016/J.WASMAN.2016.07.041.

Pita, F. and Castilho, A. (2018). Separation of copper from electric cable waste based on mineral processing methods: a case study. *Minerals* 8 (11): 517. https://doi.org/10.3390/min8110517.

Pongstabodee, S., Kunachitpimol, N., and Damronglerd, S. (2008). Combination of three-stage sink–float method and selective flotation technique for separation of mixed post-consumer plastic waste. *Waste Management* 28 (3): 475–483. https://doi.org/10.1016/J.WASMAN.2007.03.005.

Quan, C., Li, A., and Gao, N. (2012). Study on characteristics of printed circuit board liberation and its crushed products. *Waste Management & Research* 30 (11): 1178–1186. https://doi.org/10.1177/0734242X12457119.

Ruan, J., Dong, L., Zheng, J. et al. (2017). Key factors of eddy current separation for recovering aluminum from crushed e-waste. *Waste Management* 60: 84–90. https://doi.org/10.1016/J.WASMAN.2016.08.018.

Ruan, J. and Xu, Z. (2011). A new model of repulsive force in eddy current separation for recovering waste toner cartridges. *Journal of Hazardous Materials* 192 (1): 307–313. https://doi.org/10.1016/J.JHAZMAT.2011.05.025.

Ruj, B., Pandey, V., Jash, P., and Srivastava, V.K. (2015). Sorting of plastic waste for effective recycling. *Journal of Applied Sciences and Engineering Research* 4. https://doi.org/10.6088.ijaser.04058 (4): 564–571.

Saeki, M. (2008). Triboelectric separation of three-component plastic mixture. *Particulate Science and Technology* 26 (5): 494–506. https://doi.org/10.1080/02726350802367936.

Saisinchai, S. (2014). Separation of PVC from PET/PVC mixtures using flotation by calcium lignosulfonate depressant. *Engineering Journal* 18 (1): 45–54. https://doi.org/10.4186/ej.2014.18.1.45.

Saitoh, K., Nagano, I., and Izumi, S. (1976). New separation technique for waste plastics. *Resource Recovery and Conservation* 2 (2): 127–145. https://doi.org/10.1016/0304-3967(76)90004-4.

Salerno, D., Jordão, H., La Marca, F., and Carvalho, M.T. (2018). Using factorial experimental design to evaluate the separation of plastics by froth flotation. *Waste Management* 73: 62–68. https://doi.org/10.1016/j.wasman.2017.12.001.

Sarvar, M., Salarirad, M.M., and Shabani, M.A. (2015). Characterization and mechanical separation of metals from computer Printed Circuit Boards (PCBs) based on mineral processing methods. *Waste Management* 45: 246–257. https://doi.org/10.1016/J.WASMAN.2015.06.020.

Schubert, G. and Hoberg, H. (1997). *Comminution Techniques for the Recycling of Wastes*. Littleton, CO: Society for Mining, Metallurgy and Exploration, Inc.

SCHUTTE Hammer Mill. (2019). *E-Scrap Grinding Hammer Mill*. https://www.hammermills.com/equipment/e-cycler-series-hammer-mill/#tab-description (accessed 18 March 2021).

SELFRAG. (2012). *Application area | Recycling*. http://www.selfrag.com/recycling-examples.php (accessed 18 March 2021).

Senden, M. M. G. (1980). Performance of zigzag classifiers at low particle concentrations: a study of the effect of stage geometry variations. *ASME 1980 National Waste Processing Conference*.

Serranti, S. and Bonifazi, G. (2019). Techniques for separation of plastic wastes. In: *Use of Recycled Plastics in Eco-Efficient Concrete* (eds. F. Pacheco-Torgal, J. Khatib, F. Colangelo and R. Tuladhar), 9–37. https://doi.org/10.1016/B978-0-08-102676-2.00002-5. Woodhead Publishing.

Shariati, S., Ramadi, A., and Salsani, A. (2015). Beneficiation of low-grade phosphate deposits by a combination of calcination and shaking tables: Southwest Iran. *Minerals* 5 (3): 367–379. https://doi.org/10.3390/min5030367.

Shen, H., Forssberg, E., and Pugh, R.J. (2001). Selective flotation separation of plastics by particle control. *Resources, Conservation and Recycling* 33 (1): 37–50. https://doi.org/10.1016/S0921-3449(01)00056-8.

Shen, H., Pugh, R.J., and Forssberg, E. (2002). Floatability, selectivity and flotation separation of plastics by using a surfactant. *Colloids and Surfaces A: Physicochemical and Engineering Aspects* 196 (1): 63–70. https://doi.org/10.1016/S0927-7757(01)00706-3.

Shi, F., Krishnan, N., von der Weid, F., van der Wielen, K., Zuo, W., & Manlapig, E. (2014). A potential application of high voltage pulse technology in a gold-copper ore grinding circuit. *XXVII International Mineral Processing Congress*, 68–77.

Shibata, J., Matsumoto, S., Yamamoto, H. et al. (1996). Flotation separation of plastics using selective depressants. *International Journal of Mineral Processing* 48 (3–4): 127–134. https://doi.org/10.1016/S0301-7516(96)00021-X.

Singh, B.P. (1998). Wetting mechanism in the flotation separation of plastics. *Filtration & Separation* 35 (6): 525–527. https://doi.org/10.1016/S0015-1882(98)80027-X.

Smith, Y.R., Nagel, J.R., and Rajamani, R.K. (2019). Eddy current separation for recovery of non-ferrous metallic particles: a comprehensive review. *Minerals Engineering* 133: 149–159. https://doi.org/10.1016/J.MINENG.2018.12.025.

Sun, Z.H.I., Xiao, Y., Sietsma, J. et al. (2015). Characterisation of metals in the electronic waste of complex mixtures of end-of-life ICT products for development of cleaner recovery technology. *Waste Management* 35: 227–235. https://doi.org/10.1016/J.WASMAN.2014.09.021.

Takoungsakdakun, T. and Pongstabodee, S. (2007). Separation of mixed post-consumer PET–POM–PVC plastic waste using selective flotation. *Separation and Purification Technology* 54 (2): 248–252. https://doi.org/10.1016/J.SEPPUR.2006.09.011.

Tripathy, S.K., Banerjee, P.K., and Suresh, N. (2014). Separation analysis of dry high intensity induced roll magnetic separator for concentration of hematite fines. *Powder Technology* 264: 527–535. https://doi.org/https://doi.org/10.1016/j.powtec.2014.05.065.

UNTHA Shredding Technology. (2018). *Electronic scrap*. https://www.untha.com/en/applications/electronic-scrap_a12 (accessed 18 March 2021).

Wielen, K. Van Der, Weh, A., Giese, H., & Käppeler, J. (2014). High voltage breakage: a review of theory and applications. *XXVII International Mineral Processing Congress*, 78–87.

Vega-Garcia, D., Brito-Parada, P.R., and Cilliers, J.J. (2018). Optimising small hydrocyclone design using 3D printing and CFD simulations. *Chemical Engineering Journal* 350: 653–659. https://doi.org/10.1016/J.CEJ.2018.06.016.

Veit, H.M., Diehl, T.R., Salami, A.P. et al. (2005). Utilization of magnetic and electrostatic separation in the recycling of printed circuit boards scrap. *Waste Management* 25 (1): 67–74. https://doi.org/10.1016/J.WASMAN.2004.09.009.

Veit, H.M., Bernardes, A.M., Ferreira, J.Z. et al. (2006). Recovery of copper from printed circuit boards scraps by mechanical processing and electrometallurgy. *Journal of Hazardous Materials* 137 (3): 1704–1709. https://doi.org/10.1016/J.JHAZMAT.2006.05.010.

Veit, H.M., de Juchneski, N.C., F., and Scherer, J. (2014). Use of gravity separation in metals concentration from printed circuit board scraps. *Rem: Revista Escola de Minas* 67 (1): 73–79. https://doi.org/10.1590/S0370-44672014000100011.

Ventura, E., Futuro, A., Pinho, S.C. et al. (2018). Physical and thermal processing of waste printed circuit boards aiming for the recovery of gold and copper. *Journal of Environmental Management* 223: 297–305. https://doi.org/10.1016/J.JENVMAN.2018.06.019.

Vesilind, P.A. (1980). The Rosin–Rammler particle size distribution. *Resource Recovery and Conservation* 5 (3): 275–277. https://doi.org/10.1016/0304-3967(80)90007-4.

Vidyadhar, A., & Das, A. (2012). Kinetics and efficacy of froth flotation for the recovery of metal values from pulverized printed circuit boards. *XXVI International Mineral Processing Congress*, 236–243.

Vidyadhar, A. and Das, A. (2013). Enrichment implication of froth flotation kinetics in the separation and recovery of metal values from printed circuit boards. *Separation and Purification Technology* 118: 305–312. https://doi.org/10.1016/J.SEPPUR.2013.07.027.

Wang, H., Wang, C., and Fu, J. (2013). Floatability of polymer materials modulated by frothers. *Waste Management* 33 (12): 2623–2631. https://doi.org/10.1016/j.wasman.2013.09.003.

Wang, H., Wang, C., Fu, J., and Gu, G. (2014). Flotability and flotation separation of polymer materials modulated by wetting agents. *Waste Management* 34 (2): 309–315. https://doi.org/10.1016/J.WASMAN.2013.11.007.

Wen, X., Duan, C., Jiao, H., Zhao, Y., Zhou, X., & Song, S. (2005a). Study on metals recovery from discarded printed circuit boards by physical methods. *IEEE International Symposium on Electronics and the Environment*, 121–128. https://doi.org/10.1109/isee.2005.1437005

Wen, X.F., Zhao, Y.M., Pan, Y.J. et al. (2005b). Metals recovery from fine printed circuit boards using falcon SB concentrator. *Journal of China University of Mining and Technology* 15 (4): 322–328.

Wills, B.A. and Finch, J.A. (2016a). Crushers. In: *Wills' Mineral Processing Technology*, 8e (eds. B.A. Wills and J.A. Finch), 123–146. Butterworth-Heinemann https://doi.org/https://doi.org/10.1016/B978-0-08-097053-0.00006-6.

Wills, B.A. and Finch, J.A. (2016b). Classification. In: *Wills' Mineral Processing Technology*, 199–221. Butterworth-Heinemann https://doi.org/10.1016/B978-0-08-097053-0.00009-1.

Wills, B.A. and Finch, J.A. (2016c). Magnetic and electrical separation. In: *Wills' Mineral Processing Technology*, 381–407. Butterworth-Heinemann https://doi.org/10.1016/B978-0-08-097053-0.00013-3.

Wills, B.A. and Finch, J.A. (2016d). Sensor-based ore sorting. In: *Wills' Mineral Processing Technology*, 409–416. Butterworth-Heinemann https://doi.org/10.1016/B978-0-08-097053-0.00014-5.

Wu, C., Shi, G., and Yan, C. (2012). Study on wet treatment process of waste printed circuit boards. *Advanced Materials Research* 518–523: 3621–3626. https://doi.org/10.4028/http://www.scientific.net/AMR.518-523.3621.

Xia, M.-C., Wang, Y.-P., Peng, T.-J. et al. (2017). Recycling of metals from pretreated waste printed circuit boards effectively in stirred tank reactor by a moderately thermophilic culture. *Journal of Bioscience and Bioengineering* 123 (6): 714–721. https://doi.org/10.1016/J.JBIOSC.2016.12.017.

Xiao, C., Allen III, L., Biddle, M. B., & Fisher, M. M. (2003). *Electrostatic Separation and Recovery of Mixed Plastics.*

Yang, J., Wang, H., Zhang, G. et al. (2019). Recycling organics from non-metallic fraction of waste printed circuit boards by a novel conical surface triboelectric separator. *Resources, Conservation and Recycling* 146: 264–269. https://doi.org/10.1016/J.RESCONREC.2019.03.008.

Yenial, Ü., Burat, F., Yüce, A.E. et al. (2013). Separation of PET and PVC by flotation technique without using alkaline treatment. *Mineral Processing and Extractive Metallurgy Review* 34 (6): 412–421. https://doi.org/10.1080/08827508.2012.702705.

Yoo, J.-M., Jeong, J., Yoo, K. et al. (2009). Enrichment of the metallic components from waste printed circuit boards by a mechanical separation process using a stamp mill. *Waste Management* 29 (3): 1132–1137. https://doi.org/10.1016/J.WASMAN.2008.06.035.

Yue-min, Z., Xue-feng, W., Hong-xia, S. et al. (2006). Study on metals recovery from −0.074 mm printed circuit boards by enhanced gravity separation. *Chinese Journal of Process Engineering* 6 (2): 201–204.

Zelmat, M.E.M., Rizouga, M., Tilmatine, A. et al. (2013). Experimental comparative study of different tribocharging devices for triboelectric separation of insulating particles. *IEEE Transactions on Industry Applications* 49 (3): 1113–1118. https://doi.org/10.1109/TIA.2013.2251991.

Zhang, G., Wang, H., He, Y. et al. (2017). Triboelectric separation technology for removing inorganics from non-metallic fraction of waste printed circuit boards: influence of size fraction and process optimization. *Waste Management* 60: 42–49. https://doi.org/10.1016/j.wasman.2016.08.010.

Zhang, S. and Forssberg, E. (1999). Intelligent liberation and classification of electronic scrap. *Powder Technology* 105 (1–3): 295–301. https://doi.org/10.1016/S0032-5910(99)00151-5.

Zhang, S., Forssberg, E., Arvidson, B., and Moss, W. (1998). Aluminum recovery from electronic scrap by High-Force® eddy-current separators. *Resources, Conservation and Recycling* 23 (4): 225–241. https://doi.org/10.1016/S0921-3449(98)00022-6.

Zhang, Y., He, Y., Zhang, T. et al. (2018). Application of falcon centrifuge in the recycling of electrode materials from spent lithium ion batteries. *Journal of Cleaner Production* 202: 736–747. https://doi.org/10.1016/J.JCLEPRO.2018.08.133.

Zhao, Y., Yang, S., Wen, H. et al. (2019). Adsorption behavior and selectivity mechanism of flotation reagents applied in ternary plastic mixtures. *Waste Management* 87: 565–576. https://doi.org/10.1016/J.WASMAN.2019.02.044.

Zhao, Y.M., Duan, C.L., Wu, L.L. et al. (2012). The separation mechanism and application of a tapered diameter separation bed. *International journal of Environmental Science and Technology* 9 (4): 719–728. https://doi.org/10.1007/s13762-012-0105-z.

Zhou, C., Pan, Y., Lu, M., and Yang, C. (2016). Liberation characteristics after cryogenic modification and air table separation of discarded printed circuit boards. *Journal of Hazardous Materials* 311: 203–209. https://doi.org/10.1016/J.JHAZMAT.2016.03.008.

Zhou, X., Wang, X., Zhang, X., and Luo, X. (2012). The efficiency change of spiral classifier with different overflow discharge methods. *Advanced Materials Research* 412: 412–453. https://doi.org/10.4028/http://www.scientific.net/AMR.412.453.

Zhu, X., Nie, C., Wang, S. et al. (2020). Cleaner approach to the recycling of metals in waste printed circuit boards by magnetic and gravity separation. *Journal of Cleaner Production* 248: 119235. https://doi.org/10.1016/J.JCLEPRO.2019.119235.

Zhu, X., Nie, C., Zhang, H. et al. (2019). Recovery of metals in waste printed circuit boards by flotation technology with soap collector prepared by waste oil through saponification. *Waste Management* 89: 21–26. https://doi.org/10.1016/J.WASMAN.2019.03.061.

7

Pyrometallurgical Processes for Recycling Waste Electrical and Electronic Equipment

Jean-Philippe Harvey[1], *Mohamed Khalil*[2], *and Jamal Chaouki*[2]

[1] *Polytechnique Montréal, Centre for Research in Computational Thermochemistry (CRCT), Department of Chemical Engineering, C.P. 6079, succ. Centre-Ville, Montreal, Quebec H3C 3A7, Canada*
[2] *Polytechnique Montréal, Process Engineering Advanced Research Lab (PEARL), Department of Chemical Engineering, C.P. 6079, succ. Centre-Ville, Montreal, Quebec H3C 3A7, Canada*

7.1 Introduction

In the last few decades, the high demand for emerging electronic technologies combined with the short life-in-service of obsolete appliances has led to an exponential rise in waste electrical and electronic equipment (WEEE), which resulted in severe environmental issues: when incinerated, WEEE releases greenhouse gases and toxic emissions; when disposed of in landfills, it contaminates groundwater (Needhidasan et al., 2014). These terrible environmental consequences directly impact developing countries such as Ghana, the world's largest e-waste landfill (Zvezdin et al. 2020; Álvarez-de-los-Mozos et al., 2020; Vaccari et al. 2019). The United Nations (UN) reported that over 50 million metric tonnes (Mt) of WEEE were produced worldwide in 2019 and that this volume is expected to increase by over 70 million metric tonnes (Mt) over the next few years (Franzolin 2020; Adusei et al. 2020; Aboughaly and Gabbar 2020).

Pyrometallurgy is one of the most viable extractive metallurgy techniques for recycling high volumes of WEEE (Ma 2019b; Willner et al. 2014). One significant advantage of many pyrometallurgical processes is the generation of high-temperature liquid phases (such as matte in the case of copper smelting and slag in the case of lead smelting) which are perfect reactive media to melt/dissolve WEEE. Pyrometallurgical processes often involve the simultaneous presence of multiple liquid solutions such as slag, matte, metallic melt, and molten salt in which valuable metals and impurities can be partitioned and eventually discarded (in the case of impurities) (Espinosa et al. 2015; Anindya et al. 2013).

Pyrometallurgy includes many extractive metallurgical operations such as smelting (performed in high-temperature furnaces [Hageluken, 2006; Theo 1998]) and refining (which includes chemical and electrochemical processes [Rudnik and Bayaraa 2016]). High-temperature electrochemical processes use electrical work to produce metals from the reduction of metallic cations present in electrolytes (such

Electronic Waste: Recycling and Reprocessing for a Sustainable Future, First Edition.
Edited by Maria E. Holuszko, Amit Kumar, and Denise C. R. Espinosa.

as molten salts [Flandinet et al., 2012]) and can also be considered in this branch of extractive metallurgy.

In terms of recovered metals and sustainability practices, pyrometallurgical processes significantly enhance recycling efficiency as they operate at high temperatures and involve liquid reactive media. In addition to copper and precious metals, modern smelters are integrated into the recycling route of other valuable metals such as selenium, cadmium, tellurium, and nickel (Ma 2019a). Another positive outcome of such operations is the energy savings from the exothermic plastic waste oxidation/combustion during the smelting operations. Pyrometallurgical processes, therefore, close many metal loops such as for iron (via basic oxygen furnace and electric arc furnace), copper and lead (via smelting operations), as well as aluminum (via tilting or stationary melting furnaces). They are considered an essential part of the recycling chain. This chapter describes these processes to recover metals from WEEE, such as waste printed circuit boards (WPCBs). Furthermore, we discuss their advantages, limitations, and associated challenges.

7.2 Printed Circuit Boards

Printed circuit boards (PCBs) are the most valuable and common component of consumer electronics (i.e. one of the most essential categories of WEEE). They constitute approximately 3 wt% of WEEE (Zhu et al. (2020); Chatterjee (2012); Luda (2011)). Their recycling has recently drawn considerable attention from both an economic and an environmental perspective (Ippolito et al. 2021; Wang et al. 2021a, b; Rao et al. 2021; Tanísalí et al., 2021). The presence of many valuable elements, which can be recovered (i.e. the concept of urban mining), combined with their scarcity in the earth's crust is at the heart of this waste valorization. On the other hand, the toxic nature of many of their components (which may contain lead, chromium, and brominated compounds) is a critical aspect to consider when designing eco-friendly recycling processes.

Each PCB type has its own overall chemical composition, as well as mechanical and physicochemical properties. This leads to recycling challenges as each WPCB type will react differently to the unit operations of a given process (such as shredding, leaching, and smelting) Charles et al. (2020) presented an elemental mapping of the valuable metals that can be recovered from PCB (see Figure 7.1). It includes copper, gold, silver, palladium, yttrium, cobalt, antimony, and iridium, just to identify a few. More specifically, a typical PCB consists of approximately 70 wt% of plastics and other materials such as brominated flame retardants (BFRs), ceramics, and metals (Charitopoulou et al. 2020; Jandric et al. 2020).

Copper (20 wt%), solders alloys made of tin and lead (5 wt%), iron (3 wt%), as well as precious metals (about 200 ppm of gold, 800 ppm of silver, and 100 ppm palladium) represent the most common metals to be recovered in these waste materials. As seen in the previous chapters of this book, the wide variety of metallic elements to be recovered and the unique chemical nature of PCBs naturally lead to the design of many distinct recycling processes, including physical

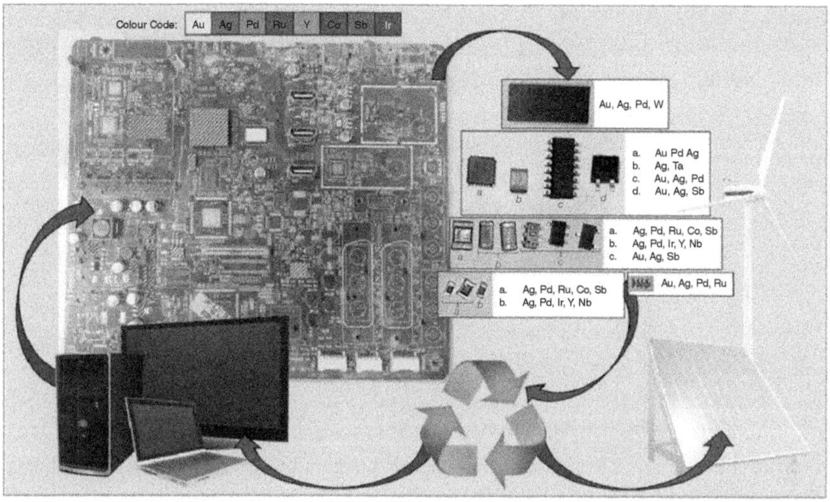

Figure 7.1 Map identifying the rare and precious metals distribution in a PCB. Source: Charles et al. (2020). From an open access article permitting unrestricted use of the original work.

separation, hydrometallurgical, biometallurgical, and pyrometallurgical processes (Moosakazemi et al. 2020; Cui and Anderson 2020; Harikrushnan et al. 2016). We explore this in detail in this pyrometallurgical processes chapter.

7.3 Pyrometallurgical Processes

A pyrometallurgical process involves using ores, concentrates, or waste materials to produce valuable metals ultimately. In the primary metal production route, these metals are in one of their oxidation states due to the ore genesis conditions (i.e. oxidizing conditions). Therefore, a metal production will require a reduction step, which is often performed using a carbon-based reducing agent such as coke or methane. In the case of WEEE, it is interesting to note that these metals are already in their fully reduced state, which has to be considered when designing an optimal recycling strategy. Moreover, basic thermodynamic considerations justify the use of high temperatures in most pyrometallurgical processes since carboreduction is energetically favored at high temperatures for many metals. Also, high-temperature liquid phases are often formed in these processes as they act as excellent reactive media. Another important consideration that promotes primary metal pyrometallurgical processes to recycle WEEE is that they can already process concentrates that naturally contain most of these metallic elements to be recovered. Therefore, these pyrometallurgical technologies can naturally process WEEE without any significant modification. A perfect example of such good process compatibility is the smelting of copper sulfide concentrate (such as chalcopyrite, $CuFeS_2$) in the Noranda process, which can integrate up to about 14 wt% of shredded electronic waste in its feed as reported by Cui and Zhang (2008). In this case, smelting operations performed between 1200 and 1300 °C, which are designed to remove sulfur and iron from the system via the gas and the slag phase, respectively, are used to produce liquid copper, which also entrained noble metals (such as gold, silver, and platinum-group metals – PGMs). They are collected in the low-temperature anode mud of the electrolytic refining process, which is required to obtain pure copper that qualifies for electrical applications (Vegliò and Birloaga, 2018).

7.3.1 Smelting

The initial steps in the processing of e-wastes generally involve their breaking and shredding followed by some physical separation operations (such as magnetic separation, eddy current separation, air current separation, and corona electrostatic separation as reported by Zhang and Xu (2016)) to sort different fractions, which have distinct chemistry. The nonferrous metallic-rich fraction is then processed in one of the available smelting technologies. It mainly contains common base metals such as copper, nickel, tin, zinc, lead, and aluminum, and some residual ferrous components, and plastics (which cannot be removed entirely from the physical separation steps). Depending on the major base metal (i.e. Cu or Pb), the metallic fraction can be processed in either a copper or a lead smelter, which acts as a recycling unit to recover these elements as well as noble metals (gold, silver, and platinum-group metals) following further hydrometallurgical and electrochemical processes. Most smelters also simultaneously process important fractions of sulfide concentrates, which explain the formation of a matte phase (sulfur-rich liquid) and an oxide-rich slag phase (induced by the oxygen blowing and the addition of fluxes) along with a metallic phase at some point during the process. The evolution of an SO_2-rich gaseous phase, which is collected to produce sulfuric acid, is also typical for such smelting operations. Further details on the copper (Cu)- and lead (Pb)-smelting strategies used to recycle e-wastes are presented in the Sections 7.3.1.1, 7.3.1.2, and 7.3.1.3.

7.3.1.1 Copper-Smelting Processes – Sulfide Route

The general principle of copper smelting is to remove sulfur and iron (i.e. the two main impurities of copper sulfide ores) using oxygen. Figure 7.2 shows the predominance diagram for the CuFeS-SiO_2-O_2 system at a temperature of 1200 °C. Silica is added to the system (using a 2 : 1 $CuFeS_2$:SiO_2 molar ratio) to produce a slag phase upon ferrous oxide (FeO) formation during the O_2-enriched air blowing. At the beginning of the smelting operation, a (Cu,Fe)-rich matte and solid silica are present in the reactor. As oxygen is introduced into the system, it generates SO_2(g) and ferrous oxide. The obtention of the adequate SiO_2:FeO ratio leads to the formation of fayalite slag. When the (Cu-rich matte + slag) meta-equilibrium is reached (gray zone #1 on Figure 7.2), the smelting process is completed. The copper matte smelting can be summarized using the following overall reaction:

$$2CuFeS_2(s) + 4O_2(g) + SiO_2(s) \xrightarrow{1200\,^\circ C} Fe_2SiO_4(slag) + Cu_2S(matte) + 3SO_2(g)$$

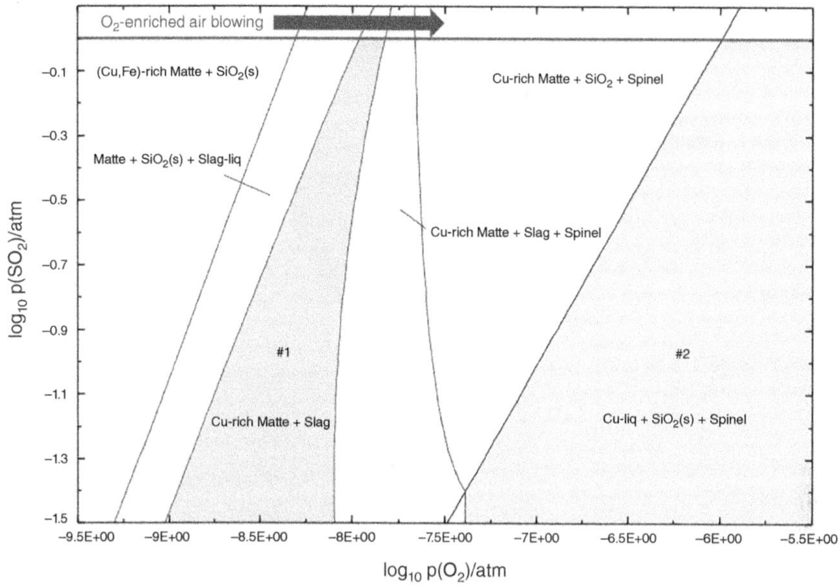

Figure 7.2 $CuFeS_2$-SiO_2-O_2 predominance diagram (using a 2 : 1 $CuFeS_2$:SiO_2 molar ratio) at 1200 °C calculated using the FactSage software. The arrow shows the evolution of the partial pressure of O_2 as air is blown in the system.

In theory, it is thermodynamically possible to identify operating conditions that lead to the direct production of metallic copper called direct-to-copper smelting by further oxidizing the matte (see the grey zone #2 in Figure 7.2). However, these operating conditions lead to significant copper oxidation and solid magnetite formation (spinel). Other fluxes such as CaO need to be used to lower the liquidus temperature of the slag. More importantly, copper is transferred to the slag phase. To make the process economically viable, the copper-rich slag phase needs to be treated with coke to recover the valuable metal. The Cu-rich matte is then converted into liquid copper using a similar strategy.

The primary production of copper from sulfides is performed via one of the following smelting technologies (Schlesinger et al. (2011c)):

- Submerged tuyere smelting: Oxygen-enriched air is the primary reactant, which is blown into the liquid matte via submerged tuyeres to produce SO_2 and FeO (which is subsequently fluxed into a fayalite slag synthetically produced by adding silica). The concentrate can be injected into the system via either the submerged tuyere (in which case the concentrate needs to be dry and has to meet specific particle size requirements) or from the top of the reactor as in the Noranda technology. The latter feeding strategy is more suitable for charging a scrap of various sizes into the smelter. Silica is also added to the system to flux the oxidized iron.
- Flash smelting: This technology values a solid (concentrate) -gas (O_2-enriched air) combustion-like reaction to promote the simultaneous formation of liquid matte and slag droplets, which eventually reach the matte–slag pool at the bottom of the

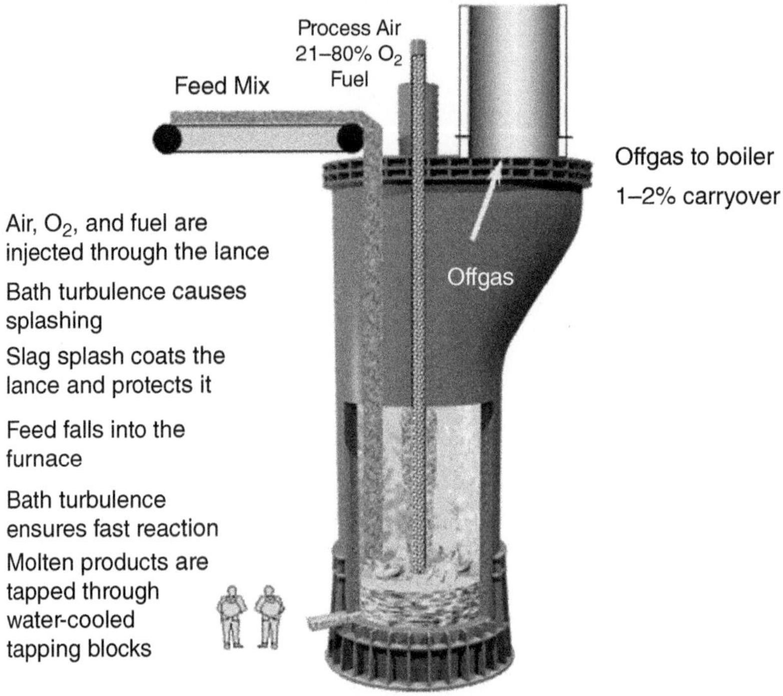

Figure 7.3 Umicore's IsaSmelt furnace. Source: Bakker et al. (2011).

reactor. This two-phase liquid system needs to settle in the furnace to promote efficient liquid–liquid phase separation and transfer of the copper from the slag phase to the matte. One drawback of this approach is that the concentrate feed needs to be calcined and being injected as fine particles.

- Vertical-lance smelting (ISASMELT): This technology developed in Australia is based on the use of a submerged gas-cooled lance, which blows a mixture of (oxygen-enriched + natural gas) blast into the liquid matte (see Figure 7.3). The vertical reactor is fed from the top and can tolerate various feed sizes, which ease the introduction of scrap. Magnetite, which is formed by the matte's oxidation, is a key phase in this process. Firstly, it protects the lance from the matte as it solidifies around it. It also plays a central role in the production of $SO_2(g)$ as it reacts with the matte phase via a solid–liquid reaction. Alvear Flores et al. (2014) presented an exhaustive study about the importance of this compact technology for recycling. It is also a central unit operation of the Umicore process detailed by Hageluken (2006).
- Direct-to-copper smelting: As explained previously, the general idea behind this smelting strategy is to severely oxidize the matte until it produces metallic copper. According to Schlesinger et al. (2011a), about 25 wt% of Cu is inevitably transferred to the slag phase. It will be recovered by reducing the copper oxide in the slag using coke.

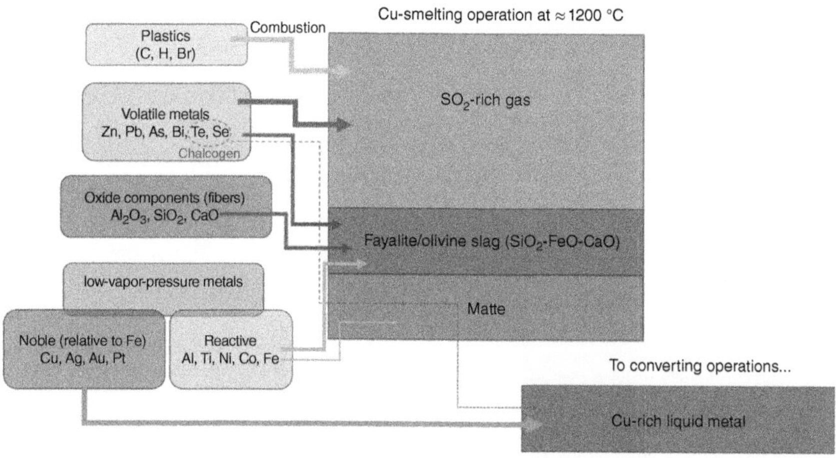

Figure 7.4 Partitioning of WPCBs during the Cu-smelting operations.

Copper smelters clearly offer an opportunity to partition valuable metallic elements and other impurities that constitute WPCBs in various phases that form during the process (such as the matte, the slag, the metallic melt, and the exhaust gas-dust). One fundamental aspect to understand here is the fate of each element and contaminant under these smelting operating conditions. Figure 7.4 presents a schematic representation of the partitioning of various elements/component present in WPCBs. The first group called plastics is incinerated in situ. The products of this incineration are transferred to the gas phase. The second group called volatile metals includes zinc, lead, arsenic, bismuth, tellurium, and selenium. Apart from tellurium and selenium (which end up in the liquid copper matte as they are chalcogen elements), all these elements are mostly transferred to the SO_2-rich exhaust gas. This is potentially problematic as they may exceed regulated concentrations when released to the atmosphere and become a danger for the environment. The third group is associated to the presence of glass fibers in WPCBs. It contains oxide components like Al_2O_3, SiO_2, and CaO. These oxide components are easily fluxed in the slag phase during smelting. The last group is called low-vapor-pressure metals. It is subdivided into two categories: (i) the noble metals relative to iron (such as Cu, Ag, Au, and Pt), which will be carried in the Cu-rich liquid metal, and (ii) the reactive metals (such as Al, Ti, Ni, Co, and Fe), which will mostly be oxidized and transferred to the slag phase. A converting step is performed on the copper matte to produce the blister copper, which is refined using natural gas to remove the dissolved monatomic oxygen.

Finally, it is to be noted that many authors (Khaliq et al. (2014); Cui and Zhang (2008); Ma (2019a); Zhang and Xu (2016)) have reviewed the principal industrial e-waste recycling practices involving copper-smelting unit operations. Their studies reported the Noranda process in Canada, the Aurubis's Kayser recycling system in Germany, the Boliden Ronnskar Smelter in Sweden, and the Umicore's precious metal refining process in Belgium (see the process flowsheet in Figure 7.5). It is to be noted that the smelting operations at Boliden and Umicore are integrated into a

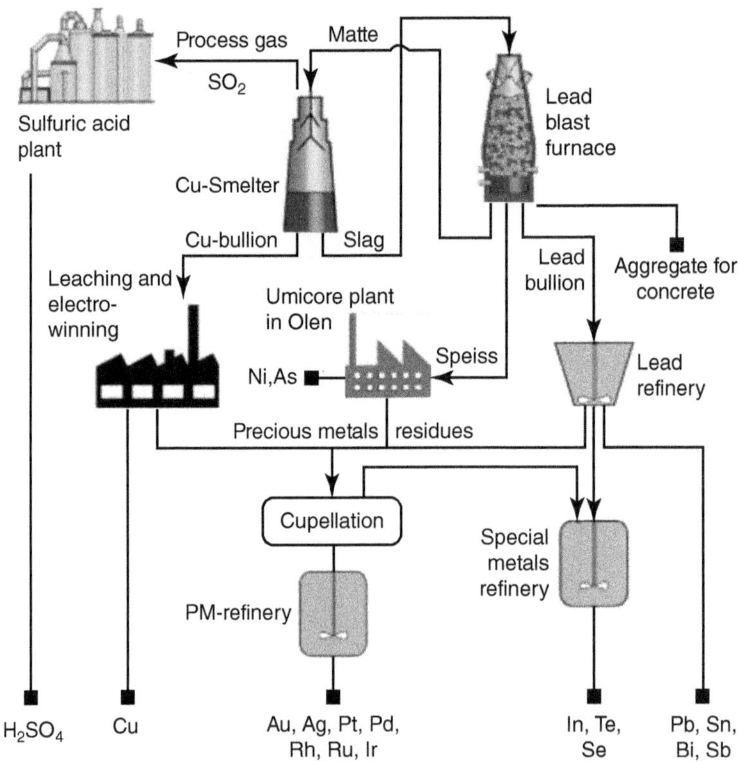

Figure 7.5 Flowsheet of Hoboken integrated smelter and refinery plant of Umicore. Source: Tuncuk et al. (2012).

much larger process scheme that also involves lead-smelting operations, which will be explained in the Section 7.3.1.3.

7.3.1.2 Copper-Smelting Processes – Secondary Smelters

There are two types of secondary smelters to process copper-rich scrap, i.e. the metal smelters and the black-copper smelters (Schlesinger et al. (2011b)). Tiltable reverberatory furnaces are used to process Cu-rich (high-grade) scrap. For low-grade scrap, a more elaborate strategy (involving a reduction step using coke followed by an O_2(g) oxidizing refining step) is required. Low-grade scrap often includes shredded automobile products and dross coming from the decopperizing of lead bullion, which both contain lead. Therefore, the typical furnaces associated to the production of black copper (such as the Kaldo furnace and top submerged lance reactors like ISASMELT and AUSMELT) are often found in lead extraction operations.

7.3.1.3 Lead-Smelting Processes

Lead, like copper, is a chalcophile element which is naturally found in association with sulfur in different minerals. Lead sulfide ores such as galena (PbS) typically contain other minerals that are rich in zinc, copper, and iron (in the form of sulfates and sulfides) as well as precious metals, notably silver (Ayres et al. 2003). E-wastes such as lead-acid batteries are therefore fully compatible with this ore type as they simultaneously contain metallic lead, PbO_2, and $PbSO_4$ (Arnout et al. 2011). They can be mixed with the sulfide concentrate feeds.

Laubertova et al. (2017) presented an overview of the lead-smelting technologies. Two main routes can be used to process lead sulfide concentrates and lead-rich scrap, i.e. (i) the traditional sintering–smelting in blast furnaces (for Zn-lean concentrate) or using Imperial Smelting furnaces (for Zn-rich concentrate), and (ii) the direct smelting approach. Grant (2001) reported many direct lead-smelting approaches such as the Kaldo furnace (top-blown rotary converter), the QSL kiln, the ISASMELT and AUSMELT vertical reactor with submerged lance, and the Kivcet process. Contrary to conventional copper-smelting operations, which always require large amounts of concentrate to operate, several authors report that many lead-smelting operations run almost exclusively using scrap feed. This is explained by the high lead fraction that contains specific waste materials (such as lead-acid batteries) when compared to the WPCBs, which have smaller fractions of metallic components.

- Sintering–smelting: In this approach, the concentrate is first roasted using air to convert the sulfide into an oxide via the following reaction:

  $$2PbS(s) + 3O_2(g) \xrightarrow{400\text{–}900\,^\circ C} 2PbO(s/l) + 2SO_2(g)$$

  This reaction is important as lead oxide can be easily reduced to metallic lead by a carbon-based reducing agent (which is not the case for lead sulfide). Lead oxide is not the only phase in the sintered product. It also contains melilite, spinel, calcium silicate sulfate, as well as copper sulfide (Zhao 2013). Next, the lead oxide sinters are reduced in the blast furnace using coke. Silica and lime are also charged in the blast furnace from the top (Watt et al. (2018)). The tuyeres at the bottom of the reactor blast O_2-enriched air, which react with the excess coke to produce CO(g) via the Boudouard reaction. This carbon-monoxide-rich gas will act as the main reducing agent for this reduction process (Vanparys et al. (2020)):

  $$PbO(l) + CO(g) \xrightarrow{T>800\,^\circ C} Pb(l) + CO_2(g)$$

  The liquid lead (called lead bullion) obtained from this operation is collected at the bottom of the furnace along with a slag and a matte phase. In many cases, arsenic is present in the charge and will form speiss (i.e. iron arsenide) upon the reaction with iron. Figure 7.6 taken from the work of CHAIDEZ-FELIX et al. (2014) shows the phase assemblage obtained from a lab-scale lead blast furnace experiment at 1573 K. It consists of an equilibrium between lead bullion, speiss, matte, and slag. In the Umicore process presented in Figure 7.5, the lead bullion obtained from the blast furnace is sent to a refinery unit, the matte is sent to a copper smelter, while the speiss is processed in the Umicore Plant in Olen. The Imperial Smelting process works similar to the blast furnace and is used to process zinc-rich concentrates.
- Direction smelting: In this approach, the sulfide concentrate is directly reduced by coke in a furnace. Slag-forming fluxes are also added to the reactor. The following

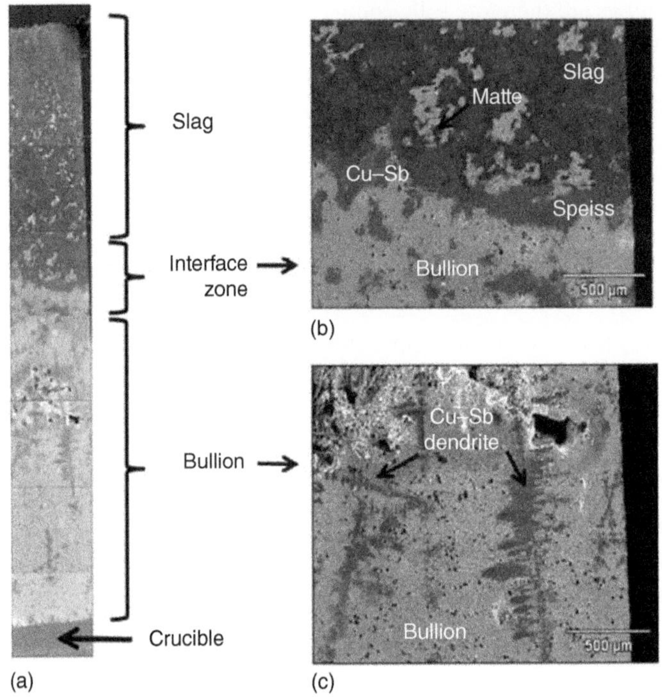

Figure 7.6 (a) longitudinal section of the solidified sample: crucible is at the bottom, followed by the bullion region, an interface zone (Cu–Sb alloy, speiss and matte) and the slag at the top of the sample. (b) magnification of the interface zone. The slag phase appears in this figure as dark globules and the lead bullion as a light zone. (c) Cu–Sb dendritic precipitation in the bulk of the bullion phase. Source: Taken from Chaidez-Felix et al. (2014)). Reprinted with permission from Transactions of Nonferrous Metals Society of China 2014.

reactions simultaneously take place:

$$PbS(s) + O_2(g) \xrightarrow{1300\,^\circ C} Pb(l) + SO_2(g)$$

$$2PbS(s) + 3O_2(g) \xrightarrow{1300\,^\circ C} 2PbO(l) + 2SO_2(g)$$

$$2PbO(l) + C(s) \xrightarrow{1300\,^\circ C} 2Pb(l) + CO_2(g)$$

As explained by Grant (2001), the direct smelting of lead sulfide concentrate to obtain a low-sulfur lead bullion and a low-lead slag in a single step is difficult to implement in practice. The $PbS\text{-}O_2\text{-}S_2\text{-}(SiO_2)_{0.4}(FeO)_{0.3}(CaO)_{0.1}$ predominance diagram calculated at 1473 K and 1 atm presented in Figure 7.7 validates this affirmation.

Figure 7.7 shows that the sulfur removal of the system performed via the O2-enriched air blowing may provide the desired (liquid lead + slag) equilibrium but inadequate compositions (see the red star on Figure 7.7). In this zone, the liquid lead activity of about 0.97 implies that it dissolved a significant amount of oxygen.

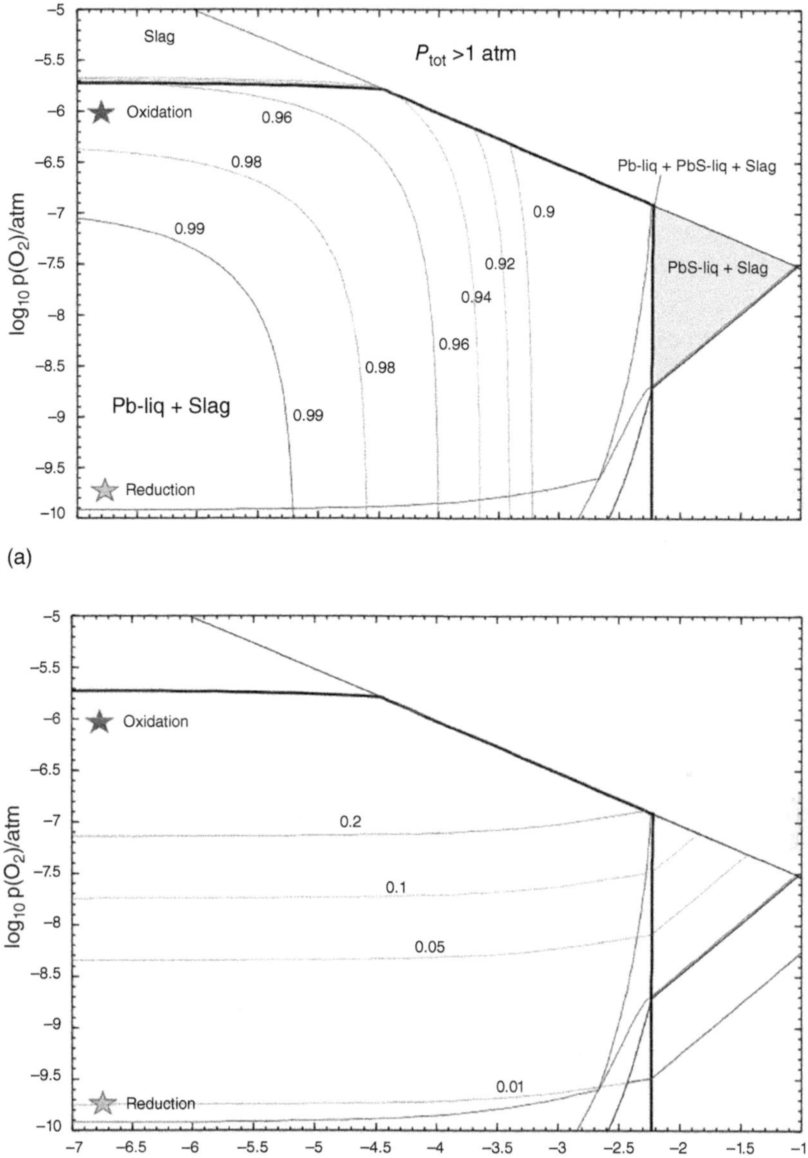

Figure 7.7 PbS-O_2-S_2-$(SiO_2)_{0.4}(FeO)_{0.3}(CaO)_{0.1}$ predominance diagram generated at 1473 K and 1 atm calculated using the FactSage software. Panel (a) shows iso-activities of pure liquid lead, while panel (b) shows iso-activities of pure liquid PbO.

Moreover, Figure 7.7b shows that the liquid PbO activity is high in this zone, which would lead to important lead losses in the slag. Therefore, the oxidation alone is not enough to smelt the concentrate. The addition of coke into the system allows a significant lowering of the oxygen chemical potential. This leads to an important reduction of the Pb losses to the slag phase (see the star in the bottom left corner in Figure 7.7b). The reduction step is also not sufficient by itself as it cannot fully control the system's sulfur chemical potential. This explains why a two-stage strategy – oxidation followed by reduction – is required (Grant, 2001). Finally, it is to be mentioned that some direct smelting furnaces can process concentrate-free feeds. Laubertova et al. (2017) reported the example of a shaft furnace battery recovery process that uses battery scrap, coke, iron scrap, and reusable slag to produce lead bullion.

7.3.1.4 Advantages and Limitations of Smelting Processes

We conclude this section by highlighting the advantages and limitation of the smelting processes to recycle electronic waste.

Advantages One major advantage of using smelting technologies to recycle waste electronics is that they were initially designed to deal with several elements' simultaneous presence in the concentrates they process. Most primary metal-smelting operations have to deal with: (i) the presence of undesired silicates/oxide compounds, which are removed via the formation of a slag phase, (ii) the presence of reactive metals and other reactive compounds such as plastics and sulfides, which can be oxidized using O_2-enriched air and then transferred to the slag phase or the exhaust gas of the process; (iii) the possibility to reduce valuable oxide components using carbon-based reducing agents; (iv) the production of metallic melts, which are then refined and tapped. Therefore, these smelting processes can tolerate the variable nature of the overall chemical composition of waste electronics. It is of prime importance to select the primary smelting process that matches the waste's major metallic component. The integration of scrap into primary metal processes is especially important for electronic wastes since their metallic fraction is not always high. Therefore, these recycling operations benefit from the formation of liquid phases during the primary smelting operations. The other advantages can be listed as follows:

- Most primary metal-smelting processes have been exploited for decades and optimized to recover all valuable metals. The perfect example is copper's primary production to process copper sulfides in Quebec (i.e. the Noranda process). Apart from copper production, the other metallurgical operations that have been appended to this process (such as the electro-refining of copper) have allowed recovering noble metals such as silver, gold, platinum, and palladium as chalcogen metalloids such as tellurium and selenium.
- Many chemical reactions involved in the smelting operations are exothermic, which lower these processes' overall energetic requirements. The in situ exothermic incineration of plastics can also be viewed as a positive outcome.

- Smelters are typically large reactors with high productivity, which allows the daily processing of large volumes of electronic waste.
- Many primary smelters are already available to process electronic waste, which lowers the investments required to recycle them.

Limitations The major limitation of primary smelting operations to process electronic waste is their relatively low selectivity. As an example, noble metals and chalcogen metalloids are alloyed with copper at the end of the smelting, converting, and refining process. Their recovery requires subsequent hydrometallurgical and electrorefining operations. Other limitations include:

- The high replacement cost of refractory and other critical components. The liquid phases that are formed during smelting are chemically aggressive and wear many critical components of the vessel.
- The initial investments to build such processes when not available.
- The need, in many cases, to simultaneously process concentrates when recycling electronic waste to ensure a proper operation of the smelter.

Finally, it is to be mentioned that the sustainability of the primary metal-smelting processes to recycle electronic waste will need to be analyzed in the future, especially in the context of the establishment of an authentic circular economy. The reduction of harmful gaseous compounds such as dioxins and brominated compounds via improved scrubbing operations is another important aspect that will need to be addressed in the future to make these smelting operations more sustainable.

7.3.2 Electrochemical Processes

As mentioned in the introduction of Section 7.3 of this chapter, one interesting characteristic of electronic waste is several metals' simultaneous presence in their fully reduced state. This contrasts with the oxidized states they naturally adopt in ores. The reduction operations required to extract metals are energy-intensive, especially for reactive elements like aluminum, magnesium, and lithium. Therefore, there is the potential for huge energy savings to be made if one can design a process that accounts for this feed specificity. Electrochemical processes fall into this opportunistic category. Su (2020) presented a review of the different electrochemical methods that exist to recycle solid and liquid waste (see Figure 7.8). It includes electrolysis, electrodialysis, electrocoagulation, and electroflotation. In this chapter, we only focus on electrolytic processes.

In a nutshell, an electrolytic process requires an electrical system constituted of a conductive anode and cathode, and an electrolyte (which can host metallic cations), and a power source. It is also important that this electrolyte possess a good ionic conductivity via highly mobile ions. Electrical work is transferred to the system to overcome the positive Gibbs free energy variation associated with the overall oxidoreduction reaction that leads to the production of metal (as well as the energy associated with anodic and cathodic overpotentials). The anode hosts the oxidation reaction (which typically leads to the production of O_2, Cl_2, or CO_2), while the

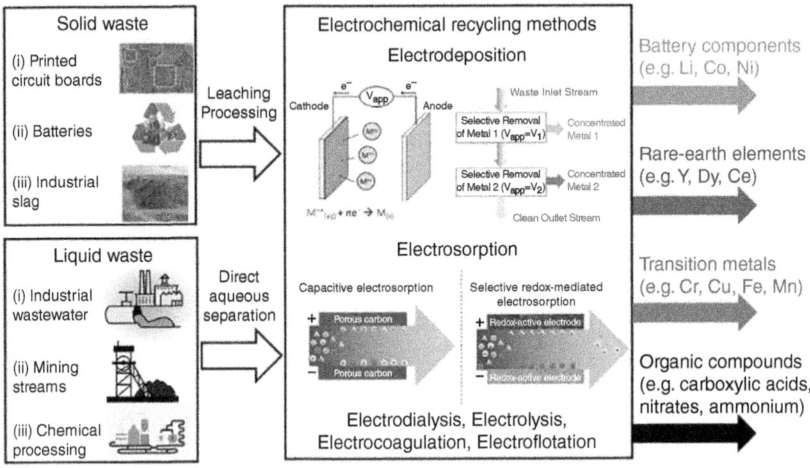

Figure 7.8 An overview of select electrochemical approaches for waste recycling and revalorization. Source: Su (2020). Reprinted with permission from The Electrochemical Society 2020.

cathode hosts the metallic reduction (the metal to be recovered is collected at the cathode). In electrorefining operations (such as in electrolytic refining of copper), the anodic reaction is the oxidation of the metal itself. As it will be presented in this section, the important selectivity of electrolytic processes (linked to the reduction potential uniqueness of each metallic ion) makes them interesting candidates for electronic waste recycling operations in the future. Electrowinning and electrorefining technologies can be divided into low- and high-temperature processes.

7.3.2.1 High-Temperature Electrolysis

In extractive metallurgy, many reactive metals can be produced using high-temperature electrolysis. This is typically performed in molten salts such as chlorides, fluorides, and chlorofluorides. Excellent review work on the subject is presented by Yan and Fray (2010). A famous example of high-temperature electrolysis is the production of aluminum via the Hall–Heroult process, which uses cryolite (Na_3AlF_6) as the electrolyte. Magnesium and lithium can also be commercially produced via high-temperature electrolysis using molten chloride electrolytes. One significant advantage of using chloride systems is that they melt at relatively low temperatures when compared to oxide-equivalent systems. It is also possible to design a molten chloride electrolyte that is not too volatile and has good ionic conductivity. Fluoride salts can dissolve oxides, which is a major advantage in primary metal production (e.g. alumina in cryolite). However, fluoride melts tend to be more aggressive and are liquid in higher temperature ranges.

In the context of recycling electronic waste, the use of high-temperature electrolysis would normally require a chlorination step to oxidize their metallic components. This step could be done via a roasting strategy. During the molten chloride electrolysis, $Cl_2(g)$ is evolved at the anode and can be recycled in the chlorination/roasting step or used in situ to chlorinate metallic components. The main environmental

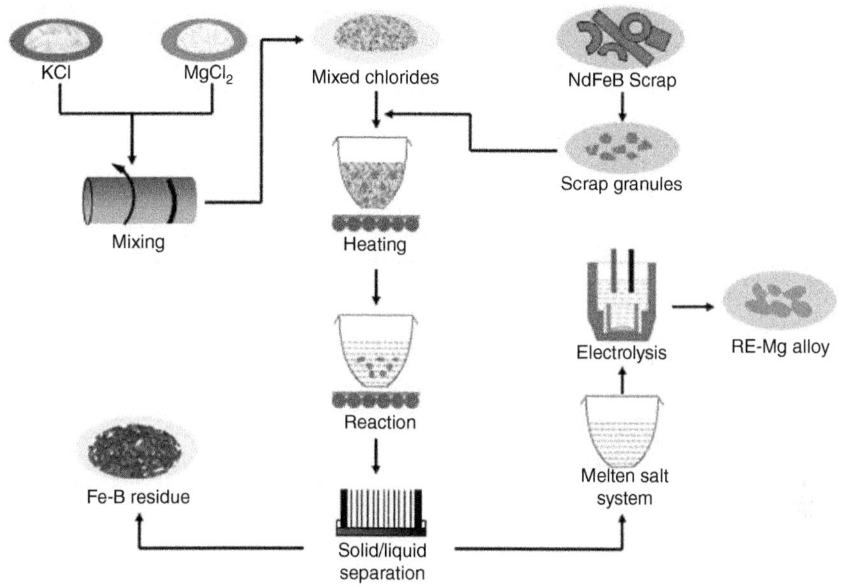

Figure 7.9 Flowsheet of the RE recovery from NdFeB scrap using molten chlorides. Source: Hua et al. (2014).

concern with this process is the accidental release of chlorine in the atmosphere as a result of a process failure. The material selection to building the equipment for handling chlorination operations is also critical and challenging. Applications of molten chloride electrolysis to recover rare-earth elements from electronic waste were recently reviewed by Xi et al. (2020). They reported Abbasalizadeh et al. (2016)'s work in which $AlCl_3$ is used as the chlorination agent to transfer Nd and Dy (via an exchange reaction) into the eutectic LiCl-KCl-NaCl molten chloride. Nd and Dy's transfer in the salt is possible as their respective chloride ($NdCl_3$ and $DyCl_3$) are more thermodynamically stable than $AlCl_3$.

Interestingly, this chlorination strategy will not transfer iron and boron in the molten chloride since their chlorides are less stable than $AlCl_3$. As a result, the electrodeposited rare-earth metals are free of iron. Another strategy to chlorinate the rare-earth elements from magnet scrap is to use a molten $MgCl_2$-KCl bath (Hua et al. (2014)). In this case, the electrolysis leads to an Mg-Nd alloy's electrodeposition, which is inevitable because of the small reduction potential differences between these two elements. The proposed flowsheet of Hua et al. (2014) to process NdFeB scrap is presented in Figure 7.9.

7.3.2.2 Low-Temperature Electrolysis

Even though low-temperature electrolytic operations using aqueous solution should not be considered as pyrometallurgical processes, their technological importance in refining primary metals and in recycling e-waste motivated an overview of these processes in this section. Low-temperature electrolytic processes typically use sulfuric, hydrochloric, and boric acid solutions (Jin and Zhang (2020)) as well as ionic liquids. It is interesting to note that conventional hydrometallurgical processes to recover valuable metals from electronic waste use similar aqueous solution to perform the chemical leaching (Su 2020), i.e. the first step of the extraction. The subsequent purification and recovery steps in extractive hydrometallurgy are based on completely different strategies such as precipitation and cementation. Here are some recent examples of low-temperature electrolytic processes to recycle electronic waste.

- Copper recovery from e-waste: There is rich literature on the recovery of copper, which is the main metallic element contained in WPCBs using low-temperature electrolytic processes. In their work, Fogarasi et al. (2013) presented the two main electrolytic routes that can be taken to perform the copper recovery (Figure 7.10). In route A (sulfuric acid electrolyte), copper's direct electrochemical oxidation is performed, followed by the electrodeposition of pure copper at the cathode. In route B, the Fe^{3+}/Fe^{2+} redox pair is used to achieve a faster dissolution of copper (Cu). Fogarasi et al. (2013) reported that route A leads to the extraction of a 98% purity of copper for the energy consumption of 1.06 kWh per kg, while route B produces a 99% purity of copper energy consumption of 1.75 kWh per kg. Here are other examples of copper extraction based on similar processes.

Song et al. (2021) were able to electrochemically extract 95 wt% of the copper dissolved in a multimetal sulfuric acid leaching solution of waste liquid crystal display panels that contained indium, copper, aluminum, and molybdenum. Barragan et al. (2020) proposed a new strategy for the efficient recovery of highly pure copper and antimony (in the form of Sb_2O_3) from WPCB by leaching the waste in hydrochloric acid solution that also contained ferric chloride (to accelerate the copper dissolution), followed by a precipitation step and an electrowinning process. Zhang

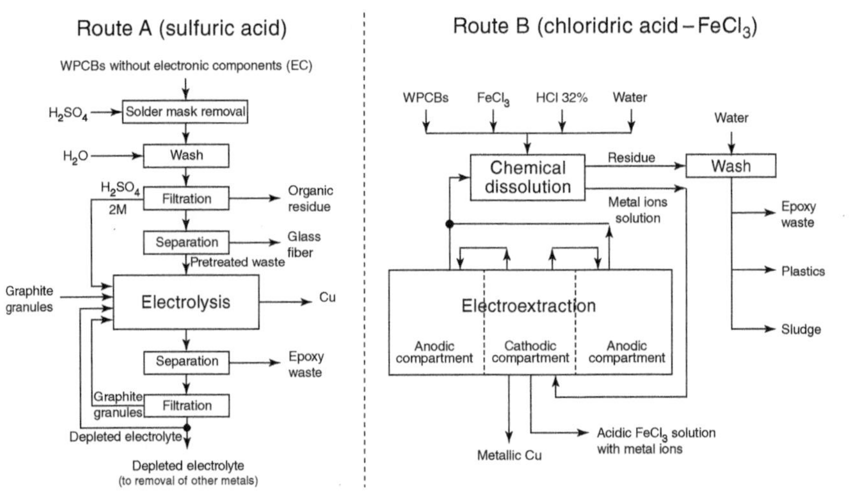

Figure 7.10 Block diagram of the copper recovery process from WPCBs by mediated electrochemical oxidation and cathodic deposition using sulfuric acid electrolyte (Route A) and hydrochloric-acid-containing $FeCl_3$ electrolyte (Route B). Source: Fogarasi et al. (2013).

et al. (2018) proposed a greener ionic liquid electrolyte that contains *N*-butyl sulfonate pyridinium hydrosulfate, i.e. $[BSO_3HPy]HSO_4$, to replace H_2SO_4 in a slurry electrolytic system made of a graphite anode and a titanium cathode. The optimal replacement of 10% H_2SO_4 by this ionic liquid allowed to obtain a copper recovery of 90.94% with a purity of 81.69%. Finally, it is worth mentioning the technoeconomic assessment for the recovery of metals from WEEE reported by Diaz and Lister (2018), which compared the low-temperature electrolytic recycling technology to the black copper-smelting operations (which also include an electrolytic step to produce pure copper). As explained previously, the black copper-smelting route of low-grade scrap starts with its melting and subsequent reduction using coke. An oxidizing step is then performed to remove less noble elements like iron, lead, zinc, and tin. The more noble metals (Ni, Ag, Au, and platinum groups) are collected in the anode mud of the electrorefining process. In their study, Diaz and Lister (2018) noted a competitive alternative to process electronic waste via the low-temperature electrolytic recycling technology alone when compared to black copper smelting. A significantly lower capital investment can be achieved with the electrochemical process (i.e. 2.9 kg e-waste per dollar of capital investment) compared to the black copper-smelting process (i.e. 1.3 kg per dollar of capital investment). This is not a surprising conclusion and validates the conventional industrial practices that use leaching and electrolytic refining to extract copper from copper oxide ores. In this case, the main difference is the much higher quantity of impurities in the e-waste (compared to the oxide concentrate), which will require larger electrolyte purification units.

- Noble metal and rare-earth recovery: The acidic sulfate electrolytic solution typically used to electrorefine copper is not strong enough to dissolve noble metals, which end up in the anode mud. Because of that, a stronger electrolyte would be required to extract them via a subsequent electrolytic process. Lister et al. (2014) reported such a two-stage electrorecycling strategy (Figure 7.11). In the first stage of the process, iron, nickel, tin, copper, silver, and rare-earth elements are dissolved in an acidic sulfate solution. Copper and silver are then plated at the cathode. Rare-earth elements dissolved in the solution could be potentially recovered at this stage by precipitation using sulfate double salts. In stage 2, pallidum and gold could be dissolved in an HCl solution (via the anodic production of Cl_2) and finally plated at the cathode.

It is also to be noted that the reduction potential of copper and gold is significantly different (i.e. +0.34 V vs +1.83 V relative to the standard hydrogen electrode). Therefore, it is interesting to evaluate the possibility to electrodeposit each metal from a single electrolytic solution selectively. This is a particularly attractive strategy if one can find a more environment-friendly electrolyte than cyanide solution that can efficiently dissolve gold such as ammoniacal thiosulfate solutions. Kasper et al. (2018) studied the electrochemical behavior of gold and copper simultaneously present in such a solution. They performed electrowinning tests using graphite electrodes to conclude that it is feasible to use such an electrolyte to extract copper and gold at two distinct potentials. Unfortunately, the purity of the individual metal deposits was not reported in their study.

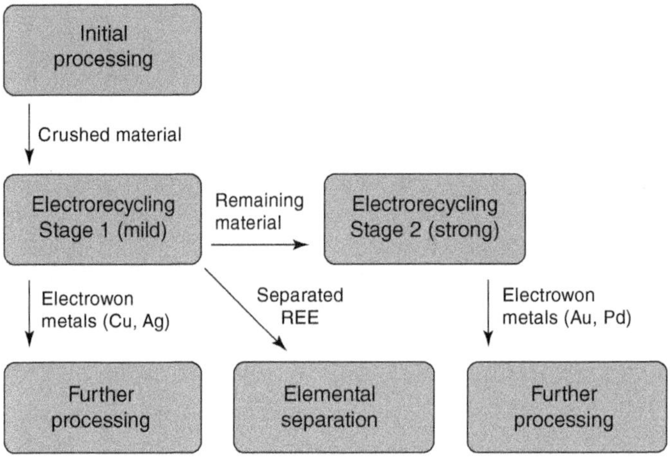

Figure 7.11 Proposed flow sheet for a mobile electronic recycling by Lister et al. (2014). Source: Lister et al. (2014).

7.3.3 Other Pyrometallurgical Operations Used in Electronic Waste Recycling

7.3.3.1 Roasting

Roasting is a pyrometallurgical treatment that can be used during the recycling process of electronic waste. It consists of heating a system to a specific temperature in some reactants' presence (which can be solid, liquid, or gas). Solid–gas reactions are often promoted in these reactors. In the case of electronic waste, Panda et al. (2020) recently proposed a low-temperature chlorination roasting step of pyrolyzed WPCB using ammonia chloride (NH_4Cl) in air at temperatures between 200–325 °C (Figure 7.12).

The thermal decomposition of NH_4Cl leads to the production of H_2(g) and HCl(g). The latter gaseous compound acts as the chlorination agent via a chemical reaction of the following type (e.g. copper):

$$Cu(s) + 2HCl(g) + 0.5O_2(g) \xrightarrow{300\,^{\circ}C} CuCl_2(s) + H_2O(g)$$

This approach provides a recovery of 93% for Cu and 100% for Ni, Zn, and Pb using a roasting temperature of 300 °C in the form of metallic chlorides. These metallic chlorides could in turn be processed using a molten chloride electrolytic method as detailed previously. Roasting strategies have also been used in the recycling of permanent magnets. Yoon et al. (2014) used oxidation roasting at 600 °C to convert Nd into Nd_2O_3, which was then processed via leaching.

7.3.3.2 Molten Salt Oxidation Treatment

Molten salt oxidation treatment is another pyrometallurgical strategy that can be used to recycle electronic waste. It relies on the introduction of electronic waste in a molten salt bath (which can be a eutectic carbonate or hydroxide mixture). As explained by Flandinet et al. (2012), these molten salts are specifically selected for

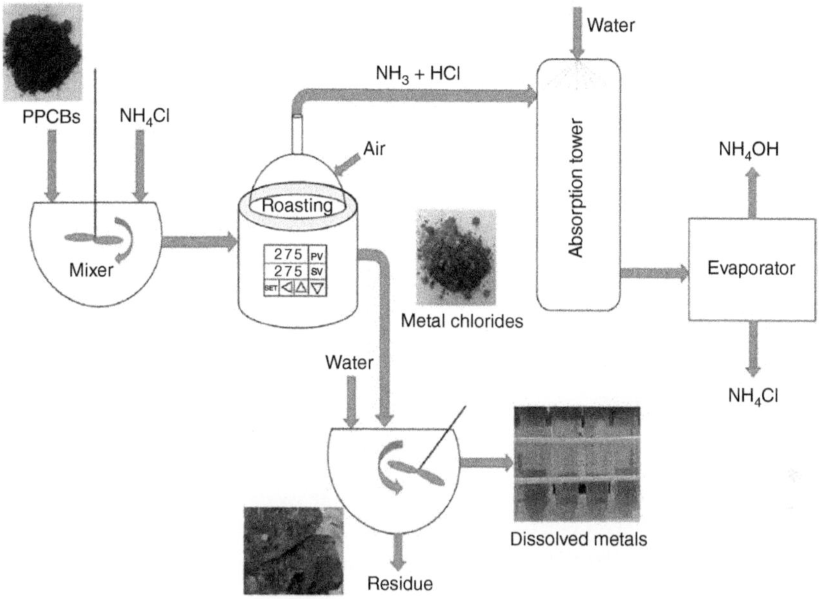

Figure 7.12 Flowsheet of the recycling process of pyrolyzed printed circuit boards using a roasting strategy. Source: Panda et al. (2020). Reprinted with permission from Journal of Hazardous Materials 2020.

their ability to dissolve solid undesirable e-waste components such as oxides, glasses, and plastics, as well as gaseous compounds (CO_2 and halogenated compounds). It is also important to prevent the dissolution of metals in the molten salt to recover them more easily, it is not always desired to generate a more severe oxidizing environment (using air injection in the reactor for example) as it could ultimately lead to the transfer of valuable metals into the salt. In their work, Flandinet et al. (2012) used a eutectic KOH-NaOH molten salt at 300 °C to treat WPCBs. They were able to virtually recover all the metallic fraction of the WPCBs. This approach also prevented the release in the gas phase of most of the halogenated compounds, which were trapped in the molten salt.

Lin et al. (2017) used a different approach based on the use of a molten $(Li,Na,K)_2CO_3$ eutectic carbonate salt reactor operated at temperatures between 550 and 700 °C. They also performed air injection to promote the oxidation reactions. Overall, they were able to recover 95% of the copper available in the WPCB even though they expected more copper oxidation (which should have lowered its recovery efficiency). Other molten salts have been studied in the literature, such as the NaOH-Na_2CO_3-$NaNO_3$ salt (Liu et al. (2016)) and LiCl–KCl eutectic mixtures (see for example Riedewald and Sousa-Gallagher (2015)) (Figure 7.13).

7.3.3.3 Distillation

As discussed previously, some metals (like cadmium, zinc, and lead) contained in e-waste are more volatile than others (like copper and tin). One can take advantage of this property to separate and purify some metallic fractions by distillation.

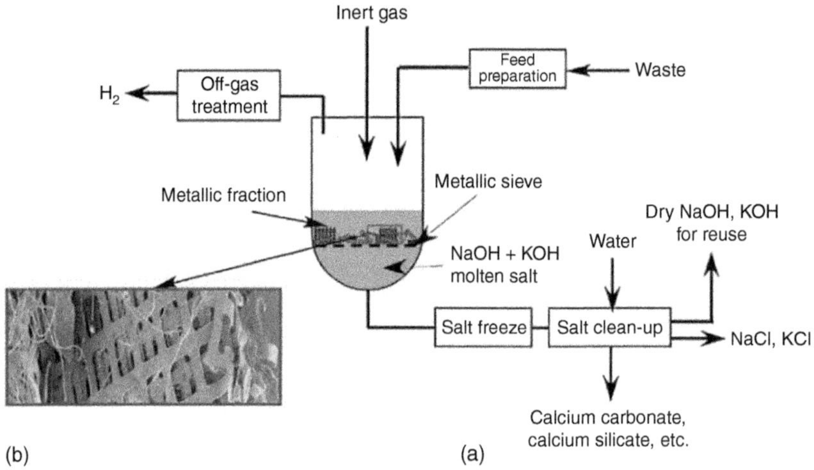

Figure 7.13 Flowsheet of a molten NaOH-KOH eutectic salt oxidation process (a) to recover the metallic fraction of WPCb (b). Source: Flandinet et al. (2012). Reprinted with permission from Journal of Hazardous Materials 2012.

Figure 7.14 shows the evolution of the vapor pressure (expressed in $\log_{10}$(Patm)) of some pure metals calculated using the FactSage software. Lead chloride is also presented in Figure 7.14 to see the impact of chlorination on the lead volatility. The boiling point at 1 atm for Cd (766.5 °C), Zn (908.3 °C) and $PbCl_2$ (950.8 °C) is also clearly visible in Figure 7.14 (i.e. the temperature at which the vapor pressure

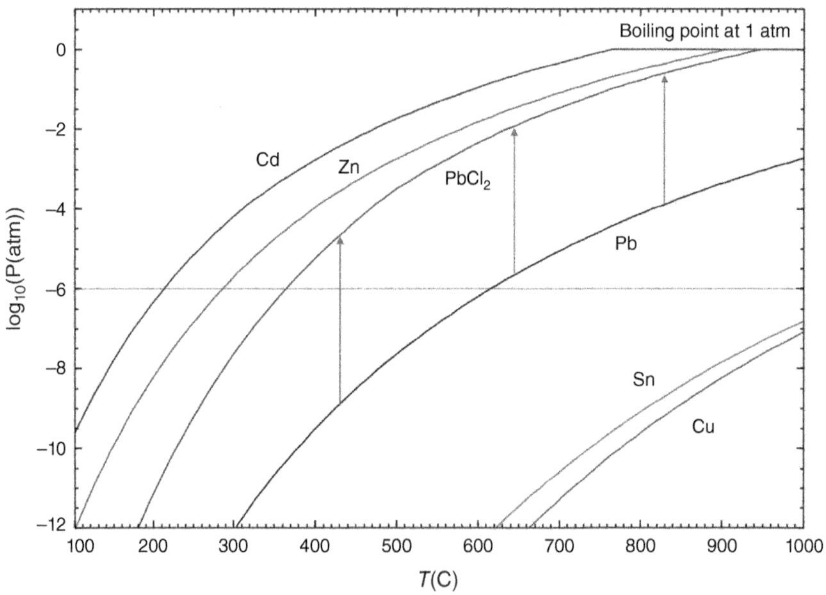

Figure 7.14 Evolution of the vapor pressure of different metallic elements as a function of temperature as calculated by the FactSage software.

reaches 1 atm). A threshold vapor pressure of 10^{-6} atm is also drawn in Figure 7.14. It represents a typical vapor pressure above which volatilization becomes experimentally non-negligible in open systems.

Zhan and Xu (2008) explored the possibility to remove zinc from Cu-rich fractions via vacuum separation at 1123 K. The application of a vacuum lowers the total pressure of the system, which increases the volatility of all the metals at a given temperature. Using such an approach, they were able to purify copper particle obtained from WPCBs from and original purity of 90.68 wt% to a final purity of 99.84 wt% by volatilizing zinc. These authors presented a series of experimental studies on the purification of copper-rich particle using vacuum separation as reported by Ma (2019a).

7.3.3.4 Pyrolysis

Pyrolysis is another important recycling operation, which allows an efficient and more environment-friendly separation of organic and volatile components from the metallic fraction of WPCB. It is defined as the thermal degradation of organic materials under vacuum or inert conditions to produce value-added products such as oil, gas, and carbon black (Xiong et al. 2020; Jadhao et al. 2020; Huang and Lo 2020). The use of conventional or microwave pyrolysis in recycling electronic plastics such as high-impact polystyrene (HIPS) and acrylonitrilebutadiene styrene (ABS) widely emerged in the last decade due to the several advantages it offers: (i) it closes the loop of waste plastic stream, (ii) it produces value-added products at low operating costs and temperature, and (iii) it potentially prevents the release of toxic gases when compared to combustion and incineration processes(Czégény et al., 2012; Miskolczi et al. 2008).

The pioneering work of de Marco et al. (2008) provides fundamental data about the pyrolysis products of electronic wastes. According to these authors, the solids that remain after the pyrolysis are composed of metals and ceramics mixed with about 3–34 wt% of char-like carbonaceous compounds. The collected condensable gases produce a low-viscosity liquid composed of aromatic compounds (in the case of WPCB). The noncondensable pyrolysis gases are composed of light hydrocarbon molecules as well as significant amounts of CO and CO_2. It is to be noted that one major challenge that limits the use of e-waste pyrolysis byproducts is the presence of halogenated flame retardants, mostly BFRs such as tetrabomobisphenol A (TBBPA). These brominated compounds contaminate the collected liquids and gaseous products by producing bromophenols and release toxic emissions such as hydrogen bromide (HBr) (Das et al. 2020; Ma and Kamo 2019).

Pyrolysis is typically an endothermic process that requires energy to proceed. Conventional pyrolysis methods of e-waste plastics or WPCB include fixed-bed reactors (Hall and Williams 2007), fluidized-bed reactors (Zhang et al., 2018; Guo et al. 2010), and rotating reactors (Ma et al. 2018). The performance of all these pyrolysis strategies is well documented in the literature. The use of microwave irradiation to perform the pyrolysis of WPCB is an interesting alternative to these conventional methods because of the following advantages (Huang et al., 2016): (i) it is a volumetric heating method (and therefore more energy efficient), (ii) it is a selective-heating method; (iii) it is a high-heating-rate method, and (iv) it is an easy-to-control (start/stop) method. Zhang et al. (2021) recently proposed a microwave-assisted catalytic process to pyrolyze WPCB.

Other scientific advances and progress in the field of e-waste pyrolysis in recent years mainly focus on three distinct aspects, which are:

- Fundamental description of the degradation mechanisms and kinetics of WPCBs: The reaction mechanisms and pathways for the degradation of WPCB and e-waste plastics (as well as their corresponding kinetics) have been extensively studied in the literature (see for example Yao et al. (2020);Alenezi and Al-Fadhli (2018); Quan et al. (2009); Cai and Shao-hong (2006)). Thermal analysis methods such as the thermogravimetric analysis (TGA) and the differential scanning calorimetry (DSC) have been used to identify reaction mechanisms at the laboratory-milligram scale. Larger-scale experiments typically performed in tubular furnaces are also frequently used in the literature to analyze and characterize the different pyrolysis products (i.e. the solid residue, the condensate, and the volatile fraction).

  Studies on the pyrolysis pathways of WPCB reported that hydrogen bromide is produced during the initial step of the process, followed by the depolymerization of epoxy resins, random rupture, and reactions of free radical to form the different pyrolysis products (Gao et al. 2021; Gao and Xu 2019; Ma and Kamo, 2018). The effect of different process parameters (such as the heating rate, the carrier gas, and the operating temperature) on the degradation mechanisms and the composition of the pyrolysis products have also been reported by many authors (Diaz et al. 2018; Evangelopoulos et al. 2017; XIE et al. 2016).

  Interestingly, the presence of copper is of prime importance during the pyrolysis of WPCB as it acts as a catalyst to promote some chemical reactions, especially in the presence of brominated compounds (Gao et al. 2021; Liu et al. 2018; Ma and Kamo 2018). More specifically, copper changes the pyrolysis mechanism of WPCB by promoting the conversion of organic bromides to Br_2 and HBr, while reducing the apparent activation energy (E_a) of the overall pyrolysis reaction (Gao et al. 2021). Ma and Kamo (2018) also noted that the simultaneous addition of iron and nickel particles in a two-stage pyrolysis bed reactor leads to a synergistic effect on the product yields of WPCB. Their presence promotes the formation of phenol and other aromatic compounds, such as benzene and toluene.
- Recovery of metals from the solid residue: Pyrolysis of WPCB also proved to be a promising technique to efficiently separate the volatile organic fraction from the valuable metallic components. The solid residue that contains these valuable metals (along with oxides and carbonaceous compounds) is then processed via hydrometallurgical or electrometallurgical methods to selectively recover each metal. Jadhao et al. (2020) proposed a new approach to recover the metallic fraction of WPCB using a chemical-free green ultrasonication technology of the solid residue obtained after a mild-temperature pyrolysis process performed between 200 and 600 °C. Many other approaches have been proposed to improve individual metal recovery from solid pyrolysis residue. It includes: (i) low-temperature alkaline smelting coupled with liquid–liquid phase

separation (Chen et al. (2020), (ii) combination of physical separation methods such as sieving, gravity separation, and magnetic separation (Wang et al. 2020), and (iii) the use of higher pyrolysis temperatures (i.e. 850 °C) to obtain distinct copper-rich, lead-tin-rich, and slag-rich residues to promote more efficient partitioning of rare-earth elements (Khanna et al. 2018).

- Capture and stabilization of brominated compounds: Copyrolysis of e-waste plastics or WPCB in the presence of alkali and alkaline earth-based additives (such as CaO and $CaCO_3$) significantly enhance the quality of the pyrolytic liquid products (Gao and Xu 2019; Xie et al. 2017; Kumagai et al. 2017; Jung et al. 2012; Zuo et al., 2011; Bhaskar et al. 2004). These additives stabilize the hydrogen bromide emissions in the form of brominated salt compounds via a reaction such as the following:

$$CaO(s) + 2HBr(g) \xrightarrow{T>300\,^{\circ}C} CaBr_2(s) + H_2O(g)$$

As a result, bromine is not transferred to the gas phase, which prevents the contamination of the collected pyrolytic oil and gas. There are still several important scientific elements to understand regarding the identification and thermal stability of the brominated compounds formed during the pyrolysis, which may impact the recovery of valuable metals.

References

Abbasalizadeh, A., Teng, L., Seetharaman, S. et al. (2016). Chapter 24 – Rare earth extraction from NdFeB magnets and rare earth oxides using aluminum chloride/fluoride molten salts. In: *Rare Earths Industry* (eds. I.B. De Lima and W.L. Filho), 357–373. Elsevier ISBN 9780128023280, https://doi.org/10.1016/B978-0-12-802328-0.00024-3.

Aboughaly, M. and Gabbar, H.A. (2020). Recent technologies in electronic-waste management. In: *E-waste Recycling and Management*, 63–80. Cham: Springer.

Adusei, A., Arko-Mensah, J., Dzodzomenyo, M. et al. (2020). Spatiality in health: the distribution of health conditions associated with electronic waste processing activities at Agbogbloshie, Accra. *Annals of Global Health* 86 (1): 31. https://doi.org/10.5334/aogh.2630.

Alenezi, R.A. and Al-Fadhli, F.M. (2018). Thermal degradation kinetics of waste printed circuit boards. *Chemical Engineering Research and Design* 130: 87–94.

Álvarez-de-los-Mozos, E., Rentería-Bilbao, A., and Díaz-Martín, F. (2020). WEEE recycling and circular economy assisted by collaborative robots. *Applied Sciences* 10 (14): 4800.

Anindya, A., Swinbourne, D.R., Reuter, M.A., and Matusewicz, R.W. (2013). Distribution of elements between copper and FeOx–CaO–SiO_2 slags during pyrometallurgical processing of WEEE: Part 1–Tin. *Minerals Processing and Extractive Metallurgy* 122 (3): 165–173.

Arnout, S., Nagels, E., and Blanpain, B. (2011). Thermodynamics of lead recycling. In: *Proceedings – European Metallurgical Conference, EMC 2011*, vol. 2, 363–372.

Ayres, R.U., Ayres, L.W., and Råde, I. (2003). Lead, zinc and other byproduct metals. In: *The Life Cycle of Copper, Its Co-Products and Byproducts*, 101–147. Dordrecht: Springer.

Bakker, M.L., Nikolic, S., and Mackey, P.J. (2011). ISASMELT™ TSL – applications for nickel. *Minerals Engineering* 24 (7): 610–619.

Barragan, J.A., Ponce de León, C., Alemán Castro, J.R. et al. (2020). Copper and antimony recovery from electronic waste by hydrometallurgical and electrochemical techniques. *ACS Omega* 5 (21): 12355–12363.

Bhaskar, T., Uddin, A., Kaneko, J. et al. (2004). Pyrolysis of polypropylene/polyethylene/polystyrene and polyvinylchloride mixed plastics using $CaCO_3$. *Progress in Rubber, Plastics and Recycling Technology* 20 (2): 163–170.

Cai, G.G.C.L.Q. and Shao-hong, M.Z.P. (2006). Pyrolysis kinetics of waste print circuit board in vacuum condition. *Journal of South China University of Technology* 3.

Chaidez-Felix, J., Romero-Serrano, A., Hernandez-Ramirez, A. et al. (2014). Effect of copper, sulfur, arsenic and antimony on silver distribution in phases of lead blast furnace. *Transactions of Nonferrous Metals Society of China* 24 (4): 1202–1209.

Charitopoulou, M.A., Kalogiannis, K.G., Lappas, A.A., and Achilias, D.S. (2020). Novel trends in the thermo-chemical recycling of plastics from WEEE containing brominated flame retardants. *Environmental Science and Pollution Research*: 1–24. https://doi.org/10.1007/s11356-020-09932-5.

Charles, R.G., Douglas, P., Dowling, M. et al. (2020). Towards increased recovery of critical raw materials from WEEE – evaluation of CRMs at a component level and pre-processing methods for interface optimisation with recovery processes. *Resources, Conservation and Recycling* 161: 104923.

Chatterjee, S. (2012). Sustainable electronic waste management and recycling process. *American Journal of Environmental Engineering* 2 (1): 23–33.

Chen, B., He, J., Sun, X. et al. (2020). Separating and recycling metal mixture of pyrolyzed waste printed circuit boards by a combined method. *Waste Management* 107: 113–120.

Cui, H. and Anderson, C. (2020). Hydrometallurgical treatment of waste printed circuit boards: Bromine leaching. *Metals* 10 (4): 462.

Cui, J. and Zhang, L. (2008). Metallurgical recovery of metals from electronic waste: a review. *Journal of Hazardous Materials* 158 (2): 228–256.

Czégény, Z., Jakab, E., Blazsó, M. et al. (2012). Thermal decomposition of polymer mixtures of PVC, PET and ABS containing brominated flame retardant: formation of chlorinated and brominated organic compounds. *Journal of Analytical and Applied Pyrolysis* 96: 69–77.

Das, P., Gabriel, J.C.P., Tay, C.Y., and Lee, J.M. (2020). Value-added products from thermochemical treatments of contaminated e-waste plastics. *Chemosphere*: 129409.

De Marco, I., Caballero, B., Chomon, M. et al. (2008). Pyrolysis of electrical and electronic wastes. *Journal of Analytical and Applied Pyrolysis* 82 (2): 179–183.

Diaz, L.A. and Lister, T.E. (2018). Economic evaluation of an electrochemical process for the recovery of metals from electronic waste. *Waste Management* 74: 384–392.

Diaz, F., Flerus, B., Nagraj, S. et al. (2018). Comparative analysis about degradation mechanisms of printed circuit boards (PCBS) in slow and fast pyrolysis: the influence of heating speed. *Journal of Sustainable Metallurgy* 4 (2): 205–221.

Espinosa, D.C.R., Moraes, V.T., and Tenorio, J.A.S. (2015). Pyrometallurgical processing. In: *Electronic Waste. Topics in Mining, Metallurgy and Materials Engineering* (eds. H. Veit and A. Moura Bernardes), 81–85. Cham: Springer https://doi.org/10.1007/978-3-319-15714-6_8.

Evangelopoulos, P., Kantarelis, E., and Yang, W. (2017). Experimental investigation of pyrolysis of printed circuit boards for energy and materials recovery under nitrogen and steam atmosphere. *Energy Procedia* 105: 986–991.

Flandinet, L., Tedjar, F., Ghetta, V., and Fouletier, J. (2012). Metals recovering from waste printed circuit boards (WPCBS) using molten salts. *Journal of Hazardous Materials* 213: 485–490.

Flores, G.R.A., Nikolic, S., and Mackey, P.J. (2014). ISASMELT™ for the recycling of E-scrap and copper in the US case study example of a new compact recycling plant. *JOM* 66 (5): 823–832.

Fogarasi, S., Imre-Lucaci, F., Ilea, P., and Imre-Lucaci, A. (2013). The environmental assessment of two new copper recovery processes from waste printed circuit boards. *Journal of Cleaner Production* 54: 264–269.

Franzolin, C.J. (2020). Planned obsolescence resulting from electrical and electronic equipment: waste rights and brazil's national solid waste policy. In: *Sustainable Consumption* (eds. A. Amaral Junior, L. Almeida and L. Klein Vieira), 463–477. Cham: Springer https://doi.org/10.1007/978-3-030-16985-5_25.

Gao, R. and Xu, Z. (2019). Pyrolysis and utilization of nonmetal materials in waste printed circuit boards: debromination pyrolysis, temperature-controlled condensation, and synthesis of oil-based resin. *Journal of Hazardous Materials* 364: 1–10.

Gao, R., Liu, B., Zhan, L. et al. (2021). Catalytic effect and mechanism of coexisting copper on conversion of organics during pyrolysis of waste printed circuit boards. *Journal of Hazardous Materials* 403: 123465.

Grant, R. (2001). Lead production. In: *Encyclopedia of Materials: Science and Technology* (eds. K.J. Buschow, R.W. Cahn, M.C. Flemings, et al.), 4439–4442. Ox: Elsevier.

Guo, Q., Yue, X., Wang, M., and Liu, Y. (2010). Pyrolysis of scrap printed circuit board plastic particles in a fluidized bed. *Powder Technology* 198 (3): 422–428.

Hageluken, C. (2006). Recycling of electronic scrap at umicore's integrated metals smelter and refinery. *World of Metallurgy – ERZMETALL* 59 (3): 152–161.

Hall, W.J. and Williams, P.T. (2007). Separation and recovery of materials from scrap printed circuit boards. *Resources, Conservation and Recycling* 51 (3): 691–709.

Harikrushnan, B., Shreyass, G., Hemant, G., and Pandimadevi, M. (2016). Recovery of metals from printed circuit boards (PCBS) using a combination of hydrometallurgical and biometallurgical processes. *International Journal of Environmental Research* 10 (4): 511–518.

Hua, Z., Wang, J., Wang, L. et al. (2014). Selective extraction of rare earth elements from ndfeb scrap by molten chlorides. *ACS Sustainable Chemistry & Engineering* 2: 2536–2543.

Huang, Y.-F. and Lo, S.-L. (2020). Energy recovery from waste printed circuit boards using microwave pyrolysis: product characteristics, reaction kinetics, and benefits. *Environmental Science and Pollution Research* 27 (34): 43274–43282.

Huang, Y.-F., Chiueh, P.-T., and Lo, S.-L. (2016). A review on microwave pyrolysis of lignocellulosic biomass. *Sustainable Environment Research* 26 (3): 103–109.

Ippolito, N.M., Medici, F., Pietrelli, L., and Piga, L. (2021). Effect of acid leaching pre-treatment on gold extraction from printed circuit boards of spent mobile phones. *Materials* 14 (2): 362.

Jadhao, P.R., Ahmad, E., Pant, K., and Nigam, K. (2020). Environmentally friendly approach for the recovery of metallic fraction from waste printed circuit boards using pyrolysis and ultrasonication. *Waste Management* 118: 150–160.

Jandric, A., Part, F., Fink, N. et al. (2020). Investigation of the heterogeneity of bromine in plastic components as an indicator for brominated flame retardants in waste electrical and electronic equipment with regard to recyclability. *Journal of Hazardous Materials* 390: 121899.

Jin, W. and Zhang, Y. (2020). Economic evaluation of an electrochemical process for the recovery of metals from electronic waste. *ACS Sustainable Chemistry & Engineering* 8: 4693–4707.

Jung, S.-H., Kim, S.-J., and Kim, J.-S. (2012). Fast pyrolysis of a waste fraction of high impact polystyrene (HIPS) containing brominated flame retardants in a fluidized bed reactor: the effects of various ca-based additives (CAO, $CA(OH)_2$ and oyster shells) on the removal of bromine. *Fuel* 95: 514–520.

Kasper, A.C., Veit, H.M., Garcıa-Gabaldon, M., and Herranz, V.P. (2018). Electrochemical study of gold recovery from ammoniacal thiosulfate, simulating the PCBS leaching of mobile phones. *Electrochimica Acta* 259: 500–509.

Khaliq, A., Rhamdhani, M.A., Brooks, G., and Masood, S. (2014). Metal extraction processes for electronic waste and existing industrial routes: a review and Australian perspective. *Resources* 3 (1): 152–179.

Khanna, R., Ellamparuthy, G., Cayumil, R. et al. (2018). Concentration of rare earth elements during high temperature pyrolysis of waste printed circuit boards. *Waste Management* 78: 602–610.

Kumagai, S., Grause, G., Kameda, T., and Yoshioka, T. (2017). Thermal decomposition of tetrabromobisphenol – a containing printed circuit boards in the presence of calcium hydroxide. *Journal of Material Cycles and Waste Management* 19 (1): 282–293.

Laubertova, M., Piroskova, J., and Dociova, S. (2017). The technology of lead production from waste. *World of Metallurgy - ERZMETALL* 70 (1): 47–54.

Lin, C., Chi, Y., and Jin, Y. (2017). Experimental study on treating waste printed circuit boards by molten salt oxidation. *Waste and Biomass Valorization* 8 (7): 2523–2533.

Lister, T.E., Wang, P., and Anderko, A. (2014). Recovery of critical and value metals from mobile electronics enabled by electrochemical processing. *Hydrometallurgy* 149: 228–237.

Liu, J., Guo, X., Liu, Y. et al. (2016). Effects of alkali-salt fusion process on recovery of amphoteric metals from waste printed circuit boards. *Minerals Processing and Extractive Metallurgy* 125 (4): 211–215.

Liu, W., Xu, J., Han, J. et al. (2018). Kinetic and mechanism studies on pyrolysis ofprinted circuit boards in the absence and presence of copper. *ACS Sustainable Chemistry & Engineering* 7 (2): 1879–1889.

Luda, M.P. (2011). Recycling of printed circuit boards. In: *Integrated Waste Management - Volume II* (ed. S. Kumar). Inte-chOpen, https://doi.org/10.5772/17220. Available from: https://www.intechopen.com/books/integrated-waste-management-volume-ii/recycling-of-printed-circuit-boards.

Ma, E. (2019a). Chapter 11 – recovery of waste printed circuit boards through pyrometallurgy. In: *Electronic Waste Management and Treatment Technology* (eds. M.N.V. Prasad and M. Vithanage), 247–267. Butterworth-Heinemann, ISBN 9780128161906, https://doi.org/10.1016/B978-0-12-816190-6.00011-X.

Ma, E. (2019b). Recovery of waste printed circuit boards through pyrometallurgy. In: *Electronic Waste Management and Treatment Technology*, 247–267.

Ma, C. and Kamo, T. (2018). Two-stage catalytic pyrolysis and debromination of printed circuit boards: effect of zero-valent fe and ni metals. *Journal of Analytical and Applied Pyrolysis* 134: 614–620.

Ma, C. and Kamo, T. (2019). Enhanced debromination by Fe particles during the catalytic pyrolysis of non-metallic fractions of printed circuit boards over ZSM-5 and Ni/Sio2-Al_2O_3 catalyst. *Journal of Analytical and Applied Pyrolysis* 138: 170–177.

Ma, H., Du, N., Lin, X. et al. (2018). Experimental study on the heat transfer characteristics of waste printed circuit boards pyrolysis. *Science of the Total Environment* 633: 264–270.

Miskolczi, N., Hall, W.J., Angyal, A. et al. (2008). Production of oil with low organobromine content from the pyrolysis of flame retarded hips and ABS plastics. *Journal of Analytical and Applied Pyrolysis* 83 (1): 115–123.

Moosakazemi, F., Ghassa, S., Soltani, F., and Mohammadi, M.R.T. (2020). Regeneration of sn-pb solder from waste printed circuit boards: a hydrometallurgical approach to treating waste with waste. *Journal of Hazardous Materials* 385: 121589.

Needhidasan, S., Samuel, M., and Chidambaram, R. (2014). Electronic waste – an emerging threat to the environment of urban indi. *Journal of Environmental Health Science and Engineering* 12: 36.

Panda, R., Jadhao, P.R., Pant, K.K. et al. (2020). Eco-friendly recovery of metals from waste mobile printed circuit boards using low temperature roasting. *Journal of Hazardous Materials* 395: 122642.

Quan, C., Li, A., and Gao, N. (2009). Thermogravimetric analysis and kinetic study on large particles of printed circuit board wastes. *Waste Management* 29 (8): 2353–2360.

Rao, M.D., Singh, K.K., Morrison, C.A., and Love, J.B. (2021). Recycling copper and gold from e-waste by a two-stage leaching and solvent extraction process. In: *Separation and Purification Technology*, vol. 263, ISSN 1383-5866, https://doi.org/10.1016/j.seppur.2021.118400, 118400. https://www.sciencedirect.com/science/article/pii/S1383586621001027.

Riedewald, F. and Sousa-Gallagher, M. (2015). Novel waste printed circuit board recycling process with molten salt. *Methods X* 2: 100–106.

Rudnik, E. and Bayaraa, E. (2016). Electrochemical dissolution of smelted low-grade electronic scraps in acid sulfate-chloride solutions. *Hydrometallurgy* 159: 110–119.

Schlesinger, M.E., King, M.J., Sole, K.C., and Davenport, W.G. (2011a). Chapter 10 – direct-to-copper flash smelting. In: *Extractive Metallurgy of Copper*, 5e (eds. M.E. Schlesinger, M.J. King, K.C. Sole and W.G. Davenport), 179–190. Oxford: Elsevier.

Schlesinger, M.E., King, M.J., Sole, K.C., and Davenport, W.G. (2011b). Chapter 19 – chemical metallurgyof copper recycling. In: *Extractive Metallurgy of Copper*, 5e (eds. M.E. Schlesinger, M.J. King, K.C. Sole and W.G. Davenport), 389–396. Oxford: Elsevier.

Schlesinger, M.E., King, M.J., Sole, K.C., and Davenport, W.G. (2011c). Chapter 5 – matte smelting fundamentals. In: *Extractive Metallurgy of Copper*, 5e (eds. M.E. Schlesinger, M.J. King, K.C. Sole and W.G. Davenport), 73–88. Oxford: Elsevier.

Song, Q., Liu, Y., Zhang, L., and Xu, Z. (2021). Selective electrochemical extraction of copper from multi-metal e-waste leaching solution and its enhanced recovery mechanism. *Journal of Hazardous Materials* 407: 124799.

Su, X. (2020). Electrochemical separations for metal recycling. *Electrochemical Society Interface* 29 (3): 55–61.

Tanísalí, E., Ozer, M., and Burat, F. (2021). Precious metals recovery from waste printed circuit boards by gravity separation and leaching. *Mineral Processing and Extractive Metallurgy Review* 42 (1): 24–37.

Theo, L. (1998, May). Integrated recycling of non-ferrous metals at Boliden Ltd. Ronnskar smelter. In: *Proceedings of the 1998 IEEE International Symposium on Electronics and the Environment*. ISEE-1998 (Cat. No. 98CH36145), 42–47. IEEE.

Tuncuk, A., Stazi, V., Akcil, A. et al. (2012). Aqueous metal recovery techniques from e-scrap: hydrometallurgy in recycling. *Minerals Engineering* 25 (1): 28–37.

Vaccari, M., Vinti, G., Cesaro, A. et al. (2019). Weee treatment in developing countries: environmental pollution and health consequences – an overview. *International Journal of Environmental Research and Public Health* 16 (9): 1595.

Vanparys, R., Brooks, G., Rhamdhani, M.A., and Crivits, T. (2020). Reduction of lead-rich slags with coke in the lead blast furnace. In: *PbZn 2020: 9th International Symposium on Lead and Zinc Processing* (eds. A. Siegmund, S. Alam, J. Grogan, et al.), 173–185. Cham: Springer International Publishing.

Vegliò, F. and Birloaga, I. (2018). *Waste Electrical and Electronic Equipment Recycling: Aqueous Recovery Methods*. Woodhead Publishing.

Wang, X., Jiao, F., Qin, W. et al. (2020). Combination of pyrolysis and physical separation to recover copper and tin from waste printed circuit boards. *JOM* 72 (9): 3179–3185.

Wang, C., Sun, R., and Xing, B. (2021a). Copper recovery from waste printed circuit boards by flotation-leaching process optimized using response surface methodology. *Journal of the Air & Waste Management Association*.

Wang, R., Zhang, C., Zhao, Y. et al. (2021b). Recycling gold from printed circuit boards gold-plated layer of waste mobile phones in “mild aqua regia” system. *Journal of Cleaner Production* 278: 123597.

Watt, W., Hidayat, T., Shishin, D., and Jak, E. (2018). Advanced thermochemical fundamental and applied research to improve the integrity of the steel water jacketed furnace at port Pirie. In: *Extraction 2018*, 229–239. Cham: Springer.

Willner, J., Fornalczyk, A., Cebulski, J., and Janiszewski, K. (2014). Preliminary studies on simultaneous recovery of precious metals from different waste materials by pyrometallurgical method. *Archives of Metallurgy and Materials* 59 (2) https://doi.org/10.2478/amm-2014-0136.

Xi, X.-L., Feng, M., Zhang, L.-W., and Nie, Z.-R. (2020). Applications of molten salt and progress of molten salt electrolysis in secondary metal resource recovery. *International Journal of Minerals, Metallurgy, and Materials* 27: 1599–1617.

Xie, Y., Sun, S., and Zeng, J. (2016). Effect of pyrolysis conditions on vacuum pyrolysis gaseous product from waste printed circuit board. *Environmental Science & Technology*: 02.

Xie, Y., Sun, S., Liu, J. et al. (2017). The effect of additives on migration and 35 transformation of gaseous pollutants in the vacuum pyrolysis process of waste printed circuit boards. *Waste Management & Research* 35 (2): 190–199.

Xiong, J., Yu, S., Wu, D. et al. (2020). Pyrolysis treatment of nonmetal fraction of waste printed circuit boards: focusing on the fate of bromine. *Waste Management & Research* 38 (11): 1251–1258.

Yan, X.Y. and Fray, D.J. (2010). Molten salt electrolysis for sustainable metals extraction and materials processing: a review. In: *Electrolysis: Theory, Types and Applications* (eds. S. Kuai and J. Meng), 255–302. Nova Science.

Yao, Z., Xiong, J., Yu, S. et al. (2020). Kinetic study on the slow pyrolysis of nonmetal fraction of waste printed circuit boards (NMF-WPCBS). *Waste Management & Research* 38 (8): 903–910.

Yoon, H.-S., Kim, C.-J., Chung, K.W. et al. (2014). Leaching kinetics of neodymium in sulfuric acid from e-scrap of ndfeb permanent magnet. *Korean Journal of Chemical Engineering* 31: 706–711.

Zhan, L. and Xu, Z. (2008). Application of vacuum metallurgy to separate pure metal from mixed metallic particles of crushed waste printed circuit board scraps. *Environmental Science & Technology* 42 (20): 7676–7681.

Zhang, L. and Xu, Z. (2016). A review of current progress of recycling technologies for metals from waste electrical and electronic equipment. *Journal of Cleaner Production* 127: 19–36.

Zhang, G., He, Y., Feng, Y. et al. (2018). Recovery of residual metals from fine non-metallic fractions of waste printed circuit boards using a vibrated gas-solid fluidized bed. *Separation and Purification Technology* 207: 321–328.

Zhang, T., Mao, X., Qu, J. et al. (2021). Microwave-assisted catalytic pyrolysis of waste printed circuit boards, and migration and distribution of bromine. *Journal of Hazardous Materials* 402: 123749.

Zhao, B. (2013). *Lead and zinc sintering*, 165–199. Rijeka: Sintering applications. InTech.

Zhu, X.-N., Zhang, Y.-K., Zhang, Y.-Q. et al. (2020). Flotation dynamics of metal and non-metal components in waste printed circuit boards. *Journal of Hazardous Materials* 392: 122322.

Zuo, X., Damoah, L.N., Zhang, L. et al. (2011). Green pyrolysis of used printed wiring board powders. In: *Recycling of Electronic Waste II* (eds. L. Zhang and G.K. Krumdick). https://doi.org/10.1002/9781118086391.ch3.

Zvezdin, A., Di Mauro, E., Rho, D. et al. (2020). En route toward sustainable organic electronics. *MRS Energy & Sustainability* 7 (16) https://doi.org/10.1557/mre.2020.16.

8

Recycling Technologies – Hydrometallurgy

Denise C. R. Espinosa, Rafael P. de Oliveira, and Thamiris A. G. Martins

University of São Paulo. Polytechnic School, Department of Chemical Engineering, Av. Prof. Luciano Gualberto, 380 – Butantã, São Paulo, SP 05508-010, Brazil

8.1 Background

Electronic waste (e-waste) has in its composition polymers, ceramics, and metals. The metallic fraction can contain critical metals, such as rare-earth-elements (REE), and valuable metals such as silver and gold. Furthermore, e-waste can also contain hazardous materials, such as lead, cadmium, polybrominated diphenyl ethers, and arsenic. Therefore, recycling end-of-life (EoL) electronic devices has economic and environmental benefits (Chauhan et al. 2018; EASAC 2016). E-waste utilization as a secondary raw material is a complementary route of metal production included in urban mining. These activities also promote necessary techniques to a circular economy, through recovery of metals, which were lost in waste discharge or landfilling. However, a challenge to close the loop in life cycle of products is the collection of e-waste. Problems are the storage of small electronic devices and the lack of collection strategy. People must realize that it is better to dispose of the old smartphone in a collection point for recycling instead of storing it (Chauhan et al. 2018; Tesfaye et al. 2017).

One challenge with e-waste recycling is related to technological advancement and the programmed obsolescence of this equipment that generate an increasing amount of residues with a heterogeneous composition (Babu et al. 2007; Raele et al. 2017; Wang et al. 2017). Although unit operations for secondary source processing are similar to primary sources, recycling of e-waste has advantages, such as reduction of extraction of raw materials from nonrenewable resources, a higher metal content when compared with natural ores and the collaboration to waste management, reducing contamination of environment with inappropriate release of metals, and energy savings due to lower-energy requirements to process waste instead of using natural ores (Tesfaye et al. 2017).

Currently, e-waste recycling is done initially by pyrometallurgical route, which provides viable recovery of metals in industrial scale. However, the following hydrometallurgical process is necessary for a possible total separation and recovery

Electronic Waste: Recycling and Reprocessing for a Sustainable Future, First Edition.
Edited by Maria E. Holuszko, Amit Kumar, and Denise C. R. Espinosa.

of metals. Thus, hybrid routes of e-waste processing are more common than pyrometallurgical exclusive route. In this context, Umicore can be mentioned. Located in Belgium, Umicore is a company that industrially recovers metals from waste electrical and electronic equipment (WEEE). Initially, electronic scrap undergoes pretreatment (preprocessing or mechanical processing), pyrometallurgical route, and, later, recovery/refining by hydrometallurgical process (Cui and Anderson 2016).

Steel recycling uses entirely pyrometallurgy to recover iron from large household appliances as refrigerators. This overview indicates that there is a technological opportunity to search for sustainable hydrometallurgical e-waste recycling. Tailing generation, toxic and expensive reagents are difficulties to be overcome. If these problems are solved, hydroprocessing can be applied industrially for recovering different metals from the heterogeneous e-waste composition (Chauhan et al. 2018; Tesfaye et al. 2017).

Hydrometallurgical techniques consist of chemical reactions in aqueous medium. Extraction of metals is performed by chemical leaching, usually followed by stages of purification and recovery of the pregnant leaching solution (PLS) (Gupta 2003; Perez et al. 2019; Tesfaye et al. 2017). In the leaching process, the transfer of metals to the liquid phase occurs. The employed leaching solution may be acidic (e.g. H_2SO_4, HNO_3, and HCl solution), alkaline (e.g. NaOH, NH_3, and Na_2CO_3 solution), saline (e.g. $CuCl_2$), or even water (Gupta and Mukherjee 1990; Jackson 1986; Swain and Mishra 2019). Other lixiviants (e.g. cyanide, halide, thiourea, and thiosulfate) are significant, due to efficacy in precious metal recovery (Namias 2013). The purification and recovery steps can occur by selective chemical precipitation/cementation, solvent extraction, ion-exchange resins, and electrorefining methods (Gupta and Mukherjee 1990; Perez et al. 2019; Silvas et al. 2015; Tesfaye et al. 2017).

The first step in hydrometallurgical processing before the leaching of metals from electronic is the dismantling, separating, and comminuting. Dismantling and manual separation can concentrate metals in e-waste, thereby reducing the use of chemical reagents during leaching. Magnetic separation can separate iron before leaching, thereby reducing the use of reagents and facilitates purification of leaching solution (Chauhan et al. 2018; Tesfaye et al. 2017; Valix et al. 2017). Polymers or ceramic materials present in e-waste could hinder contact of metals with leaching solution, as presented by Valix et al. (2017).

Hydrometallurgical processing of e-waste , in contrast to pyrometallurgy, can be controlled to selectively recover different metals, even those present in ppm concentration. Aqueous processing also reduces gas emitted and requires less energy than smelting metals in furnaces (Gupta 2003; Swain and Mishra 2019).

The optimal condition for leaching metals from e-waste depends on pulp density, temperature, time, particle size, percentage of magnetic fraction, and mixing velocity. Polymer coating hinders the contact with leaching solution, also influencing hydroprocessing, and is related with parameters such as particle size, solubility in the solution, and degradation. Some authors show a desoldering step that facilitates the leaching of copper, gold, and base metals from electronics. Desoldering is made by leaching with fluoroboric acid, which removes lead and tin, and separates the nonmetallic components. The residue is then leached to recover copper, zinc, nickel, silver, palladium, and gold (Ashiq et al. 2019; Kamberović et al. 2009; Namias 2013; Valix et al. 2017).

Several studies have been carried out on the recycling of WEEE through hydrometallurgical routes and/or hybrid routes. Even in hybrid routes, the hydrometallurgical technique is used in the metal recovery and purification stage, as it manages to treat several metals present in solution, even in low concentrations (on a ppm scale). Thus, this technique is used for recycling WEEE such as printed circuit boards (PCBs), photovoltaic modules (PV), batteries, and light-emitting diodes (LEDs). Due to the constant change in the composition and amount of waste generated each year, hydrometallurgical processing must always be in constant adaptation and study. This technique allows the recovery of several materials present at the same time in solution, in different concentrations.

8.2 Waste Printed Circuit Boards (WPCBs)

PCBs are present in almost all electronics products, such as air conditioners, televisions, computers, mobile phones, and printers (Ning et al. 2017; Zhou and Qiu 2010). They are nonconductive substrates that electrically connect the circuit components, and they can be made up of about 30% polymeric materials, 30% ceramic materials, and up to 40% metals (Li et al. 2004; Kaya 2016; Silvas et al. 2015). Studies estimate that the fraction of PCBs in e-waste (by mass) can vary from 3% to 8% (Cucchiella et al. 2015; Raele et al. 2017).

Several studies report routes for characterization and recovery/concentration of metals in waste printed circuit boards (WPCBs) such as mobile phones (Camelino et al. 2015; Jing-ying et al. 2012; Petter et al. 2014), digital video discs (DVD) players and vacuum cleaners (VCs) (Kumar et al. 2015), printers (Silvas et al. 2015), and computers (Veit et al. 2005; Yamane et al. 2011).

Table 8.1 shows the mass percentage of the main metals in different WPCBs. Due to this complexity and heterogeneity of the composition of different types of existing WPCBs, the release and recovery of the metals present have become a challenge for the existing studies. Different routes hydrometallurgical can be adopted depending on the materials of interest (Cui and Anderson 2016).

Rocchetti et al. (2018) presented a study with various patents on recycling WPCBs. The survey and search for these patents were updated until July 2017. The main processes based on the analysis of these patents are shown in Figure 8.1.

Figure 8.1 presents a flowchart illustrating a general outline based on the study that was carried out on existing and updated patents until 2017 (Rocchetti et al. 2018). The main processes that can be included in the recycling of WPCBs have been shown. However, these steps can be modified, depending on the metal of interest, the available technologies, and the type of WPCB. Generally, metal recycling/recovery in WPCBs starts with a pretreatment that is directed toward the removal of components (e.g. capacitors and transistors) from WPCBs through physical and manual processes (Silvas et al. 2015). Pretreatment may lead to the

Table 8.1 Mass percentage of main metals in different WPCBs.

WPCB sample	Yamane et al. (2011)	Silvas et al. (2015)	Kasper et al. (2011)	Guo et al. (2011)	Behnamfard et al. (2013)	Yamane et al. (2011)	Kumar et al. (2015)	Kumar et al. (2015)
	Mobile phones	Printer	Mobile phones (other brand)	Computer mother-boards	Computer	Personal computer	DVD	Vacuum cleaner
Ag	0.21	0.31	0.05	0.0026	0.0704	0.16	–	–
Pd	–	–	–	–	0.0028	–	–	–
Al	0.26	3.73	0.61	1.32	4.011	5.7	10.10	3.26
Au	0.00	0.004	0.09	0.006	0.013	0.13	–	–
Cu	34.49	32.50	37.81	27.5	19.187	20.19	17.80	7.08
Fe	10.57	1.42	4.85	1.6	1.133	7.33	5.51	4.40
Ni	2.63	0.34	2.54	0.4	0.165	0.43	0.36	0.26
Pb	1.87	0.00	1.23	2.91	0.385	5.53	3.33	3.71
Sn	3.39	0.96	2.55	3.43	0.689	8.83	2.57	2.75
Zn	5.92	0.64	1.82	2.86	0.840	4.48	1.99	5.54

recovery of electronic components and the manufacture of Sn (Rocchetti et al. 2018).

Processing and/or concentration are stages that may include physical–mechanical processes such as comminution; granulometric classification; and magnetic, electrostatic, and density separation. These processes are performed for size reduction and particle separation, and they are often a prerequisite for additional treatments for metal extraction techniques (hydrometallurgical, pyrometallurgical, biohydrometallurgical, and electrohydrometallurgical, in addition to hybrid or combined routes) (Veit et al. 2006 2014; Zhang and Xu 2016). Dismantling and physical–mechanical processes should be used when there is a need to release and separate metallic components from WPCB to expose metals to subsequent chemical processes (Cui and Anderson 2016).

Kumar et al. (2015) reported the use of physical processing for the generation of a concentrate enriched with the metals of interest for subsequent selective extraction through hydrometallurgical processing. The authors processed WPCBs of DVD and VCs by dismantling, comminution, homogenization, pneumatic separation, and froth flotation. Metal concentrations in DVD WPCBs obtained in waste characterization were 43 (wt%), 88 (wt%) in concentrates from froth flotation, and 75 (wt%) from pneumatic separation. For VC WPCBs, through applications, these methods, the total quantity of the metals in characterization was 30 (wt%), and in concentrates from froth flotation and pneumatic separation was 90 and 65 (wt%) total metals. Therefore, this work obtained an enriched metal concentrate according to studied conditions, and this material can help the stages of extraction, recuperation, and purification of the metals.

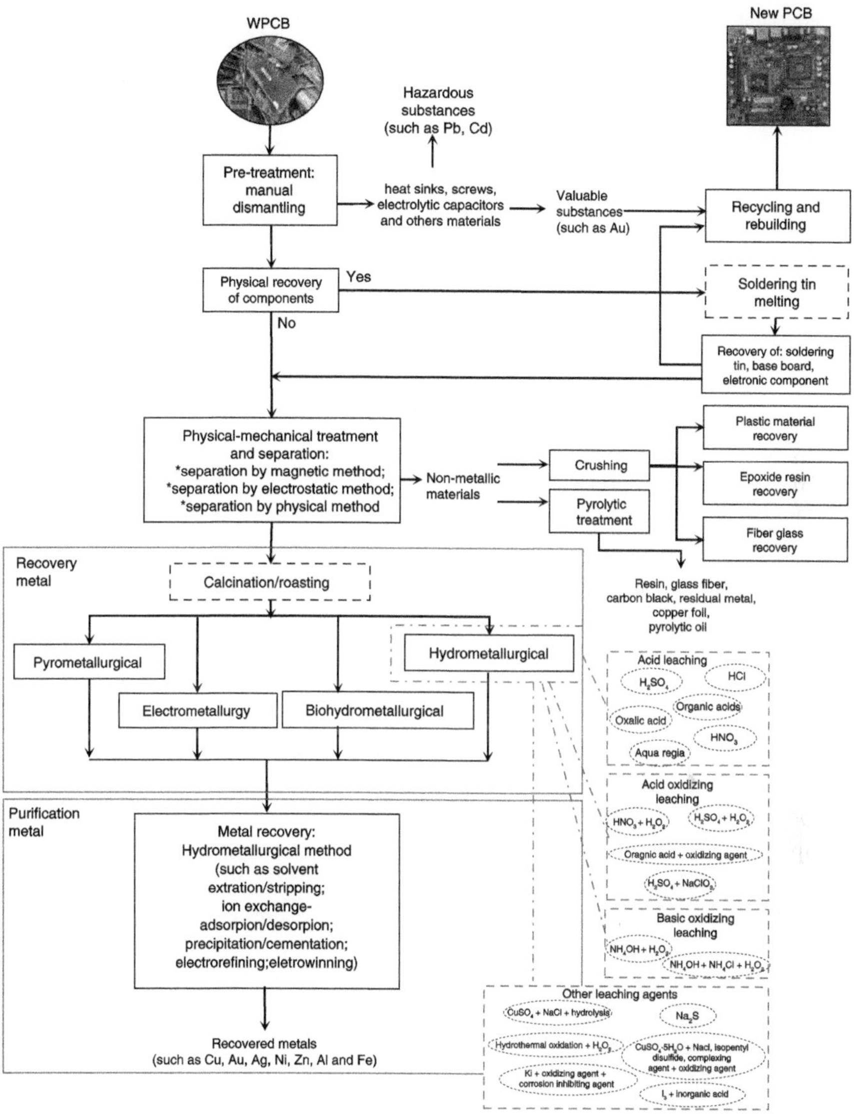

Figure 8.1 General flowchart illustrating the whole recycling value chain of the WPCB. Source: Adapted from Rocchetti et al. (2018).

The recovery/purification stage is carried out to recover elements of interest through hydrometallurgical techniques. This stage can occur by selective chemical precipitation or cementation, solvent extraction, ion-exchange resins, and electrorefining/electroplating methods (Kaya 2016; Perez et al. 2019; Silvas et al. 2015; Tesfaye et al. 2017).

The extraction and recovery of metals can be carried out using various leaching agents. Awasthi and Li (2017) presented a revision with different leaching agent studies for extraction, recovery, and purification of WPCBs metals. Some of them reached recoveries of more than 90% of Au, Ag, Cu, Pd, and Zn. Some of these leaching agents checked were sulfuric, nitric, and hydrochloric acid, ammonia, and thiosulfate. These reagents were also applied in combinations intended to be synergetic for metal extraction. The authors also compared the leaching rates, cost, toxicity, corrosion behavior, and reliabilities of lixiviants. Thiourea was well assessed in all categories, and cyanide had the disadvantage of toxicity. Chloride and aqua regia were badly rated because they have aggressive corrosive behavior. Therefore, the application of chloride and aqua regia can be a problem in industries.

Table 8.2 presents a compilation of results for the extraction and recovery of WPCBs by leaching processes that have been reported in several studies. Jing-ying et al. (2012) studied the effects of particle size on the extraction of metals (Au and Ag) from WPCBs of mobile phones. They also verified the best conditions for the concentration of leaching agents (thiourea and Fe^{3+}) and temperature. The study was able to obtain, in two hours of reaction, with particles of 100 mesh, in a leach solution with 24 g/l of thiourea and 0.6% of Fe^{3+} and room temperature, about 90% of Au and 50% of Ag.

Silvas et al. (2015) described a hydrometallurgical extraction process for copper recovery from printer WPCBs. The process was performed in two subsequent leaching steps: the first in sulfuric medium and the second in oxidant sulfuric medium. The metallic fraction present in the waste was 44.0 wt%, whose main metal was copper 32.5 wt%. In the sulfuric leaching, 90 wt% Al, 40 wt% Zn, and 8.6 wt% Sn were extracted. In the oxidative leaching, the extraction percentage was 100 wt% Cu, 60 wt% Zn, and 10 wt% Al. Thus, at the end of the proposed hydrometallurgical processing, the extraction results were 100% Cu. The recovery was 98.46%, which would correspond to 32 kg of Cu for 100 kg of residues of these WPCBs.

Complementing the studies about Cu leaching techniques, Correa et al. (2018) studied a route for purification and recovery of this metal of interest. This route was composed of an extraction step (oxidative sulfuric leaching) and another subsequent solvent extraction step for purification and recovery of Cu. In the extraction step, the nonmagnetic fraction of the WPCBs was ground and leached, according to Table 8.2. This step was similar to the study of Silvas et al. (2015). Extraction results were 60 wt% Al; 94 wt% Cu; Zn, Ni, and residual Fe. Subsequently, the solvent extraction step (applying D2EHPA) was performed in two steps: (i) separation of residual Zn, Al, and Fe; and (ii) copper separation. In the first stage, metals were removed from the solution 100 wt% Zn, Fe, and Al. pH 3.5, 2 : 1 aqueous/organic ratio (A/O), 10% (v/v) D2EHPA were used. In the second stage, 100% of the copper was extracted at pH 3.5, 1 : 1 A/O, 20% (v/v) D2EHPA.

Wang et al. (2019) performed the separation and recovery of Cu from PCB extraction solution because the composition of the leach solution (mainly Fe) could affect this recovery. The solvent extraction technique also was used, in which the concentration of the organic extractor Acorga M5640, pH of the aqueous phase, phase ratio (O/A), contact time, and the concentration of H_2SO_4 as a stripping reagent were studied for recovery Cu. Thus, in the extraction solution containing Cu (6.166 g/l) and Fe (57.5 g/l), under defined ideal extraction conditions (16% M5640, pH = 1.1, O/A = 1/1, contact time three minutes and 25 °C) and stripping (2.5 mol/l H_2SO_4,

Table 8.2 Results compilation for extraction and recovery from WPCBs by leaching processes.

References	WPCB sample	Leaching medium	Conditions	Extraction (wt%)
Correa et al. (2018)	WPCBs fraction non-magnetic	H_2SO_4 (1 M) + H_2O_2 (10 ml were added every 30 min, total added 120 ml)	1 : 10 solid/liquid rate; 75 °C; for 6 h	60 Al, 94 Cu, 76 Zn; 50 Ni; and residual iron
Silvas et al. (2015)	Printer	1st leaching: H_2SO_4 (1 M); 2nd leaching: H_2SO_4 (1 M) + H_2O_2 (30 ml were added every 30 min);	1:10 solid/liquid rate; 75 °C; 1st for 4 h; 2nd for 4 h	100 Al; 100 Cu; 8.6 Sn; 100 Zn
Camelino et al. (2015)	Mobile phones (particle size <2 mm)	$H_8N_2O_3S_2$ (0.13 M)	10 pH; 20 °C; 180 rpm	70 Au
Jing-ying et al. (2012)	Mobile phones (particle size <100 mesh)	24 g/l of thiourea +0.6% of Fe^{3+}	Room temperature; for 2 h;	90 Au; 50 Ag
Petter et al. (2014)	Mobile phones (particle size <1 mm)	(i) for leaching Ag: HNO_3 (1/3 v/v);	1 : 20 solid/liquid rate; 25 °C and 60 °C; for 2 h	100 Ag
		(ii) for leaching Au: commercial cyanide (Galvastripper concentrate);	1 : 20 solid/liquid rate; 25 °C; for 2 h and 4 h; pH ~ 12.5	60 Au
Havlik et al. (2010)	Personal computer (after thermal pretreatment)	400 ml HCl (1 M)	Sample 3 g; 80 °C; for 180 min	98 Cu
Neto et al. (2016)	Motherboards <0.250 mm	HNO_3 (2 mol/l)	1 : 10 solid/liquid rate; 50 °C; por 210 min; 200 rpm	59 Cu
Birloaga and Vegliò (2016)	Computers	1.7 M H_2SO_4 (98wt./vol%) + 17% (v/v) H_2O_2 (30 wt./vol%),	1 : 15 solid/liquid rate; room temperature; 200 rpm; por 1 h	98 Cu
Behnamfard et al. (2013)	Computers	1st leaching: H_2SO_4 (2 M/l) + H_2O_2 (35%) 20 V%	1 : 10 solid/liquid rate; ~25 °C; 200 rpm; por 3 h	85.76 Cu 0.86 Ag
		2nd leaching: H_2SO_4 (2 M/l) + H_2O_2 (35%) 20 V%	1 : 10 solid/liquid rate; ~25 °C; 200 rpm; por 3 h	13.99 Cu 11.30 Ag
		3rd leaching: Thiourea (20 g/l) + Ferric Iron (6 g/l) + H_2SO_4 (10 g/l)	1:10 solid/liquid rate; ~25 °C; 200 rpm; por 3 h	84.31 Au 71.36 Ag 2.13 Pd
		4th leaching: HCl (5 M/l) + H_2O_2 (1%V) + NaClO (10%V)	1 : 10 solid/liquid rate; ~63 °C; 300 rpm; por 3 h	97.87 Pd 6 Au 16.48 Ag

O/A = 1/1, contact time three minutes, 25 °C and five stages), more than 88.6% Cu and less than 2.4% Fe were transferred to the recovery solution for these e-waste.

Different from Correa et al. (2018) and Wang et al. (2019), Neto et al. (2016) studied extraction, recovery, and purification of Cu in motherboards with ion-exchange resins. Initially, they obtained a multielement leaching solution with 78% of the total amount of Cu of the residue. After this leaching, a bispicolylamine resin was used to recover 59.0% Cu, and a final Cu solution with high purity (99.0%) was obtained after eluting, with H_2SO_4 (4 mol/l) in the column.

Kasper et al. (2011) also recovered Cu from mobile phone WPCBs. They used mechanical processing techniques for concentration of iron in the magnetic fraction, copper in the conductive fraction, and another fraction with polymers and ceramics. Magnetic separation concentrated 60% of Cu. This metal fraction was extracted through leaching in aqua regia (solid/liquid ratio of 1 : 20, under magnetic stirring, heating at 60 °C, for two hours). Subsequently, the solution resulting from the extraction step was electrowinned for Cu recovery. The electrolytic cell was assembled with a Pt anode and a Cu cathode, and different variations of current densities (1, 3, and 6 A/dm^2) and deposition times (30, 60, 90, 120, and 180 minutes) were studied. Thus, 92.8% of the dissolved copper was recovered.

Table 8.2 also presents the study of Behnamfard et al. (2013) that proposes a new hydrometallurgical process for the selective recovery of Cu, Ag, Au, and Pd from PCBs. They propose a hybrid route that is composed of leaching processes and precipitation/cementation (Figure 8.2).

As shown in Figure 8.2, about 99% of the Cu content was extracted after the two consecutive leaching steps in an oxidizing sulfuric medium. The solid residue from the second leaching step was treated (acid thiourea + ferric iron as an oxidizing agent), and an extraction of 85.76% Au and 71.36% Ag was obtained. The precipitation of Au and Ag from this third lixiviation was investigated in which sodium borohydride (SBH) was used as a reducing agent. The leaching of Pd and gold (from the solid residue of the third stage) was carried out in the leaching system $NaClO–HCl–H_2O_2$. The total amount of Pd and Au resulting from the chloride leachate was precipitated using 2 g/l of SBH. Thus, they proposed a process flow chart for the recovery of Cu, Ag, Au, and Pd from the PCB.

These studies show that the quantity and diversity of metals present in WPCBs can also vary according to the type and year of manufacture of the original equipment (Kasper et al. 2011; Veit et al. 2006). This metal fraction can transform PCB scrap into an interesting raw material according to the environmental and economic points of view, assigning high-added value to the e-waste through reusing it in the manufacture of other products (Hadi et al. 2015; Mdlovu et al. 2018).

8.3 Photovoltaic Modules (PV)

Since the 2000s, the photovoltaic module technology has been extensively used due to the increase in cell efficiency and the decrease in production costs. The total world production of photovoltaic energy rose from 278 MW (megawatt) in 2000

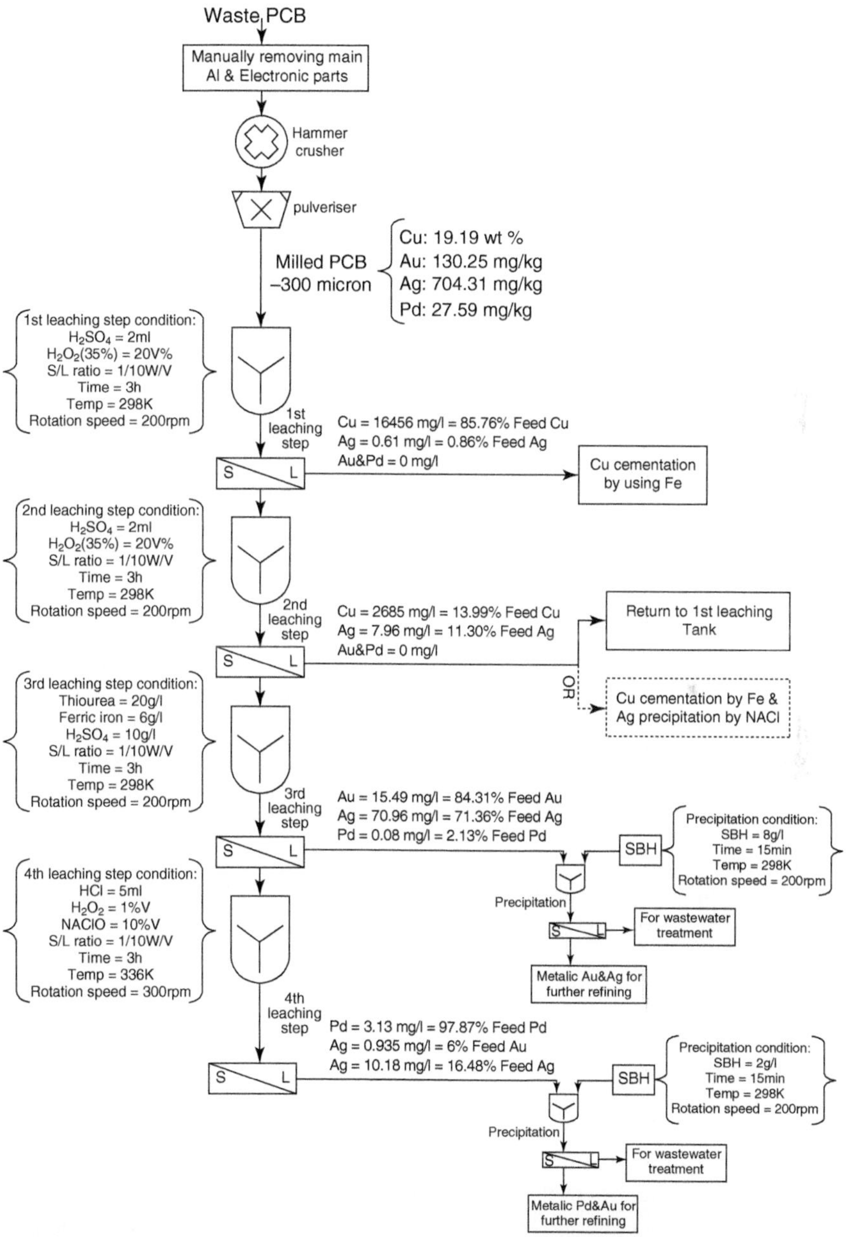

Figure 8.2 The proposed route for extraction of Cu, Ag, Au, and Pd from WPCB of computers. Source: Behnamfard et al. (2013).

to 56 000 MW in 2012 (Klugmann-Radziemska and Kuczynska-Łazewska 2020; Tammaro et al. 2015). These PV modules have a life span of around 25–30 years; hence, it is estimated that the amount of waste will increase from 2030 (Klugmann-Radziemska & Kuczynska-Łazewska, 2020; Tammaro et al., 2015; Dias et al., 2016b).

Types of PV modules can differ according to production technologies and semiconductor materials. PV modules can be classified as first, second, or third generation, depending on the production technology. The first generation is composed of crystalline silicon (c-Si), which can be monocrystalline, and polycrystalline or ribbon sheets (approximately 80% of world production). Second generation is formed by thin-film amorphous silicon (a-Si), cadmium telluride (CdTe), multijunction cells (a-Si-lc-Si), copper indium gallium diselenide (CIGS), and copper indium diselenide (CIS). Third generation is composed of photovoltaic concentrator (CPV) and emerging technologies (Dias et al. 2016a; Tammaro et al. 2015; Tao and Yu 2015).

Industrial-scale recycling processes silicon photovoltaic modules crystalline (c-Si) at SolarWorld, First Solar, and ANTEC Solar GmbH companies were cited in Tao and Yu (2015) studies. SolarWorld recycles silicon through calcination (600 °C) of these modules (for thermal decomposition of ethylene-vinyl acetate [EVA]), and, subsequently, the solids are manually separated into fractions with glass, Si-cells, and metals. Then, the semiconductor is purified by etching, and recovering the glass and semiconductor. First Solar is a manufacturer that recycle CdTe thin-film modules. The modules are comminuted (shredding + hammer milling), and, later, they are leached in oxidizing sulfuric medium in stainless steel drum. The company recovers 90% of glass, 95% of semiconductors, and 80% of tellurium. ANTEC Solar GmbH has a pilot plant in Germany for recycling CdTe and CdS modules. The initial step is also physical disintegration, which is followed by calcination at 300 °C in an oxygen atmosphere. An etching step occurs when exposing the resulting fragments in a chlorine gas atmosphere and the condensation of $CdCl_2$ and $TeCl_4$ occurs, and these compounds can be precipitated by cooling.

Latunussa et al. (2016b) summarize the SASIL S.p.A. project in which it was called full recovery end-of-life photovoltaic (FRELP). SASIL S.p.A. created a pilot-scale plant for recycling c-Si modules in which FRELP uses mechanical and chemical separation steps. This project became an industrial-scale plant with a processing capacity of 1 t/h up to 8000 t/year of crystalline-silicon waste PV modules (Latunussa et al. 2016a). Figure 8.3 presents a proposal for the recovery process of silicon photovoltaic waste in FRELP.

The photovoltaic waste recycling process proposed/developed by the SASIL photovoltaic waste treatment project (Figure 8.3) consists of a sequence of processes (transport of waste PV modules, unloading of the waste panels, disassembly, cable treatment, incineration of cable polymers, glass separation and refinement, cutting of modules, incineration of encapsulation, and back-sheet layer with energy recovery, sieving, acid leaching, filtration, electrolysis, neutralization, and filter press) that are described in detail in Latunussa et al. 2016a. SASIL can recover about 94% Ag, 99% Cu, and 99% Al (Latunussa et al. 2016a, 2016b).

Studies of Wang et al. (2012) have proposed a thermal method to separate materials (such as silicon, glass, and metal) from conventional crystalline silicon modules.

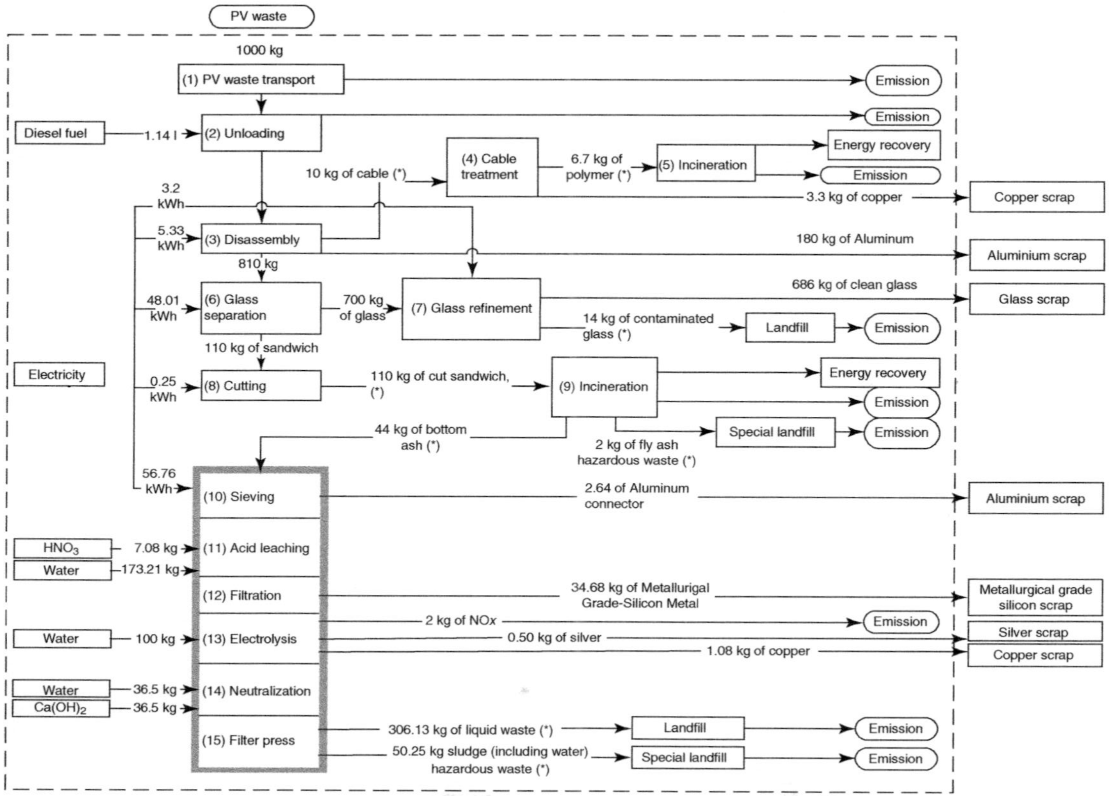

Figure 8.3 Recycling process of c-Si modules proposed by the FRELP method of SASIL. Source: Latunussa et al. (2016a).

Two steps of thermal processing were carried out, and the EVA was burnt. The glass plate was recovered without breaking, and the copper could be recovered (85%) in further acid treatment.

Dias et al. (2016a) studied characterization of PV waste modules (c-Si modules), and they checked if Ag could be leached by hydrometallurgical procedures. To obtain the semiconductor material, the modules were cut and immersed, for two days, in H_2SO_4 (95%), at room temperature and constant agitation. This dry and ground material was leached with HNO_3 (64%), for two hours at room temperature, and later analyzed chemically. In these modules, the Ag content was 600 g/t.

The estimated amount of silver is 10 g of silver/m^2 of photovoltaic panel produced. Silver scrap recycling plays an important role in the silver market as it accounts for a third of the total market (Latunussa et al. 2016b). Even so, in the literature consulted, few published works addressing the recovery of silver from solar modules were found.

The extraction of silver from crystalline silicon modules was performed at 20 °C and 40 °C by Yi et al. (2014). H_2SO_4, HCl, and HNO_3 (the concentrations varied from 1 to 3 mol/l) were used as leaching agents with a solid/liquid ratio of 1 : 5 for two hours. Nitric acid extracted about 100% of the available silver, while sulfuric and hydrochloric acids recovered 10% less silver in two hours, even at a concentration of 3 mol/l.

The recycling of several metals present in PV modules, such as Ag, Cu, Al, Ga, In, Ge, and Te is not yet explored. Latunussa et al. (2016b) report that glass, Al, and Cu make up most of the PV mass. However, despite the untapped economic potential, these materials are lost in the EoL of crystalline PV modules. Yi et al. (2014) affirm that Ag, Cu, and Al will be the metals of interest in PV recycling between 2030 and 2050 and that these metals correspond to US$ 0.54–1.70/W of the cost of crystalline silicon modules. Hence, in future, recycling strategies will no longer focus mainly on Si recovery, but on metal recovery. In this context, hydrometallurgy routes for recovering these metals from PV modules should be further studied and explored.

8.4 Batteries

Batteries are present in many electronic devices to supply energy, thereby allowing autonomy. Mobile phones, laptops, tablets, cameras, and recently electric cars require batteries, which use critical materials and generate a great amount of waste. Batteries can cause a potential environmental impact due to their composition of hazardous metals, such as cadmium, lead, and mercury (Bernardes et al. 2004). There are two main groups of batteries: primary and rechargeable. The rechargeable batteries include lead–acid, nickel–cadmium, nickel–metal hydride (NiMH), and lithium-ion (Zeng et al. 2014).

The recycling processes for batteries can be applied through pyrometallurgical routes, hydrometallurgy, or a combination of these two extractive methodologies. Examples of hydrometallurgical processes for batteries are TNO process for NiCd batteries; BATENUS process for most batteries types; ZINCEX process for

Zn-bearing materials; and RECUPYL for most batteries types. UMICORE process for Li-ion and NiMH batteries is a hybrid process (Goodship and Stevels 2012).

TNO process initiates sorting batteries and selecting NiCd to comminution and hydrometallurgical step. The fraction with granulometry above 3 mm is Fe rich, then is submitted to magnetic separation, and the magnetic and non-magnetic portions are washed in HCl to remove Cd. The washing solution is used to leach comminuted fraction (with granulometry below 3 mm) at 90 °C. Leaching liquor is purified with solvent extraction to remove Cd and then Fe is removed by precipitation. The final purified leaching solution is used to recover Ni by electrolysis process. Stripping solution of Cd solvent extraction is also submitted to electrolysis to recover this metal (Erkel 1995).

In BATENUS process, batteries pass through a sorting step with sieving. Button batteries are separated and sent to be treated in another process. The remaining are shredded, sieved, and magnetically separated. Fine fraction is leached and solution purified to recover Zn, Hg, Cu, Ni, Cd, and Mn (Fröhlich and Sewing 1995).

Hydrometallurgical processing is commonly executed after the discharging and dismantling of spent batteries. In the case of lithium-ion, dismantled parts are composed of separators, cathodes, anodes, and shells; electrolyte salts and cathodes are the lithium-containing components; anodes are made of a carbon material, copper, active substances, and organic binder. Cathode-active materials are commonly leached by acids to selectively extract metals, such as cobalt, nickel, manganese, and lithium. However, some studies indicate the use of NaOH leaching during dismantling step to dissolve cathode aluminum foil. The powder collected after filtration is then heated to burn organics, and exhausted gas is managed due to its hazardous content. The calcined material is grounded and leached by acid to extract metals. Organic acids are preferable in environmentally sustainable methods, such as lactic, gluconic, and tartaric acids (Ashiq et al. 2019; Nayaka et al. 2019; Roshanfar et al. 2019; Siqi et al. 2019).

Existing industrial recycling processes for spent lithium batteries, which involves hydrometallurgy, are the Toxco process, which was developed to avoid risks with Li explosion, through cryogenic treatment; and Umicore VAL'EAS process, which is a hybrid pyro–hydro process of spent Li-ion and NiMH batteries to recover Ni, Co, Cu, Fe, and Mn, with exception of Al and Li that are lost (Georgi-Maschler et al. 2012).

Li et al. (2019) presented a review on clean recovery of lead–acid battery, based on alkaline or acid leaching to extract lead. These batteries represented more than 50% market share in 2015. Some processes involve organic acid leaching–calcination, which is the main technology published for batteries. CLEANLEAD , PLACID, and PLINT processes are examples of Pb recycling from spent batteries. PLACID and PLINT processes consist of leaching battery pastes, purification of leaching solution, and metallic lead production, which differ in electrowinning for PLACID process and precipitation for PLINT. CLEANLEAD just produces desulfurized pastes through leaching for further processing (Andrews et al. 2000). Discussion suggests that that hydrometallurgical process is preferably applied, if compared with pyrometallurgy, due to its environment-friendly characteristics. Other technologies are reagent leaching–electrowinning and alkaline leaching–crystallization (Li et al. 2019).

Provazi et al. (2011) studied the separation of metals from a solution composed of a mixture of the main types of household batteries. This solution was prepared by grinding several batteries, and the reduction and elimination of volatile metals took place in an oven. The extraction step was performed by acid leaching (solid/liquid ratio 1 : 10, H_2SO_4–1 M, at room temperature, for 24 hours). Thus, two hydrometallurgical routes were applied: selective precipitation with sodium hydroxide (NaOH) and solvent extraction using Cyanex 272 acid. In the selective precipitation step, a solution of NaOH (2 M) was added to the liquor resulting from the sulfuric acid leaching step to increase the pH of this solution, so selective precipitation can occur. For solvent extraction, Cyanex 272 (0.6 M-extractor) was used and kerosene was used as a diluent. In this comparative study, solvent extraction obtained the highest recovery rates: 99% for Zn (pH 2.5); above 95% Fe, 90% Ce, 88% Mn, Cr, and Co (pH 7.0); more than 85% Ni (pH 3.0); and more than 80% Cd and La (pH 3.5).

8.5 Light-Emitting Diodes (LEDs)

LEDs are dominating lighting market, with expectations of 74% of market share in 2030, and are present in various types of equipment (displays from smartphones, TVs, laptops, tablets, and lamps), due to their energy efficiency (Mizanur Rahman et al. 2017).

Hydrometallurgical processing is also expected for metal recovery in LED waste. Elements such as yttrium, cerium, europium, and gallium are the focus, because they are critical and gold is a precious metal. As a lamp phosphor LED rare earth elements (REE) may have a distinct leaching behavior. Yttrium and europium from fluorescent lamps are leached easily when compared with other rare-earth elements. In these cases, leaching efficiency is more significant than leaching rate. Other benefits from diverse leaching kinetics are the selectivity. Leaching of REE may be performed with the duration of more than 24 hours with separation of rare-earth phosphors in different leaching steps (Jha et al. 2016; Tunsu et al. 2015). Contribution of Fang et al. (2018) assessed criticality and recyclability of LEDs in different colors and products, based on cost for processing, supply risk, environmental impact, technology availability, composition, and total amount of materials.

Ruiz-Mercado et al. (2017) proposed a process of metal recovery from LEDs, which used hydrochloric acid and ammonia as leachate. At the initial leaching with HCl, the REE were dissolute and the solids were releached with NH_3 to separate silver from indium and gallium nitrides. The solution from initial leaching precipitated cerium oxide and in the remaining solution rare-earth metals are recovered by solvent extraction. Aqueous strip solution containing Eu and Y was separated by selective reduction and precipitation, producing respective oxalates, which were calcined to oxides. Figure 8.4 shows schematic flowchart with proposed route to recover rare-earth metals from the LED.

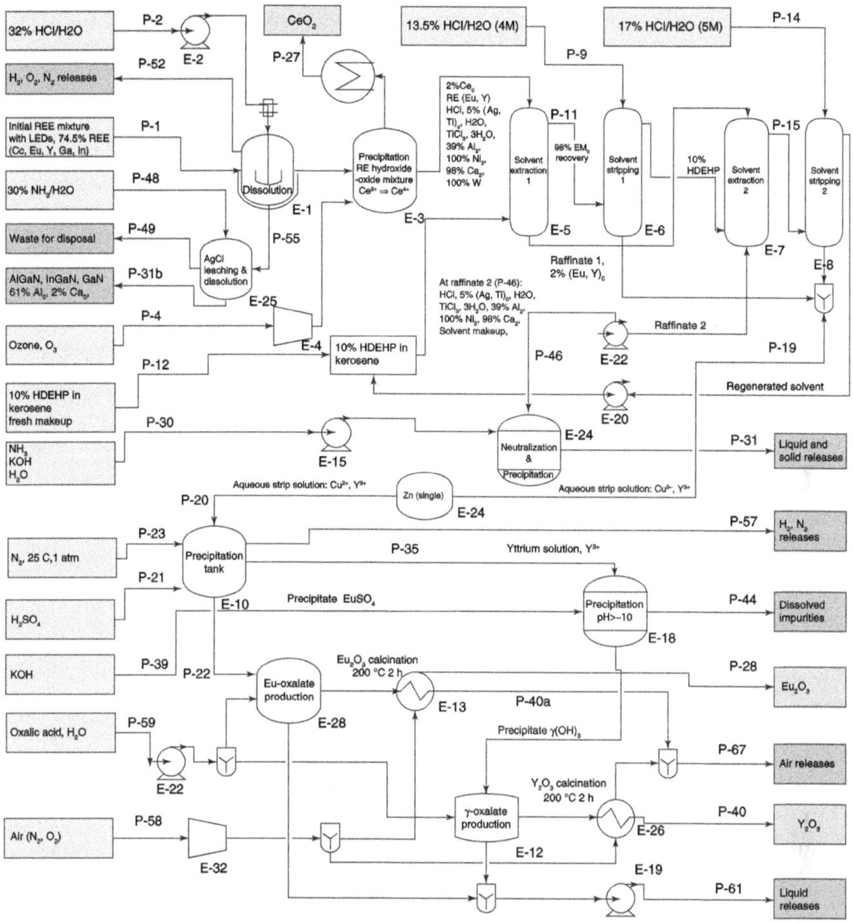

Figure 8.4 Recycling flowchart of REE recovery from LED waste . Source: Ruiz-Mercado et al. (2017).

Maarefvand et al. (2020) studied the recovery of gallium from LED waste based on a process of oxidation and subsequent leaching. Gallium in LED is present as nitride. However, leaching gallium oxide is preferable due to refractory characteristics of gallium nitride. Oxidation step at 1100 °C facilitates leaching, which reaches 91.4% of Ga recovery using 4 M hydrochloric acid, at 93 °C. Increasing oxidation temperature may unprivilege leaching, because surface area decreases and grain size increases, which reduces the contact with acid. Zhou et al. (2019) treated the LED waste through a pyrolysis step at 460 °C before leaching. Escaping from inorganic acids, leaching was performed with oxalic, citric, and DL-malic acids, oxalic acid being a better option due to 83.42% of Ga recovery, which was higher and with modifications reached 90.36% of efficiency at optimal conditions (90 °C, pulp density of 10 g/l, 0.7 M of acid concentration and particles size between 48 and 150 μm). In the same study, a possible mechanism of gallium reaction with oxalic acid and precipitation of iron oxalate was proposed (Maarefvand et al. 2020; Zhou et al. 2019).

Leaching can also be performed under pressure in autoclaves, reduce the consumption of acid, and give better leaching efficiencies. Studies affirmed an increase of extraction in pressurized leaching, such as the recovery of yttrium and europium from spent fluorescent lamp tubes during four hours of leaching, at 125 °C and 5 MPa (Jha et al. 2016), and gallium leaching from GaN waste of LED industry, at 23 atm, 220 °C, in four hours (Chen et al. 2018).

Murakami et al. (2015) studied the recovery of gold by ion-exchange method from LED leaching solution, obtained by leaching with aqua regia in an autoclave, at 80 °C, 1 atm, during 24 hours. The adsorption resin needed to be used in low aqua regia concentrations to avoid decomposition of the amine group and improve the recovery efficiency. Elution was made by thiourea solution, and gold was precipitated from this solution using sodium borohydride to reduce gold. Gold was produced with a purity of 100%.

8.6 Trends

E-waste processing keeps changing as long as there are technological advances of electronic products. New equipment, such as electric vehicles batteries, compact PCBs, organic light-emitting diodes (OLEDs), and flexible smartphones, will require renewed recycling processes. Hybrid car batteries, as an example, may contain cobalt, nickel, and rare earths in a NiMH battery, which has an average weight of 50 kg, and is necessary to develop a metal recovery process (Petranikova et al. 2017; Sethurajan et al. 2019).

A novel and potentially greener technology applied to extract and recover metals from electronic wastes and their leaching solution is the use of ionic liquid (IL). ILs have low volatility, low combustibility and high extractability, and thermal stability, making them environmental friendly (Chauhan et al. 2018). These liquids have been used in studies to separate REE and transitional metals also applied in WEEE. Common functional groups in IL constitution are imidazolium, pyridinium, pyrrolidinium, ammonium, sulfonium, and phosphonium, which can functionalize IL to extract specific metals (Huang et al. 2014; Sethurajan et al. 2019; Tunsu et al. 2015; Zhou et al. 2018).

Supercritical fluid technology is a recent method to recycle metals and deteriorate polymers in e-waste. Supercritical fluid is an efficient extraction media due to its properties of low viscosity, low surface tension, high mass transport coefficient, high solubility and diffusivity for organics. These fluids are pressurized in 25–40 MPa pressures with temperatures varying from 350 to 550 °C. Studies with the recovery of metals from WPCBs using supercritical fluid show metals concentrating on solid products of the treatment. Another variation of this process is the combination of supercritical water oxidation and electrodeposition process, which decomposes organics with oxidative environment and turns metals into oxides; these oxides are soluble in chloride media and deposition occurs in electrochemical cells (Ram et al. 2014; Zhang and Xu 2016).

Magnetic nanohydrometallurgy is a potential technique for metal recovery from waste electric and electronic equipment. This novel process applies magnetic nanoparticles, which are designed with functionalized organic compounds to extract specific metals from leaching solution. These particles ensure selectivity, being capable of recovering the metal through electrowinning combined with its magnetic attraction. When deposition is completed, nanoparticles are recomposed, being liberated to close the loop (Condomitti et al. 2012).

Magnetic nanoparticles are versatile, with many benefits such as the use of compact reactors, low-energy demand, non-aggressive conditions of use, reduced consumption of solvents, multiple design possibilities giving selectivity to capture metals of interest. It is important to highlight that magnetic nanoparticles may separate rare-earth elements, such as lanthanum and neodymium, in three extraction stages (Almeida and Toma 2016; Condomitti et al. 2018).

References

Almeida, S.D.N. and Toma, H.E. (2016). Neodymium(III) and lanthanum(III) separation by magnetic nanohydrometallurgy using DTPA functionalized magnetite nanoparticles. *Hydrometallurgy* 161: 22–28. https://doi.org/10.1016/j.hydromet.2016.01.009.

Andrews, D., Raychaudhuri, A., and Frias, C. (2000). Environmentally sound technologies for recycling secondary lead. *Journal of Power Sources* 88 (1): 124–129. https://doi.org/10.1016/S0378-7753(99)00520-0.

Ashiq, A., Kulkarni, J., and Vithanage, M. (2019). Hydrometallurgical recovery of metals from E-waste. *Electronic Waste Management and Treatment Technology*: 225–246. (Issue March). https://doi.org/10.1016/B978-0-12-816190-6.00010-8.

Awasthi, A.K. and Li, J. (2017). An overview of the potential of eco-friendly hybrid strategy for metal recycling from WEEE. *Resources, Conservation and Recycling* 126: 228–239. https://doi.org/10.1016/j.resconrec.2017.07.014.

Babu, B.R., Parande, A.K., and Basha, C.A. (2007). Electrical and electronic waste: a global environmental problem. *Waste Management and Research* 25: 307–318. https://doi.org/10.1177/0734242X07076941.

Behnamfard, A., Mehdi, M., and Veglio, F. (2013). Process development for recovery of copper and precious metals from waste printed circuit boards with emphasize on palladium and gold leaching and precipitation. *Waste Management* 33: 2354–2363.

Bernardes, A.M., Espinosa, D.C.R., and Tenório, J.A.S. (2004). Recycling of batteries: a review of current processes and technologies. *Journal of Power Sources* 130 (1–2): 291–298. https://doi.org/10.1016/j.jpowsour.2003.12.026.

Birloaga, I. and Vegliò, F. (2016). Study of multi-step hydrometallurgical methods to extract the valuable content of gold, silver and copper from waste printed circuit boards. *Journal of Environmental Chemical Engineering* 4: 20–29. https://doi.org/10.1016/j.jece.2015.11.021.

Camelino, S., Rao, J., Padilla, L.R., and Lucci, R. (2015). Initial studies about gold leaching from printed circuit boards (PCB's) of waste cell phones metals recycling from PCB's and ion-lithium batteries view project. *Procedia Materials Science* 9: 105–112. https://doi.org/10.1016/j.mspro.2015.04.013.

Chauhan, G., Jadhao, P.R., Pant, K.K., and Nigam, K.D.P. (2018). Novel technologies and conventional processes for recovery of metals from waste electrical and electronic equipment: challenges & opportunities – a review. *Journal of Environmental Chemical Engineering* 6 (1): 1288–1304. https://doi.org/10.1016/j.jece.2018.01.032.

Chen, W.-S., Hsu, L.-L., and Wang, L.-P. (2018). Recycling the GaN waste from LED industry by pressurized leaching method. *Metals* 8 (10): 861. https://doi.org/10.3390/met8100861.

Condomitti, U., Zuin, A., Silveira, A.T. et al. (2012). Magnetic nanohydrometallurgy: a promising nanotechnological approach for metal production and recovery using functionalized superparamagnetic nanoparticles. *Hydrometallurgy* 125–126: 148–151. https://doi.org/10.1016/j.hydromet.2012.06.005.

Condomitti, U., Almeida, S.N., Silveira, A.T. et al. (2018). Green processing of strategic elements based on magnetic nanohydrometallurgy. *Journal of the Brazilian Chemical Society* 29 (5): 948–959. https://doi.org/10.21577/0103-5053.20180009.

Correa, M.M.J., Silvas, F.P.C., Aliprandini, P. et al. (2018). Separation of copper from a leaching solution of printed circuit boards by using solvent extraction with D2EHPA. *Brazilian Journal of Chemical Engineering* 35 (3): 919–930. https://doi.org/10.1590/0104-6632.20180353s20170144.

Cucchiella, F., D'Adamo, I., Lenny Koh, S.C., and Rosa, P. (2015). Recycling of WEEEs: an economic assessment of present and future e-waste streams. *Renewable and Sustainable Energy Reviews* 51: 263–272. https://doi.org/10.1016/j.rser.2015.06.010.

Cui, H. and Anderson, C.G. (2016). Literature review of hydrometallurgical recycling of printed circuit boards (PCBs). *Journal of Advanced Chemical Engineering* 6 (1): 1–11. https://doi.org/10.4172/2090-4568.1000142.

Dias, P., Javimczik, S., Benevit, M. et al. (2016a). Recycling WEEE: extraction and concentration of silver from waste crystalline silicon photovoltaic modules. *Waste Management* 57: 220–225. https://doi.org/10.1016/j.wasman.2016.03.016.

Dias, P.R., Benevit, M.G., and Veit, H.M. (2016b). Photovoltaic solar panels of crystalline silicon: characterization and separation. *Waste Management and Research* 34 (3): 235–245. https://doi.org/10.1177/0734242X15622812.

EASAC (2016). Priorities for critical materials for circular economy: Policy report 19. (Issue November).

Erkel, V. (1995). Recovery of Cd and Ni from batteries. US Patent 5,407,463.

Fang, S., Yan, W., Cao, H. et al. (2018). Evaluation on end-of-life LEDs by understanding the criticality and recyclability for metals recycling. *Journal of Cleaner Production* 182: 624–633. https://doi.org/10.1016/j.jclepro.2018.01.260.

Fröhlich, S. and Sewing, D. (1995). The BATENUS process for recycling mixed battery waste. *Journal of Power Sources* 57 (1–2): 27–30. https://doi.org/10.1016/0378-7753(95)02234-1.

Georgi-Maschler, T., Friedrich, B., Weyhe, R. et al. (2012). Development of a recycling process for Li-ion batteries. *Journal of Power Sources* 207: 173–182. https://doi.org/10.1016/j.jpowsour.2012.01.152.

Goodship, V. and Stevels, A.L.N. (2012). Waste electrical and electronic equipment (WEEE) handbook. In: *Waste Electrical and Electronic Equipment (WEEE) Handbook*, 1e (eds. V. Goodship, A. Stevels and J. Huisman), 752. Woodhead Publishing https://doi.org/10.1533/9780857096333.

Guo, C., Wang, H., Liang, W. et al. (2011). Liberation characteristic and physical separation of printed circuit board (PCB). *Waste Management* 9–10: 2161–2166. https://doi.org/10.1016/j.wasman.2011.05.011.

Gupta, C.K. (2003). *Chemical Metallurgy Principles and Practice*. Wiley-VCH https://doi.org/10.1002/anie.200385071.

Gupta, C.K. and Mukherjee, T.K. (1990). *Hydrometallurgy in Extraction Processes*, 1e, vol. 2. CRC Press https://doi.org/10.1201/9780203751404.

Hadi, P., Xu, M., Lin, C.S.K. et al. (2015). Waste printed circuit board recycling techniques and product utilization. *Journal of Hazardous Materials* 283: 234–243. https://doi.org/10.1016/j.jhazmat.2014.09.032.

Havlik, T., Orac, D., Petranikova, M. et al. (2010). Leaching of copper and tin from used printed circuit boards afterthermal treatment. *Journal of Hazardous Materials* 183: 866–873. https://doi.org/10.1016/j.jhazmat.2010.07.107.

Huang, J., Chen, M., Chen, H. et al. (2014). Leaching behavior of copper from waste printed circuit boards with Brønsted acidic ionic liquid. *Waste Management* 34: 483–488. https://doi.org/10.1016/j.wasman.2013.10.027.

Jackson, E. (1986). *Hydrometallurgical Extraction and Reclamation*. Ellis Horwood Limited.

Jha, M.K., Kumari, A., Panda, R. et al. (2016). Review on hydrometallurgical recovery of rare earth metals. *Hydrometallurgy* 161: 77–101. https://doi.org/10.1016/j.hydromet.2016.01.035.

Jing-ying, L., Xiu-li, X., and Wen-quan, L. (2012). Thiourea leaching gold and silver from the printed circuit boards of waste mobile phones. *Waste Management* 32 (6): 1209–1212. https://doi.org/10.1016/j.wasman.2012.01.026.

Kamberović, Ž., Korać, M., Ivšić, D. et al. (2009). Hydrometallurgical process for extraction of metals from electronic waste-part I: material characterization and process option selection. *Metalurgija-Journal of Metallurgy* 15 (4): 231–243. https://doi.org/10.30544/382.

Kasper, A.C., Berselli, G.B.T., Freitas, B.D. et al. (2011). Printed wiring boards for mobile phones: characterization and recycling of copper. *Waste Management* 31 (12): 2536–2545. https://doi.org/10.1016/j.wasman.2011.08.013.

Kaya, M. (2016). Recovery of metals and nonmetals from electronic waste by physical and chemical recycling processes. *Waste Management* 57: 64–90. https://doi.org/10.1016/j.wasman.2016.08.004.

Klugmann-Radziemska, E. and Kuczynska-Łazewska, A. (2020). The use of recycled semiconductor material in crystalline silicon photovoltaic modules production – a life cycle assessment of environmental impacts. *Solar Energy Materials and Solar Cells* 205, 110259 Contents. https://doi.org/10.1016/j.solmat.2019.110259.

Kumar, V., Lee, J.c., Jeong, J. et al. (2015). Recycling of printed circuit boards (PCBs) to generate enriched rare metal concentrate. *Journal of Industrial and Engineering Chemistry* 21: 805–813. https://doi.org/10.1016/j.jiec.2014.04.016.

Latunussa, C.E.L., Ardente, F., Blengini, G.A., and Mancini, L. (2016a). Life cycle assessment of an innovative recycling process for crystalline silicon photovoltaic panels. *Solar Energy Materials and Solar Cells* 156: 101–111. https://doi.org/10.1016/j.solmat.2016.03.020.

Latunussa, C. E. L., Mancini, L., Blengini, G. A., Ardente, F., Pennington, D. (2016b). Analysis of material recovery from silicon photovoltaic panels (p. 45). https://doi.org/10.2760/378123.

Li, J., Shrivastava, P., Gao, Z., and Zhang, H.-C. (2004). Printed circuit board recycling methods. *Workshop Materials on WEEE Management in Taiwan* 27: 33–42.

Li, M., Yang, J., Liang, S. et al. (2019). Review on clean recovery of discarded/spent lead-acid battery and trends of recycled products. *Journal of Power Sources* 436 (July): 226853. https://doi.org/10.1016/j.jpowsour.2019.226853.

Maarefvand, M., Sheibani, S., and Rashchi, F. (2020). Recovery of gallium from waste LEDs by oxidation and subsequent leaching. *Hydrometallurgy* 191 (June 2019): 105230. https://doi.org/10.1016/j.hydromet.2019.105230.

Mdlovu, N.V., Chiang, C.L., Lin, K.S., and Jeng, R.C. (2018). Recycling copper nanoparticles from printed circuit board waste etchants via a microemulsion process. *Journal of Cleaner Production* 185: 781–796. https://doi.org/10.1016/j.jclepro.2018.03.087.

Mizanur Rahman, S.M., Kim, J., Lerondel, G. et al. (2017). Missing research focus in end-of-life management of light-emitting diode (LED) lamps. *Resources, Conservation and Recycling* 127: 256–258. https://doi.org/10.1016/j.resconrec.2017.04.013.

Murakami, H., Nishihama, S., and Yoshizuka, K. (2015). Separation and recovery of gold from waste LED using ion exchange method. *Hydrometallurgy* 157: 194–198. https://doi.org/10.1016/j.hydromet.2015.08.014.

Namias, J. (2013). The future of electronic waste recycling in the united states: obstacles and domestic solutions. July.

Nayaka, G.P., Zhang, Y., Dong, P. et al. (2019). An environmental friendly attempt to recycle the spent Li-ion battery cathode through organic acid leaching. *Journal of Environmental Chemical Engineering* 7 (1): 102854. https://doi.org/10.1016/j.jece.2018.102854.

Neto, I.F.F., Sousa, C.A., Brito, M.S.C.A. et al. (2016). A simple and nearly-closed cycle process for recycling copper with high purity from end life printed circuit boards. *Separation and Purification Technology* 164: 19–27.

Ning, C., Sze, C., Lin, K. et al. (2017). Waste printed circuit board (PCB) recycling techniques. *Topics in Current Chemistry* 375: 43. https://doi.org/10.1007/s41061-017-0118-7.

Perez, J.P.H., Folens, K., Leus, K. et al. (2019). Progress in hydrometallurgical technologies to recover critical raw materials and precious metals from low-concentrated streams. *Resources, Conservation and Recycling* 142: 177–188. https://doi.org/10.1016/j.resconrec.2018.11.029.

Petranikova, M., Herdzik-Koniecko, I., Steenari, B.M., and Ekberg, C. (2017). Hydrometallurgical processes for recovery of valuable and critical metals from spent car NiMH batteries optimized in a pilot plant scale. *Hydrometallurgy* 171: 128–141. https://doi.org/10.1016/j.hydromet.2017.05.006.

Petter, P.M.H., Veit, H.M., and Bernardes, A.M. (2014). Evaluation of gold and silver leaching from printed circuit board of cellphones. *Waste Management* 34 (2): 475–482. https://doi.org/10.1016/j.wasman.2013.10.032.

Provazi, K., Campos, B.A., Espinosa, D.C.R., and Tenório, J.A.S. (2011). Metal separation from mixed types of batteries using selective precipitation and liquid-liquid extraction techniques. *Waste Management* 31 (1): 59–64. https://doi.org/10.1016/j.wasman.2010.08.021.

Raele, M.P., De Pretto, L.R., and Zezell, D.M. (2017). Soldering mask laser removal from printed circuit boards aiming copper recycling. *Waste Management* 68: 475–481. https://doi.org/10.1016/j.wasman.2017.07.019.

Ram, C.G., Kislik, V.S., and Vladimir, S.K. (2014). *Solvent Extraction: Classical and Novel Approaches*. Elsevier (Issue 1). https://doi.org/10.1007/s13398-014-0173-7.2.

Rocchetti, L., Amato, A., and Beolchini, F. (2018). Printed circuit board recycling: a patent review. *Journal of Cleaner Production* 178: 814–832. https://doi.org/10.1016/j.jclepro.2018.01.076.

Roshanfar, M., Golmohammadzadeh, R., and Rashchi, F. (2019). An environmentally friendly method for recovery of lithium and cobalt from spent lithium-ion batteries using gluconic and lactic acids. *Journal of Environmental Chemical Engineering* 7 (1): 102794. https://doi.org/10.1016/j.jece.2018.11.039.

Ruiz-Mercado, G.J., Gonzalez, M.A., Smith, R.L., and Meyer, D.E. (2017). A conceptual chemical process for the recycling of Ce, Eu, and Y from LED flat panel displays. *Resources, Conservation and Recycling* 126: 42–49. https://doi.org/10.1016/j.resconrec.2017.07.009.

Sethurajan, M., van Hullebusch, E.D., Fontana, D. et al. (2019). Recent advances on hydrometallurgical recovery of critical and precious elements from end of life electronic wastes – a review. *Critical Reviews in Environmental Science and Technology* 49 (3): 212–275. https://doi.org/10.1080/10643389.2018.1540760.

Silvas, F.P.C., Jiménez Correa, M.M., Caldas, M.P.K. et al. (2015). Printed circuit board recycling: physical processing and copper extraction by selective leaching. *Waste Management* 46: 503–510. https://doi.org/10.1016/j.wasman.2015.08.030.

Siqi, Z., Guangming, L., Wenzhi, H. et al. (2019). Recovery methods and regulation status of waste lithium-ion batteries in China: a mini review. *Waste Management and Research* 37 (11): 1142–1152. https://doi.org/10.1177/0734242X19857130.

Swain, N. and Mishra, S. (2019). A review on the recovery and separation of rare earths and transition metals from secondary resources. *Journal of Cleaner Production* 220: 884–898. https://doi.org/10.1016/j.jclepro.2019.02.094.

Tammaro, M., Rimauro, J., Fiandra, V., and Salluzzo, A. (2015). Thermal treatment of waste photovoltaic module for recovery and recycling: experimental assessment of the presence of metals in the gas emissions and in the ashes. *Renewable Energy* 81: 102–112. https://doi.org/10.1016/j.renene.2015.03.014.

Tao, J. and Yu, S. (2015). Review on feasible recycling pathways and technologies of solar photovoltaic modules. *Solar Energy Materials & Solar Cells* 141: 108–124.

Tesfaye, F., Lindberg, D., Hamuyuni, J. et al. (2017). Improving urban mining practices for optimal recovery of resources from e-waste. *Minerals Engineering* 111: 209–221. https://doi.org/10.1016/j.mineng.2017.06.018.

Tunsu, C., Petranikova, M., Gergorić, M. et al. (2015). Reclaiming rare earth elements from end-of-life products: a review of the perspectives for urban mining using hydrometallurgical unit operations. *Hydrometallurgy* 156: 239–258. https://doi.org/10.1016/j.hydromet.2015.06.007.

Valix, M., Loo, Y.S., Bucknell, J. et al. (2017). Effect of FR-4 decomposition in the hydrometallurgical recovery of copper from electronic waste. *Hydrometallurgy* 173: 199–209. https://doi.org/10.1016/j.hydromet.2017.08.012.

Veit, H.M., Diehl, T.R., Salami, A.P. et al. (2005). Utilization of magnetic and electrostatic separation in the recycling of printed circuit boards scrap. *Waste Management* 25 (1): 67–74. https://doi.org/10.1016/j.wasman.2004.09.009.

Veit, H.M., Bernardes, A.M., Ferreira, J.Z. et al. (2006). Recovery of copper from printed circuit boards scraps by mechanical processing and electrometallurgy. *Journal of Hazardous Materials* 137 (3): 1704–1709. https://doi.org/10.1016/j.jhazmat.2006.05.010.

Veit, H.M., Juchneski, N.C.d.F., and Scherer, J. (2014). Use of gravity separation in metals concentration from printed circuit board scraps. *Metallurgy and Materials* 67 (1): 73–79.

Wang, T. Y., Hsiao, J. C., & Du, C. H. (2012). Recycling of materials from silicon base solar cell module. *Conference Record of the IEEE Photovoltaic Specialists Conference*, 2355–2358. https://doi.org/10.1109/PVSC.2012.6318071.

Wang, F., Zhao, Y., Zhang, T. et al. (2017). Metals recovery from dust derived from recycling line of waste printed circuit boards. *Journal of Cleaner Production* 165: 452–457. https://doi.org/10.1016/j.jclepro.2017.07.112.

Wang, L., Li, Q., Sun, X., and Wang, L. (2019). Separation and recovery of copper from waste printed circuit boards leach solution using solvent extraction with Acorga M5640 as extractant. *Separation Science and Technology* 54 (8): 1302–1311. https://doi.org/10.1080/01496395.2018.1539106.

Yamane, L.H., de Moraes, V.T., Espinosa, D.C.R., and Tenório, J.A.S. (2011). Recycling of WEEE: characterization of spent printed circuit boards from mobile phones and computers. *Waste Management* 31 (12): 2553–2558. https://doi.org/10.1016/j.wasman.2011.07.006.

Yi, Y.K., Kim, H.S., Tran, T. et al. (2014). Recovering valuable metals from recycled photovoltaic modules. *Journal Ofthe Air & Waste Management Association* 64 (7): 797–807. https://doi.org/10.1080/10962247.2014.891540.

Zeng, X., Li, J., and Singh, N. (2014). Recycling of spent lithium-ion battery: a critical review. *Critical Reviews in Environmental Science and Technology* 44 (10): 1129–1165. https://doi.org/10.1080/10643389.2013.763578.

Zhang, L. and Xu, Z. (2016). A review of current progress of recycling technologies for metals from waste electrical and electronic equipment. *Journal of Cleaner Production* 127: 19–36. https://doi.org/10.1016/j.jclepro.2016.04.004.

Zhou, Y. and Qiu, K. (2010). A new technology for recycling materials from waste printed circuit boards. *Journal of Hazardous Materials* 175: 823–828. https://doi.org/10.1016/j.jhazmat.2009.10.083.

Zhou, H., Wang, Y., Guo, X. et al. (2018). The recovery of rare earth by a novel extraction and precipitation strategy using functional ionic liquids. *Journal of Molecular Liquids* 254: 414–420. https://doi.org/10.1016/j.molliq.2018.01.078.

Zhou, J., Zhu, N., Liu, H. et al. (2019). Recovery of gallium from waste light emitting diodes by oxalic acidic leaching. *Resources, Conservation and Recycling* 146 (March): 366–372. https://doi.org/10.1016/j.resconrec.2019.04.002.

9

Recycling Technologies – Biohydrometallurgy

Franziska L. Lederer and Katrin Pollmann

Helmholtz-Zentrum Dresden-Rossendorf, Helmholtz Institute Freiberg for Resource Technology, Biotechnology Division, Bautzner Landstraße 400, 01328 Dresden, Germany

9.1 Introduction

Bio*hydro*metallurgy is one of many different processes for metal recovery. As a highly interdisciplinary field, biohydrometallurgy combines microorganisms and their metabolites (–bio) in a mainly aquatic environment (–hydro) and the treatment of metal-containing materials or solutions (–metallurgy) for metal production and treatment. It is applied to many different metal-rich materials from primary mineral sources, secondary mining products, and numerous manufactured resources (Watling 2015). Biohydrometallurgy uses biological tools for the processing of primary ores for many years – especially in case of bioleaching. Besides that, special biological tools can enhance the metal recovery from manufactured resources such as technical waste products, processing wastes, industrial waste waters, and other secondary sources (Pollmann et al. 2018). In nature, multiple processes exist that influence biogeochemical cycles of elements. These microorganism-driven processes contribute to bioaccumulation, bioweathering, biomineralization, and precipitation or microbial reduction. Using these bio-inspired processes promotes biological recycling strategies as well as several clean industrial processes, bio-based materials, and bioremediation. Modern bio-based approaches that are currently being developed for the recycling of valuable elements found in technical products contribute to a "green" circular economy. Main processes in biohydrometallurgy are bioleaching, biosorption, bioflotation, and bioreduction.

9.2 Bioleaching: Metal Winning with Microbes

Bioleaching is defined as the extraction of metals by the metabolic activity of bacteria (direct bioleaching) or metabolic compounds (indirect bioleaching). It is applicable to metal extraction from low-grade ores, beneficiation of ores or coal, removal of toxic metals, and recovery of metals from waste materials.

Electronic Waste: Recycling and Reprocessing for a Sustainable Future, First Edition.
Edited by Maria E. Holuszko, Amit Kumar, and Denise C. R. Espinosa.

Many bioleaching studies concentrate on the metal extraction from ores, but the processes can be applied principally also to other sources such as industrial residues and waste materials that can be considered as an artificial ore. During the last years, numerous studies were published that describe the application of bioleaching approaches for the recovery of metals from various technical and industrial waste products, e.g. fly ash from municipal waste incineration, electronic scrap such as printed circuit boards from computers and mobile phones, spent catalysts and batteries, and others. Bioleaching approaches on industrial residues and waste materials are based on the use of chemolithoautotrophic bacteria, heterotrophic bacteria, yeasts and fungi, and cyanogenic bacteria. Most studies with chemolithoautotrophic bacteria used the acidophilic organisms *Acidithiobacillus ferrooxidans* or *A. thiooxidans* (e.g. Brombacher et al. 1998; Gholami et al. 2011; Karwowska et al. 2014; Mishra et al. 2008). These approaches obtained relatively high leaching efficiencies (in many cases of >80% of metals) but required the addition of sulfur and acidification of the cultivation media to maintain bacterial growth and solubilization process. Studies describing approaches with heterotrophic microorganisms used bacteria, fungi, or yeasts that produce diverse organic acids such as citric acid, gluconic acid, or acetic acid as lixiviant for the respective metals. Especially the fungus *Aspergillus niger* was used for a broad range of materials (Bosshard et al. 1996; Brandl et al. 2001; Qu et al. 2015). Depending on growth conditions, this organism produces huge amounts of diverse organic acids (e.g. citric acid, gluconic acid, oxalic acid). The organism is applied commercially for organic acid production and cultivation as well as its metabolism has been studied in detail. In some cases, more than 90% of metals could be mobilized by application of this organism (Brandl et al. 2001). Biogenic cyanide produced by *Chromobacter violaceum* was used for the mobilization of gold and other noble metals, copper, and nickel from shredded printed circuit boards and automotive catalysts by using direct or indirect leaching approaches (Campbell et al. 2001; Chi et al. 2011; Faramarzi et al. 2004; Shin et al. 2015b). In indirect bioleaching processes, more than 90% of metals were mobilized. These results were comparable with the usage of commercial sodium cyanide (NaCN) demonstrating the principal suitability of such approaches for commercial applications (Shin et al. 2015b).

Most studies aimed at the recovery of valuable metals such as Cu, Ni, Au, or other noble metals. These elements can be found in high concentrations in diverse electronic wastes or in residues from mineral processing, e.g. smelter dust, fly ash, or incineration slag (Auerbach et al. 2019; Bosshard et al. 1996; Brandl et al. 2001; Brandl et al. 2008; Brombacher et al. 1998; Klink et al. 2016; Oliazadeh et al. 2006; Ramanathan and Ting 2016). However, starting with the resource crises in 2009, an increasing number of studies have been published that concentrate on other valuable elements. Several research groups investigated the mobilization of rare-earth elements (REE) that are an essential component in most modern technologies, from different solid materials. Most of these studies used a variety of heterotrophic bacteria for the extraction of REE from REE-bearing minerals such as monazite (Hassanien et al. 2014; Shin et al. 2015a). However, some recent studies investigated bioleaching approaches for different REE-bearing secondary resources such as red mud from alumina production (Cizkova et al. 2019; Qu and Lian 2013a; Qu et al. 2013b) or electronic waste material, e.g. from waste phosphors (Hopfe et al. 2017, 2018; Reed et al. 2016). In most cases, REE are mobilized either by different organic acids or by enzymatic activity of phosphate solubilizing microorganisms.

As consequence of increasing electromobility, lithium and cobalt as components of energy storage devices moved into the center of attention. Some studies investigated the extraction for Li and Co from waste lithium ion batteries either by a combination of chemical treatment with citric acid and bioactivity (Dolker and Pant 2019) or by *A. niger* (Horeh et al. 2016) or *A. ferrooxidans* (Mishra et al. 2008).

Maneesuwannarat et al. (2016) used a strain of *Cellolosemicrobium funkei*, which was isolated from cadmium- and arsenic-contaminated soil for bioleaching of GaAs (Maneesuwannarat et al. 2016). It was supposed that proteins are involved in Ga mobilization, indicating a new mechanism for metal dissolution (Maneesuwannarat et al. 2019).

These studies demonstrate the great potential that microorganisms offer for the transformation of materials that can be used for new recycling routes. It can be expected that the ongoing growing demand of other elements, and development and growth of new technologies such as renewable energies will promote further studies.

Although all these studies gave a proof of principle for the application of microorganisms for metal mobilization from various waste materials, only few of these approaches have been implemented in industrial processes just yet. Major challenges are selectivity, efficiency, and economy of bioprocesses. More recent studies developed stepwise approaches by combining different chemical or biological leaching steps. For example, Pourhossein and Mousavi obtained high leaching rates of >80% for different elements (Cu, Ni, Ga) from waste light-emitting diodes (WLEDs) by applying a stepwise indirect bioleaching approach using a biogenic ferric agent (Pourhossein and Mousavi 2019). Rizki et al. (2019) combined chemical leaching using thiourea with bioleaching by a thiourea-tolerant Fe-oxidizing microorganism to extract gold from electronic waste (Rizki et al. 2019). Huang et al. (2019) reported on a bio-electro-hydrometallurgical process that combines bioleaching by using sulfur-oxidizing bacteria with an electrokinetic recovery process (Huang et al. 2019). It can be expected that such approaches overcome current barriers in using biological approaches.

9.3 Biosorption: Selective Metal Recovery from Waste Waters

Biosorption is defined as the property of biomass or certain biomolecules to bind and concentrate selected ions or other molecules from aqueous solutions (Volesky 2007). It is a passive process and independent from metabolic activities. Therefore, nutrients are not required, and processes can be performed in environments with high toxicity. Biosorption has been mainly applied for the removal of toxic metals from polluted waters, such as arsenic, chromate, cadmium, or uranium (Volesky and Holan 1995).

Another attractive application is the recovery of valuable metals such as gold, platinum, palladium, or others from solutions (Das 2010; Pollmann et al. 2006b). Conventional pyrometallurgical or hydrometallurgical methods (e.g. adsorption by ion-exchange resin, activated carbon, or minerals, solvent extraction, chemical precipitation) require high amounts of energy and addition of chemical agents, thus generating secondary wastes, or are inefficient especially for highly diluted solutions (Das 2010). Biosorption is an environment-friendly alternative to these methods because it uses biodegradable compounds that can be easily produced in high amounts. Further, biomass is considered as carbon-neutral and petrochemical-independent process as it does not emit extra carbon dioxide when burned (Maruyama et al. 2007; Ritter 2004). Various types of biomass have been reported to bind and concentrate metal ions from industrial effluents and aqueous solutions.

Metal-containing solutions such as industrial waste waters, leachates, and mining waters are often acidic with $pH < 3$, have a complex composition containing different competing elements, and contain toxic chemicals or organic compounds that influence biosorptive properties. Therefore, major challenges of biosorptive approaches are the stability of materials, selectivity, effectivity, and cost efficiency. Several approaches address these challenges. Most studies concentrated on the use of bacterial cells, fungi, yeast (Volesky and Holan 1995), algae (reviewed by He and Chen (2014)), seaweed biomass (Figueira et al. 2000), or biocomponents such as crab shells (Daubert and Brennan 2007), plant fibers (Salamun et al. 2015), etc. as biosorptive components that can be easily produced or are waste materials (e.g. in case of crab shells). Especially biopolymers such as cellulose, chitin, or chitosan materials are chemically resistant. However, these materials possess no selectivity and bind a broad range of different elements. This is a drawback for their application in metal recovery processes because these applications are intended only the concentration of metals from highly diluted solutions but also the selective recovery of metals of interest.

A different approach was propagated by Bonificio and Clarke (2016). These authors described a selective recovery of REE by biosorption on immobilized bacterial biomass followed by a selective desorption as a function of pH. This approach enabled the separation of the three heaviest lanthanides Tm, Lu, and Yb from a mixture of different lanthanides.

Other recent developments concentrate on the use of defined proteins from biomass or the direct engineering of improved microbes and enzymes. Maruyama et al. (2007) tested different model peptides, proteins, and protein-rich biomass regarding their capability to selectively bind different precious metals from model solutions, metal-refining solutions, and industrial wastes at acidic conditions (Maruyama et al. 2007). All tested biomasses as well as proteins selectively adsorb Pd and Au ions in the presence of transition elements. Further, it was possible to remove Au, Pd, and Pt from plating wastes using protein-rich chicken egg-shell membrane.

Other approaches use the metal-binding motifs of natural proteins, e.g. metallothioneins, as biosorptive component. Metallothioneins (MT) are cysteine-rich proteins that bind different metals such as Cd, Hg, Cu, and Pb. MTs from different natural sources have been expressed in *Escherichia coli* and *Pseudomonas putida* and used as biosorbent, mainly for removal of heavy metals (reviewed by Chen et al. (1999)) and Mejare and Bulow (2001). However, these proteins are also attractive for the recovery of valuable metals. Terashima et al. (2002) produced a fusion protein composed of the maltose-binding protein and human MT and immobilized it on Chitopearl resins (Terashima et al. 2002). These materials were used for binding of Cd and Ga in a concentration range of 0.2–1.0 mM. Further, the biosorbents could be used several times without loss of binding activity.

CadR, which is a Cd-binding protein first isolated from rhizobacterium *P. putida*, has been expressed on the surface of *E. coli* cells. These engineered cells show a high Cd^{2+} adsorption capacity of 19.5 μmol Cd(II) g^{-1} cells (Liu et al. 2015).

Phytochelatins (PCs) are naturally occurring metal-binding peptides, which contain multiple repeats of the γGlu-Cys moiety terminated by a Gly residue. Various researchers have expressed different synthetic PCs onto the surface of bacterial cells to improve metal uptake and biosorption. For example, recently, Tan et al. (2019) displayed the synthetic phytochelatin EC20 onto the surface of *E. coli*. The obtained constructs showed an increased biosorption of Pt(IV) accompanied by the formation of platinum nanoparticles (Tan et al. 2019).

Proteinaceous bacterial surface layers that envelope many bacterial cells are other interesting biomolecules that have been used for the binding of different elements such as U, Pd, Au, or Cu (Allievi et al. 2011; Merroun et al. 2005; Pollmann et al. 2006b). The binding of U, Pd, and Au has been investigated in more detail in case of the S-layers from *Lysinibacillus sphaericus* JG-A12 and NCTC 9602 (Fahmy et al. 2006; Jankowski et al. 2010; Merroun et al. 2005). These elements were coordinated by phosphate and carboxyl groups (Fahmy et al. 2006; Merroun et al. 2005); in case of Au(III), it was assumed that amine groups were involved in complexation (Jankowski et al. 2010). Due to their self-assembling properties that enable the formation of nanostructured protein arrays on various technical surfaces (Sleytr et al. 2014; Toca-Herrera et al. 2005; Weinert et al. 2015), S-layer proteins are attractive biomolecules for the construction of biosorptive composites (Suhr et al. 2014). For example, so-called biocers were produced by entrapping S-layer carrying cells or S-layers in porous ceramics using sol–gel technology and used for the removal of U from contaminated waters (Soltmann et al. 2002). Pollmann et al. (Pollmann and Matys 2007) constructed modified His-tagged S-layer proteins that exhibited enhanced Ni-binding capacities while self-assembling to a nanoporous protein meshwork (Pollmann and Matys 2007).

Peptides are other less complex and easily synthesizable biomolecules that have been used for the design of various biosorbents. Stair et al. (Stair and Holcombe 2005) synthesized and immobilized various peptides of different lengths composed of Gly, Asp, and Cys residues on commercial Tentagel resins and used it as biosorbents for the binding of Ni^{2+}, Cd^{2+}, Co^{2+}, and Mg^{2+} (Stair and Holcombe 2005).

All the previously discussed biomolecules are able to interact with a specific number of ions. A selectivity to the target ion is not given in the above shown approaches.

9.3.1 Biosorption Via Metal Selective Peptides

The lack of selectivity in separation processes can be solved by using a novel, very promising approach for the selection of metal-binding peptides by phage surface display. With this technique, peptides selective for several metallic surfaces or metal ions were identified (Sarikaya et al. 2003; Seker and Demir 2011). Cetinel and coworkers describe the technique appropriately as “the directed evolution of peptides with specific interactions toward technologically relevant materials” based on combinatorial bio-based libraries (Cetinel et al. 2012). The functional groups presented by individual amino acids of the identified peptides and the interaction with neighbor functional groups are responsible for the specific and strong interaction with the target material. Insertion, deletion, or exchange of one amino acid can change the peptide–target interaction drastically. The peptide–target bonding usually occurs via long-range interactions (physisorption) or short-range interactions (chemisorption) (Schwaminger et al. 2018). Fundamental knowledge of the occurring peptide–target interactions is necessary to improve and control these bio-based interactions for an optimization of separation and recycling processes.

Material-selective peptides are used currently mainly for the development of nanomaterials and composites, but they are also attractive as biosorbents. Nian et al. (2010) selected Pb^{2+}-specific peptides and identified one bacteriophage-expressed peptide (TNTLSNN) with high affinity and specificity to Pb^{2+} as proven by cross-binding assay to different metal ions (Nian et al. 2010). In a follow-up study, Nguyen et al. (2013) constructed a recombinant *E. coli* displaying the peptide on its cell surface thus obtaining a highly selective *E. coli*-based biosorbent (Nguyen et al. 2013). Similarly, Yang et al. (2015) selected Cr(III) binding phages from a phage display library (Yang et al. 2015). A phage expressing the heptameric peptide YKASLIT was immobilized on cytopore beads for Cr(III) preconcentration. Sawada et al. (2016) selected Nd(III) binding bacteriophages via phage surface display technology (Sawada et al. 2016). These phages were used as adsorbent for the selective recovery of Nd(III) from mixed solutions of Nd(III) and Fe(III), mimicking the dissolved solution of neodymium–iron–boron alloys ($Nd_2Fe_{15}B$) indicating a high potential to be applied in recycling strategies. In another approach, lanthanide oxide particles were used as target to select peptides that induce the precipitation of lanthanide hydroxides (Hatanaka et al. 2017). Three peptides (SCLWGDVSELD-FLCS, SCLYPSWSDYAFCS, SCPVWFSDVGDFMVCS) were identified that mediate the mineralization of lanthanide ions. The researchers proposed that such peptides have a potential for the separation of lanthanides via selective mineralization. Yunus et al. (Yunus and Tsai 2015) immobilized genetically engineered fusion proteins composed of palladium-binding peptides and cellulose-binding domains on cellulose materials (Yunus and Tsai 2015). These constructs were used as biosorbents for the selective binding of Pd(II). The materials were able to selectively bind

Pd(II) from a mixture of Pd(II) and Pt(IV) with a maximum adsorption capacity of 175.44 mg/g. Further, it was possible to remove the bound Pd and reuse the biosorbent several times without losing the binding capacity. The materials were working at a wide range of different pH (pH 1.8–11) and temperatures (10–40 °C); therefore, they can be applied at different conditions. Yang et al. (2018) identified arsenic (III)-binding peptides with the ability to induce the aggregation of gold nanoparticles in the absence of arsenic (III). These peptides can be applied as colorimetric detection sensors for arsenic (III) (Yang et al. 2018). Schönberger and coworkers presented several linear gallium-binding peptides identified via phage surface display and used afterwards a cysteine-scanning methodology to introduce structures in one preferred peptide. The changed binding affinity of the modified peptides were tested in subsequent biosorption experiments for the peptides future application in biorecovery approaches for gallium (Schönberger et al. 2019a, b). In 2020, Matys and coworkers identified peptides that selectively interact with nickel (CNAKHHPRC) and cobalt (CTQMLGQLC) using sol–gel coated glass–fiber fabrics for the future application in new element-specific biosorptive materials (Matys et al. 2020). Arsenic-binding peptides were identified in the study of Braun and coworkers in 2020 for the decontamination of industrial wastewater. This group developed a new combined approach of phage display and next-generation sequencing for the identification of the strongest target-binding peptides (Braun et al. 2020). These examples demonstrate that phage surface display technology is a promising strategy to identify highly selective peptides for different elements that can be used for the construction of biosorbents not only for bioremediation but also for the recovery of valuable metals.

In numerous studies, metal-binding motifs and peptides were expressed and anchored on the surface of microbial cells via fusion with outer membrane proteins. In many approaches, the peptides were anchored to the outer membrane protein LamB, thus obtaining engineered microbes that worked as an efficient adsorbent. For example, hexa-His chains were expressed on the surface of *E. coli* by construction of LamB hybrids. These cells exhibited high affinity to Ni ions (Sousa et al. 1996). Other researchers expressed metallothioneins or metal-binding peptides as fusions to membrane or membrane associated proteins in *E. coli*, *P. putida*, yeasts, or other microorganisms (Kotrba et al. 1999a, b; Nishitani et al. 2010; Sousa et al. 1998; Valls et al. 2000a, b; Valls et al. 1998). Park et al. (2016) produced fusion proteins comprising the surface (S-layer) protein of *Caulobacter crescentes* and peptides that have been used as lanthanide-binding tags for protein purification, biosensing, and nuclear magnetic resonance (NMR) spectroscopy (Liang et al. 2013; Martin et al. 2007; Nitz et al. 2003; Park et al. 2016). These hybrid proteins were expressed in the cell surface of *C. crescentes* in high density (Park et al. 2016). The engineered cells exhibited an enhanced sorption of REE and a high specificity for REE. Further, it was possible to desorb the bound REE enabling a repeatable reuse of the bioadsorbents. Li and coworkers developed a Ni-ion biosorption process based on nickel-binding peptides presented by surface engineered yeast (Li et al. 2019). Immobilized *Saccharomyces cerevisiae* EBY100 expressing three different Ni-binding peptides at the same time showed the selective biosorption of up to

68.62% of all the Ni-ions in the system. Other heavy metals like As(III), Pb(II), Cr(III), and Cd(III) were not or were in very small dimensions bound to the yeast surface. Thus, cell surface display is an attractive approach for implementation in recycling processes. Other approaches use metabolic products as complexing agents.

9.3.2 Chelators Derived from Nature

Very interesting biomolecules are siderophores. These small organic molecules are iron chelators that are produced and secreted by bacteria and are used for the uptake of iron. In addition to iron, other metals, e.g. Ga, Co, different actinides, can be complexed by the siderophores (Brainard et al. 1992; Gascoyne et al. 1991b; Gascoyne et al. 1991a; Harrington et al. 2012). These properties make them attractive for biotechnological applications. For example, Jain et al. (2019) used the siderophores desferrioxamine A and E for Ga complexation and developed a chromatography method enabling the selective recovery of Ga from industrial waste waters. Regeneration and multiple reuse of the biomolecules was possible, which is the requirement for an economic application of the technology (Jain et al. 2019). In another study, the siderophore yersinobactin, a metal-chelating peptide derived from *Yersinia pestis*, was adsorbed on a resin within a packed-bed column. With this material, it was possible to remove >80% of copper from field water mixed with copper (Ahmadi et al. 2016).

Biopolymers are another attractive group of metabolites. Bacterial poly(γ-glutamic acid) has been used for the adsorption of toxic Hg(II) (Inbaraj et al. 2009), Pb(II) (Mu et al. 2011), and Fe(III) (Bodnar et al. 2013). Varshini constructed a modified biohydrogel and used it for the removal of the rare-earth element Ce(III) from industrial effluents (Varshini et al. 2015). Different extracellular polymeric substances have been used for the removal of Co(II), Cu(II), and other elements (Dobrowolski et al. 2017; Mona and Kaushik 2015; Perez et al. 2008). Natural polysaccharides such as alginate, chitin, chitosan, starch as well as their derivatives and polysaccharide-based composites have been widely used for the removal of not only different heavy metals (reviewed by Crini (2005)) but also precious metals (Donia et al. 2007).

To enable a low-cost usage, the biocompounds should be recycled and reused after adsorption. The repeatable use of biomolecules for biosorption requires the immobilization of the molecules to an appropriate surface. The combination of biocompounds with inorganic materials brings together advantages of both materials. Soltmann et al. (2002) developed uranium-binding composites, so-called bio-ceramics (biocers), by immobilization of bacterial cells or surface (S) layer proteins via sol–gel techniques (Soltmann et al. 2002). These composites were used not only for the removal of uranium from waters but also for the binding of Pd(II) and copper (Raff et al. 2003; Pollmann et al. 2006a, b). Yunus et al. (Yunus and Tsai 2015) immobilized fusion proteins composed of palladium-binding peptides and cellulose-binding domains on cellulose (Yunus and Tsai 2015). These complexes were used for the adsorption of Pd(II) from model solutions at various conditions. In addition, it was possible to desorb the Pd(II) using 1 M thiourea, thus creating a reusable Pd(II) selective biosorbent. Other approaches entrap the biocompounds in polyvinyl alcohol, chitosan, hydrogels, or alginate (Ting and Sun 2000).

9.4 Bioflotation: Separation of Particles with Biological Means

Microbial cells, cell components, metabolites, or other biomolecules can interact with solid substrates and modify surface properties, e.g. by introducing hydrophobic properties by adhesion to the surfaces (Das et al. 1999; Patra and Natarajan 2006). These properties can be used for mineral beneficiation. For example such biocompounds have been reported as environment-friendly collectors or depressants and were applied as flotation reagents in selective mineral separation (reviewed by Behera and Mulaba-Bafubiandi (2017)). Most of these approaches concentrate on the use of bacterial cells or their products that have been described to specifically interact with minerals such as *Acidithiobacillus* or *Leptospirillum ferrooxidans* or *Rhodococcus opacus* or on model organisms (Behera and Mulaba-Bafubiandi 2017). However, newer investigations demonstrate the applicability of a much broader range of microorganisms beyond the classical bioleaching bacteria or model organisms. Luque Consuegra et al. investigated the influence of different marine bacteria on bioflotation of pyrite and chalcopyrite and identified strains of *Halobacillus* sp. and *Marinococcus* sp. to depress pyrite in artificial sea water conditions while improving the flotation of chalcopyrite (Consuegra et al. Consuegra et al. 2019). These studies prove the high potential of the application of various bacteria in diverse environments for particle separation, thus opening up not only new perspectives in mineral separation technologies but also in recycling technologies and other industrial applications.

Biopolymers and so-called extracellular polymeric substances (EPS) mediate the attachment of bacterial cells to surfaces and biofilm formation (Gehrke et al. 1998; Kinzler et al. 2003; Vu et al. 2009). They form the structure and architecture of the biofilm matrix. The EPS are composed of an undefined complex mixture of biopolymers primarily consisting not only of polysaccharides, but also lipids, proteins, humic acids, and nucleic acids (reviewed by Vu et al. (2009)). Their composition depends on type of microorganisms, age of biofilm, and environmental conditions, including surface properties (Donlan 2002). Their affinity to surfaces makes them interesting for their application in flotation processes. Consequently, several studies investigated the effect of EPS as bioreagent in mineral separation (Figure 9.1).

Besides EPS, other biomolecules have been investigated regarding their application in mineral separation. Especially biosurfactants are interesting compounds that have been applied as frothers in many flotation experiments. Biosurfactants are surface-active organic molecules that are produced by many microorganisms. They have been attributed to lowering the surface tension at the interfaces of solid, liquid, and gases. In contrast to common commercial chemical surfactants, they are less

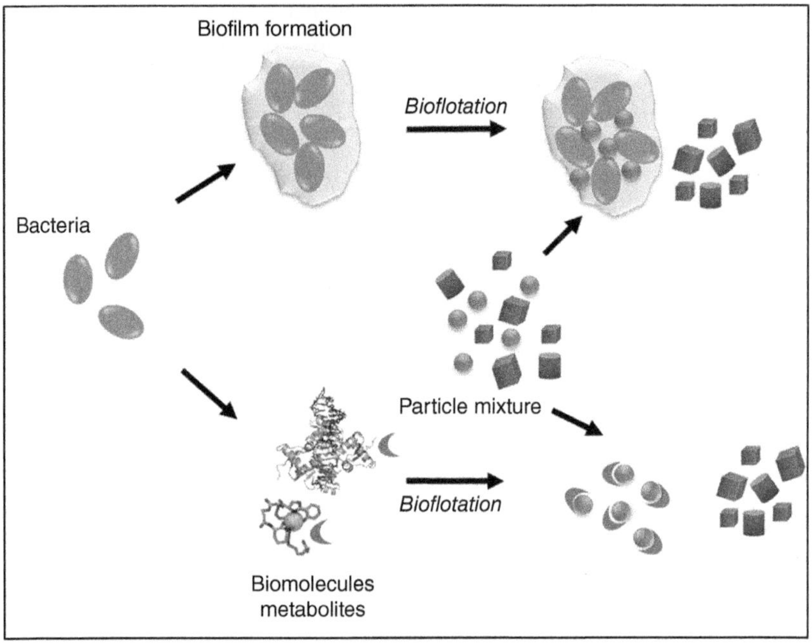

Figure 9.1 Use of extracellular polymeric substances (EPS) producing bacteria and biomolecules for particle separation.

toxic, biodegradable, and effective under extreme conditions. Consequently, there are numerous potential fields of application, including pharmaceutical industry, environmental remediation, and petroleum industry (reviewed by Saha and Rao (2017)). Amphiphilic siderophores are another class of surface-active compounds produced by microorganisms. The amphiphilic siderophore marinobactin is composed of a hydrophilic chelating head group and a hydrophobic fatty acid tail of different lengths and interacts with iron minerals. These properties make the molecules attractive as flotation agents for mineral separation in froth flotation as propagated by Schrader et al. (2017) (Schrader et al. 2017).

Hacha et al. (2018) combined innovative electroflotation processes that reduces bubble sizes, with *R. opacus* cells as bioreagent (Hacha et al. 2018). By this, it was possible to separate fine hematite particles from a mixture. The combination of different methodologies from different fields enables innovations in classical flotation procedures, thus overcoming current limitations.

All these approaches include living cells or natural biocompounds that were isolated from living cells. Principally, these developments could be transferred to the separation of fine particles that are released during recycling processes and cannot be targeted by existing processes. Lederer and coauthors described the selection of phages displaying $LaPO_4:Ce^{3+}$, Tb^{3+} (LAP), and $CeMgAl_{11}O_{19}:Tb^{3+}$ (CAT)-specific peptides on their surface (Lederer et al. 2017; Braun et al. 2018). Moreover, the researchers introduced modifications and reached $a > 5000$ fold higher binding strength to LAP in comparison to the wild type. The directed modification of individual amino acids was proven to increase or decrease the binding specificity and affinity of a peptide to the target material drastically. These phages bind to several components of compact fluorescent lamps (LAP, CAT), but not or only weak to $Y_2O_3:Eu^{3+}$ (YOX), $LaPO_4$, SiO_2, and $BaMgAl_{10}O_{17}:Eu^{2+}$ (BAM). This proof of principle shows that the researchers are able to identify perfectly fitting biomolecules for target particles by using the improved phage surface display techniques. The authors proposed an application as collectors in bioflotation processes for the separation and recycling of fluorescence phosphor components from electronic scrap (Lederer et al. 2019).

Other approaches anchored ZnO-, Au-, or TiO_2-binding peptides or organic molecules to magnetic particles and separated the respective particles from colloidal mixtures (Essinger-Hileman et al. 2013; Shen et al. 2017; Vreuls et al. 2011). Given the high number of peptides that have been described to selectively interact with various inorganic surfaces and that were mostly applied for material syntheses or sensory applications (reviewed by Care et al. (2015); Seker and Demir 2011)), one can assume that these separation technologies can be easily transferred to other materials.

9.5 Bioreduction and Bioaccumulation: Nanomaterials from Waste

Bioaccumulation and bioreduction are often accompanied by the formation of nanoparticles. These properties make the processes attractive for the creation of novel nanomaterials. Bioaccumulation describes the accumulation and enrichment of metals in the cells, relative to the environment. This mechanism can be applied for the recovery and concentration of valuable elements from diluted solutions. For example, different microorganisms have been described to accumulate Ga and were used for Ga removal (Gascoyne et al. 1991b). In this case, the accumulation is mediated by siderophores that are complexing with Ga.

The transformation of metal ions into nanoparticles is one strategy to overcome the toxic effects of the metals. Nanoparticles are formed either via bio-precipitation or biologically catalyzed metal reduction (bioreduction). Among the various described biologically produced inorganic nanoparticles, precious metals are the most interesting metals for many applications.

Precious metals of the platinum group metals such as platinum, palladium, rhodium, and ruthenium are widely used in medicine, electronics, for optical devices, and catalysis (Yong et al. 2002, 2003). Especially it is commonly used as catalyst, e.g. in automotive catalytic converters or as catalyst in chemical syntheses. Consequently, significant amounts of Pd are released during production processes, consumption and recycling processes. For minimizing loss of Pd and enabling a circular economy, Pd-efficient recycling processes while avoiding secondary waste streams of toxic chemicals as well as an efficient recovery of Pd from industrial waste waters are mandatory. The application of Pd(II)-reducing microorganisms is an attractive approach that combines the removal of Pd from waste streams, thus minimizing the loss of Pd, with the synthesis of nanocatalysts via bio-reduction and deposition of Pd-nanoparticles on biomass using nontoxic biological means (De Corte, De Corte et al. 2012). The catalysts themselves can be used for the degradation of different recalcitrant pollutants or chemical syntheses.

Different microorganisms have been described that mediate the reduction of Pd(II). In case of the extensively studied anaerobic sulfur-reducing bacteria *Desulfovibrio desulfuricans* and *Geobacter sulfurreducens* and the facultative anaerobic iron-reducing *Shewanella oneidensis*, it has been proposed that hydrogenases and cytochrome c3 are involved in bioreduction of Pd(II) and nanoparticle deposition (De Corte et al. 2012; De Windt et al. 2005; Lloyd et al. 1998; Pat-Espadas et al. 2013; Yates et al. 2013). These reactions require H_2 or formate as electron donor. In other cases (e.g. different cyanobacteria, *E. coli*), it was assumed that other enzyme systems such as nitrogenase enzyme or molybdenum-containing enzyme systems are responsible for Pd(II) reduction and deposition of Pd(0) in the medium (Foulkes et al. 2016). The majority of the formed Pd(0) particles were formed outside the cell. In other cases, Pd(0) formation is not based on enzyme activities and it was assumed that organic functional groups of the cell wall are responsible for the bioreductive process. For example, Pd(0) nanoparticles could be deposited on S-layer carrying Gram–positive bacteria as well as on the S-layer proteins following the array structure (Pollmann et al. 2006a, b; Wahl et al. 2001). Experiments with native and dead cells of *E. coli*, *S. oneidensis*, and *P. putida* and artificial systems demonstrated that the presence of amine groups mediates the reduction of Pd(II) bound to cell surfaces suggesting the use of amine-rich biomaterials rather than native cells for Pd-recovery (Rotaru et al. 2012). Consequently, De Corte et al. (2013) replaced the bacteria and used amine-functionalized surfaces as target for the synthesis of Pd(0) nanocatalysts (De Corte et al. 2012, 2013).

The application of metal-reducing bacteria for the removal of precious metals from industrial waste streams is a quite attractive alternative to conventional methods, because it requires less toxic chemicals and new products (nano-catalysts) are formed as "byproducts." A wide range of natural or genetically engineered metal-reducing bacteria were successfully applied by several authors for recovery of precious metals from synthetic solutions or scrap leachates (Creamer et al. 2006; Ito et al. 2016; Konishi et al. 2006, 2007a, b; Mabbett et al. 2006; Maes et al. 2016, 2017; Martins et al. 2013; Pat-Espadas et al. 2013). Metal reduction was accompanied by nanoparticle formation. In these approaches, recovery rates of up to 99% were obtained.

In all cases, the formed bio-Pd was catalytically active. It was especially applied to transform a wide range of pollutants, mainly by reduction (Cr(VI), ClO^{4-}) (Tuo et al. 2013; Mabbett et al. 2006; Humphries et al. 2007) or dehalogenation (e.g. printed circuit boards (PCBs), trichloroethylene, pharmaceuticals) (Baxter-Plant et al. 2004; De Windt et al. 2005; Hennebel et al. 2009a, b, 2010). Further, bio-Pd was used as a catalyst for diverse chemical syntheses, e.g. the hydrogenation of organic molecules or for coupling reactions in synthetic organic chemistry (Creamer et al. 2007). The doping of bio-Pd with other metals, e.g. Au, thus producing bimetallic catalysts, significantly enhanced catalytic activity. This relatively new approach will extend the applicability of metallic biocatalysts.

Besides catalytic applications, bio-Pd has been applied in microbial fuel cells, e.g. proton-exchange membrane fuel cells, for the generation of energy. In these approaches, biologically produced Pd(0) particles were deposited onto the anode, e.g. carbon papers, of the fuel cells and used for energy generation (Yong et al. 2009, 2010; Quan et al. 2015a). The now-modified anode possessed both electrooxidation and biodegradation capability (Quan et al. 2015a, b).

Most studies concentrated on precious metals. However, some newer publications studied the reduction and recovery of other valuable metals or used bioreduction for removal of toxic elements from industrial waste waters. For example, Lv et al. (2018) synthesized copper nanoparticles via bioreduction by a *Shewanella loihica* strain and discussed their use as antibacterial material (Lv et al. 2018). Maleke et al. (2019) described the reduction and intracellular accumulation of the REE europium by a Clostridium strain, probably mediated by active transport and intracellular precipitation (Maleke et al. 2019). The authors suggested an application for REE recovery from waste materials. Moreno-Benavides et al. (2019) used a *Bacillus cereus* strain for reduction of toxic Cr(VI) from electroplating wastewater (Moreno-Benavides et al. 2019).

9.6 Conclusion

In conclusion, many efforts have been done to recover metals from solutions by biological means ranging from the application of different biomasses, construct biosorptive composites, engineering of chelators, and use of different metabolic microbial processes. These approaches were used especially for the removal of toxic elements from waters. There are some reports on the removal of precious metals but only few studies describing the recovery of other valuable metals, e.g. REE, Ga, In. It can be assumed that many technologies developed for heavy metal removal can be transferred to other elements. Most current approaches concentrate on the usage of well-studied microorganisms such as chemolithoautotrophic sulfur-oxidizing bacteria like *A. ferrooxidans*, bioreducing bacteria such as *Shewanella* strains, or heterotrophic citric acid–producing fungi such as *A. niger*. However, there are some reports exploring the potential of novel microorganisms from diverse environments. It can be expected that especially extreme habitats such as salt lakes, volcanoes, deep sea, etc. bear many microorganisms with new metabolic properties that can be used for metal recovery also from e-wastes. Another approach is the smart design of new biomolecules, e.g. of new metal-chelating agents. For example, peptides can be designed interacting with numerous inorganic target materials. Such bioreagents can be integrated into various resource technologies such as metal extraction, fine particle flotation, and metal complexing. First results in these fields are highly promising. The combination of diverse biotechnological methods with classical resource technologies leads to new opportunities to find more environment-friendly and efficient solutions for metal extraction. Thus, a high potential for future applications also in recycling technologies can be expected. Opening up to these new multidisciplinary ideas offers new chances for a green economy.

References

Ahmadi, M.K., Ghafari, M., Atkinson, J.D. et al. (2016). A copper removal process for water based upon biosynthesis of yersiniabactin, a metal-binding natural product. *Chemical Engineering Journal* 306: 772–776.

Allievi, M.C., Florencia, S., Mariano, P.A. et al. (2011). Metal biosorption by surface-layer proteins from Bacillus species. *Journal of Microbiology and Biotechnology* 21: 147–153.

Auerbach, R., Ratering, S., Bokelmann, K. et al. (2019). Bioleaching of valuable and hazardous metals from dry discharged incineration slag. An approach for metal recycling and pollutant elimination. *Journal of Environmental Management* 232: 428–437.

Baxter-Plant, V.S., Mikheenko, I.P., Robson, M. et al. (2004). Dehalogenation of chlorinated aromatic compounds using a hybrid bioinorganic catalyst on cells of *Desulfovibrio desulfuricans*. *Biotechnology Letters* 26: 1885–1890.

Behera, S.K. and Mulaba-Bafubiandi, A.F. (2017). Microbes assisted mineral flotation a future prospective for mineral processing industries: a review. *Miner Process Extractive Metallurgy* 38: 96–105.

Bodnar, M., Hajdu, I., Rothi, E. et al. (2013). Biopolymer-based nanosystem for ferric ion removal from water. *Separation and Purification Technology* 112: 26–33.

Bonificio, W.D. and Clarke, D.R. (2016). Rare-earth separation using bacteria. *Environmental Science & Technology Letters* 3 (4): 180–184.

Bosshard, P.P., Bachofen, R., and Brandl, H. (1996). Metal leaching of fly ash from municipal waste incineration by *Aspergillus niger*. *Environmental Science & Technology* 30: 3066–3070.

Brainard, J.R., Strietelmeier, B.A., Smith, P.H. et al. (1992). Actinide binding and solubilization by microbial siderophores. *Radiochimica Acta* 58–9: 357–363.

Brandl, H., Bosshard, R., and Wegmann, M. (2001). Computer-munching microbes: metal leaching from electronic scrap by bacteria and fungi. *Hydrometallurgy* 59: 319–326.

Brandl, H., Lehmann, S., Faramarzi, M.A. et al. (2008). Biomobilization of silver, gold, and platinum from solid waste materials by HCN-forming microorganisms. *Hydrometallurgy* 94 (1–4): 14–17.

Braun, R., Bachmann, S., Schönberger, N. et al. (2018). Peptides as biosorbents – promising tools for resource recovery. *Research in Microbiology* 169: 649–658.

Braun, R., Schönberger, N., Vinke, S. et al. (2020). Application of next generation sequencing (NGS) in phage displayed peptide selection to support the identification of arsenic-binding motifs. *Viruses* 12 (12): 1360. https://doi.org/10.3390/v12121360.

Brombacher, C., Bachofen, R., and Brandl, H. (1998). Development of a laboratory-scale leaching plant for metal extraction from fly ash by Thiobacillus strains. *Applied and Environmental Microbiology* 64 (4): 1237–1241.

Campbell, S.C., Olson, G.J., Clark, T.R. et al. (2001). Biogenic production of cyanide and its application to gold recovery. *Journal of Industrial Microbiology & Biotechnology* 26: 134–139.

Care, A., Bergquist, P.L., and Sunna, A. (2015). Solid-binding peptides: smart tools for nanobiotechnology. *Trends in Biotechnology* 33: 259–268.

Cetinel, S., Dincer, S., Cebeci, A. et al. (2012). Peptides to bridge biological-platinum materials interface. *Bioinspired Biomimetic and Nanobiomaterials* 1: 143–153.

Chen, W., Bruhlmann, F., Richins, R.D. et al. (1999). Engineering of improved microbes and enzymes for bioremediation. *Current Opinion in Biotechnology* 10: 137–141.

Chi, T.D., Lee, J.C., Pandey, B.D. et al. (2011). Bioleaching of gold and copper from waste mobile phone PCBs by using a cyanogenic bacterium. *Minerals Engineering* 24: 1219–1222.

Cizkova, M., Mezricky, D., Rucki, M. et al. (2019). Bio-mining of lanthanides from red mud by green microalgae. *Molecules* 24 (7): 1356. https://doi.org/10.3390/molecules24071356.

Consuegra, G.L., Kutschke, S., Rudolph, M. et al. (2019). Halophilic bacteria as potential pyrite bio-depressants in Cu–Mo bioflotation. *Minerals Engineering* 145: 106062.

Creamer, N.J., Baxter-Plant, V.S., Henderson, J. et al. (2006). Palladium and gold removal and recovery from precious metal solutions and electronic scrap leachates by *Desulfovibrio desulfuricans*. *Biotechnology Letters* 28: 1475–1484.

Creamer, N.J., Mikheenko, I.P., Yong, P. et al. (2007). Novel supported Pd hydrogenation bionanocatalyst for hybrid homogeneous/heterogeneous catalysis. *Catalyt Tod* 128: 80–87.

Crini, G. (2005). Recent developments in polysaccharide-based materials used as adsorbents in wastewater treatment. *Progress in Polymer Science* 30 (1): 38–70.

Das, N. (2010). Recovery of precious metals through biosorption – a review. *Hydrometallurgy* 103: 180–189.

Das, A., Rao, K.H., Sharma, P. et al. (1999). Surface chemical and adsorption studies using *Thiobacillus ferrooxidans* with reference to bacterial adhesion to sulfide minerals. *Process Metallurgy* 9: 697–707. https://doi.org/10.1016/S1572-4409(99)80072-8.

Daubert, L.N. and Brennan, R.A. (2007). Passive remediation of acid mine drainage using crab shell chitin. *Environmental Engineering Science* 24: 1475–1480.

De Corte, S., Hennebel, T., De Gusseme, B. et al. (2012). Bio-palladium: from metal recovery to catalytic applications. *Microbial Biotechnology* 5: 5–17.

De Corte, S., Bechstein, S., Lokanathan, A.R. et al. (2013). Comparison of bacterial cells and amine-functionalized abiotic surfaces as support for Pd nanoparticle synthesis. *Colloid Surface B* 102: 898–904.

De Windt, W., Aelterman, P., and Verstraete, W. (2005). Bioreductive deposition of palladium (0) nanoparticles on *Shewanella oneidenis* with catalytic activity towards reductive dechlorination of polychlorinated biphenyls. *Environmental Microbiology* 7: 314–325.

Dobrowolski, R., Szczes, A., Czemierska, M. et al. (2017). Studies of cadmium(II), lead(II), nickel(II), cobalt(II) and chromium(VI) sorption on extracellular polymeric substances produced by *Rhodococcus opacus* and *Rhodococcus rhodochrous. Bioresource Technology* 225: 113–120.

Dolker, T. and Pant, D. (2019). Chemical-biological hybrid systems for the metal recovery from waste lithium ion battery. *Journal of Environmental Management* 248: 109270, ISSN 0301-4797, https://doi.org/10.1016/j.jenvman.2019.109270.

Donia, A.M., Atia, A.A., and Elwakeel, K.Z. (2007). Recovery of gold(III) and silver(I) on a chemically modified chitosan with magnetic properties. *Hydrometallurgy* 87 (3–4): 197–206.

Donlan, R.M. (2002). Biofilms: microbial life on surfaces. *Emerging Infectious Diseases* 8: 881–890.

Essinger-Hileman, E.R., Popczun, E.J., and Schaak, R.E. (2013). Magnetic separation of colloidal nanoparticle mixtures using a material specific peptide. *Chemical Communications* 49: 5471–5473.

Fahmy, K., Merroun, M., Pollmann, K. et al. (2006). Secondary structure and Pd(II) coordination in S-layer proteins from *Bacillus sphaericus* studied by infrared and X-ray absorption spectroscopy. *Biophysical Journal* 91: 996–1007.

Faramarzi, M.A., Stagars, M., Pensini, E. et al. (2004). Metal solubilization from metal-containing solid materials by cyanogenic *Chromobacterium violaceum. Journal of Biotechnology* 113: 321–326.

Figueira, M.M., Volesky, B., Ciminelli, V.S.T. et al. (2000). Biosorption of metals in brown seaweed biomass. *Water Research* 34: 196–204.

Foulkes, J.M., Deplanche, K., Sargent, F. et al. (2016). A novel aerobic mechanism for reductive palladium biomineralization and recovery by *Escherichia coli. Geomicrobiology Journal* 33 (3–4): 230–236.

Gascoyne, D.J., Connor, J.A., and Bull, A.T. (1991a). Isolation of bacteria producing siderophores under alkaline conditions. *Applied Microbiology and Biotechnology* 36: 130–135.

Gascoyne, D.J., Connor, J.A., and Bull, A.T. (1991b). Capacity of siderophore – producing alkalophilic bacteria to accumulate iron, gallium and aluminum. *Applied Microbiology and Biotechnology* 36: 136–141.

Gehrke, T., Telegdi, J., Thierry, D. et al. (1998). Importance of extracellular polymeric substances from *Thiobacillus ferrooxidans* for bioleaching. *Applied and Environmental Microbiology* 64: 2743–2747.

Gholami, R.M., Borghei, S.M., and Mousavi, S.M. (2011). Bacterial leaching of a spent Mo-Co-Ni refinery catalyst using *Acidithiobacillus ferrooxidans* and *Acidithiobacillus thiooxidans. Hydrometallurgy* 106 (1–2): 26–31.

Hacha, R.R., LeonardoTorem, M., Merma, A.G. et al. (2018). Electroflotation of fine hematite particles with *Rhodococcus opacus* as a biocollector in a modified partridge-Smith cell. *Minerals Engineering* 126: 105–115.

Harrington, J.M., Parker, D.L., Bargar, J.R. et al. (2012). Structural dependence of Mn complexation by siderophores: donor group dependence on complex stability and reactivity. *Geochim Cosmochim Ac* 88: 106–119.

Hassanien, A.G., Desouky, O.A.N., and Hussien, S.S.E. (2014). Bioleaching of some rare earth elements from Egyptian monazite using *Aspergillus ficuum* and *Pseudomonas aeruginosa*. *Walailak Journal of Science and Technology* 11 (9): 809–823.

Hatanaka, T., Matsugami, A., Nonaka, T. et al. (2017). Rationally design mineralization for selective recovery of the rare earth elements. *Nature Communications* 8: 15670. https://doi.org/10.1038/ncomms15670.

He, J.S. and Chen, J.P. (2014). A comprehensive review on biosorption of heavy metals by algal biomass: materials, performances, chemistry, and modeling simulation tools. *Bioresource Technology* 160: 67–78.

Hennebel, T., Simoen, H., De Windt, W. et al. (2009a). Biocatalytic dechlorination of trichloroethylene with bio-palladium in a pilot-scale membrane reactor. *Biotechnology and Bioengineering* 102 (4): 995–1002.

Hennebel, T., Verhagen, P., Simoen, H. et al. (2009b). Remediation of trichloroethylene by bio-precipitated and encapsulated palladium nanoparticles in a fixed bed reactor. *Chemosphere* 76 (9): 1221–1225.

Hennebel, T., De Corte, S., Vanhaecke, L. et al. (2010). Removal of diatrizoate with catalytically active membranes incorporating microbially produced palladium nanoparticles. *Water Research* 44 (5): 1498–1506.

Hopfe, S., Flemming, K., Lehmann, F. et al. (2017). Leaching of rare earth elements from fluorescent powder using the tea fungus Kombucha. *Waste Management* 62: 211–221.

Hopfe, S., Konsulke, S., Barthen, R. et al. (2018). Screening and selection of technologically applicable microorganisms for recovery of rare earth elements from fluorescent powder. *Waste Management* 79: 554–563.

Horeh, N.B., Mousavi, S.M., and Shojaosadati, S.A. (2016). Bioleaching of valuable metals from spent lithium-ion mobile phone batteries using *Aspergillus niger*. *Journal of Power Sources* 320: 257–266.

Huang, T., Liu, L.F., and Zhang, S.W. (2019). Recovery of cobalt, lithium, and manganese from the cathode active materials of spent lithium-ion batteries in a bio-electro-hydrometallurgical process. *Hydrometallurgy* 188: 101–111.

Humphries, A.C., Penfold, D.W., and Macaskie, L.E. (2007). Cr(VI) reduction by bio and bioinorganic catalysis via use of bio-H-2: a sustainable approach for remediation of wastes. *Journal of Chemical Technology and Biotechnology* 82 (2): 182–189.

Inbaraj, B.S., Wang, J.S., Lu, J.F. et al. (2009). Adsorption of toxic mercury(II) by an extracellular biopolymer poly(gamma-glutamic acid). *Bioresource Technology* 100 (1): 200–207.

Ito, R., Kuroda, K., Hashimoto, H. et al. (2016). Recovery of platinum(0) through the reduction of platinum ions by hydrogenase-displaying yeast. *AMB Express* 6: 88. https://doi.org/10.1186/s13568-016-0262-4.

Jain, R., Fan, S.Y., Kaden, P. et al. (2019). Recovery of gallium from wafer fabrication industry wastewaters by Desferrioxamine B and E using reversed-phase chromatography approach. *Water Research* 158: 203–212.

Jankowski, U., Merroun, M.L., Selenska-Pobell, S. et al. (2010). S-layer protein from *Lysinibacillus sphaericus* JG-A12 as matrix for Au(III) sorption and Au-nanoparticle formation. *International Journal of Spectroscopy* 24 (1–2): 177–181. https://doi.org/10.3233/SPE-2010-0408.

Karwowska, E., Andrzejewska-Morzuch, D., Lebkowska, M. et al. (2014). Bioleaching of metals from printed circuit boards supported with surfactant-producing bacteria. *Journal of Hazardous Materials* 264: 203–210.

Kinzler, K., Gehrke, T., Telegdi, J. et al. (2003). Bioleaching – a result of interfacial processes caused by extracellular polymeric substances (EPS). *Hydrometallurgy* 71: 83–88.

Klink, C., Eisen, S., Daus, B. et al. (2016). Investigation of *Acidithiobacillus ferrooxidans* in pure and mixed-species culture for bioleaching of Theisen sludge from former copper smelting. *Journal of Applied Microbiology* 120 (6): 1520–1530.

Konishi, Y., Tsukiyama, T., Ohno, K. et al. (2006). Intracellular recovery of gold by microbial reduction of AuCl4-ions using the anaerobic bacterium *Shewanella algae*. *Hydrometallurgy* 81 (1): 24–29.

Konishi, Y., Ohno, K., Saitoh, N. et al. (2007a). Bioreductive deposition of platinum nanoparticles on the bacterium *Shewanella algae*. *Journal of Biotechnology* 128 (3): 648–653.

Konishi, Y., Tsukiyama, T., Tachimi, T. et al. (2007b). Microbial deposition of gold nanoparticles by the metal-reducing bacterium *Shewanella algae*. *Electrochimica Acta* 53 (1): 186–192, ISSN 0013-4686, https://doi.org/10.1016/j.electacta.2007.02.073.

Kotrba, P., Doleckova, L., De Lorenzo, V. et al. (1999a). Enhanced bioaccumulation of heavy metal ions by bacterial cells due to surface display of short metal binding peptides. *Applied and Environmental Microbiology* 65: 1092–1098.

Kotrba, P., Pospisil, P., de Lorenzo, V., and Ruml, T. (1999b). Enhanced metallosorption of *Escherichia coli* cells due to surface display of beta- and alpha-domains of mammalian metallothionein as a fusion to lamb protein. *Journal of Receptor and Signal Transduction* 19 (1–4): 703–715. https://doi.org/10.3109/10799899909036681.

Lederer, F.L., Curtis, S.B., Bachmann, S. et al. (2017). Identification of lanthanum-specific peptides for future recycling of rare earth elements from compact fluorescent lamps. *Biotechnology and Bioengineering* 114: 1016–1024.

Lederer, F.L., Braun, R., Schöne, L.M. et al. (2019). Identification of peptides as alternative recycling tools via phage surface display – how biology supports geosciences. *Minerals Engineering* 132: 245–250.

Li, H., Dong, W., Liu, Y. et al. (2019). Enhanced biosorption of nickel ions on immobilized surface-engineered yeast using nickel-binding peptides. *Frontiers in Microbiology* 10: 1254.

Liang, H.H., Deng, X., Bosscher, M. et al. (2013). Engineering bacterial two-component system PmrA/PmrB to sense lanthanide ions. *Journal of the American Chemical Society* 135: 2037–2039.

Liu, Q., Yuan, F., Liang, Y. et al. (2015). Cadmium adsorption by *E. coli* with surface displayed CadR. *RSC Advances* 5: 16089–16092.

Lloyd, J.R., Yong, P., and Macaskie, L.E. (1998). Enzymatic recovery of elemental palladium by using sulfate-reducing bacteria. *Applied and Environmental Microbiology* 64: 4607–4609.

Lv, Q., Zhang, B.G., Xing, X. et al. (2018). Biosynthesis of copper nanoparticles using *Shewanella loihica* PV-4 with antibacterial activity: novel approach and mechanisms investigation. *Journal of Hazardous Materials* 347: 141–149.

Mabbett, A.N., Sanyahumbi, D., Yong, P. et al. (2006). Biorecovered precious metals from industrial wastes: single-step conversion of a mixed metal liquid waste to a bioinorganic catalyst with environmental application. *Environmental Science & Technology* 40: 1015–1021.

Maes, S., Claus, M., Verbeken, K. et al. (2016). Platinum recovery from industrial process streams by halophilic bacteria: influence of salt species and platinum speciation. *Water Research* 105: 436–443.

Maes, S., Props, R., Fitts, J.P. et al. (2017). Biological recovery of platinum complexes from diluted aqueous streams by axenic cultures. *PLoS One* 12 (1): e0169093. https://doi.org/10.1371/journal.pone.0169093.

Maleke, M., Valverde, A., Gomez-Arias, A. et al. (2019). Anaerobic reduction of europium by a *Clostridium* strain as a strategy for rare earth biorecovery. *Scientific Report* 9: 14339. https://doi.org/10.1038/s41598-019-50179-z.

Maneesuwannarat, S., Vangnai, A.S., Yamashita, M., and Thiravetyan, P. (2016). Bioleaching of gallium from gallium arsenide by *Cellulosimicrobiurn funkei* and its application to semiconductor/electronic wastes. *Process Safety and Environmental Protection* 99: 80–87. https://doi.org/10.1016/j.psep.2015.10.008.

Maneesuwannarat, S., Kudpeng, K., Yingchutrakul, Y. et al. (2019). A possible protein model involved in gallium arsenide leaching by *Cellulosimicrobium funkei*. *Minerals Engineering* 137: 207–216.

Martin, L.J., Hahnke, M.J., Nitz, M. et al. (2007). Double-lanthanide-binding tags: design, photophysical properties, and NMR applications. *Journal of the American Chemical Society* 129: 7106–7113.

Martins, M., Assuncao, A., Martins, H. et al. (2013). Palladium recovery as nanoparticles by an anaerobic bacterial community. *Journal of Chemical Technology and Biotechnology* 88: 2039–2045.

Maruyama, T., Matsushita, H., Shimada, Y. et al. (2007). Proteins and protein-rich biomass as environmentally friendly adsorbents selective for precious metal ions. *Environmental Science & Technology* 41: 1359–1364.

Matys, S., Schönberger, N., Lederer, F.L. et al. (2020). Characterization of specifically metal-binding phage clones for selective recovery of cobalt and nickel. *Journal of Environmental Chemical Engineering* 8: 103636.

Mejare, M. and Bulow, L. (2001). Metal-binding proteins and peptides in bioremediation and phytoremediation of heavy metals. *Trends in Biotechnology* 19: 67–73.

Merroun, M., Raff, J., Rossberg, A. et al. (2005). Complexation of uranium by cells and S-layer sheets of *Bacillus sphaericus* JG-A12. *Applied and Environmental Microbiology* 71: 5532–5543.

Mishra, D., Kim, D.J., Ralph, D.E. et al. (2008). Bioleaching of metals from spent lithium ion secondary batteries using *Acidithiobacillus ferrooxidans*. *Waste Management* 28 (2): 333–338.

Mona, S. and Kaushik, A. (2015). Chromium and cobalt sequestration using exopolysaccarides produced by freshwater cyanobacterium *Nostoc linckia*. *Ecological Engineering* 82: 121–125.

Moreno-Benavides, J.A., Pena-Salamanca, E.J., and Benitez-Campo, N. (2019). Reducing Cr6+ in electroplating wastewater with *Bacillus cereus* strain B1. *Universitas Scientiarum* 24 (1): 73–89.

Mu, R.M., Ma, G.X., and Zhao, X. (2011). Removal efficiency of lead(II) by a biopolymer poly-gamma-glutamic acid. *Applied Mechanics and Materials* 94–96: 995–998.

Nguyen, T.T.L., Lee, H.R., Hong, S.H. et al. (2013). Selective lead adsorption by recombinant *Escherichia coli* displaying a lead-binding peptide. *Applied Biochemistry and Biotechnology* 169: 1188–1196.

Nian, R., Kim, D.S., Thuong, N. et al. (2010). Chromatographic biopanning for the selection of peptides with high specificity to Pb2+ from phage displayed peptide library. *Journal of Chromatography. A* 1217: 5940–5949.

Nishitani, T., Shimada, M., Kuroda, K. et al. (2010). Molecular design of yeast cell surface for adsorption and recovery of molybdenum, one of rare metals. *Applied Microbiology and Biotechnology* 86: 641–648.

Nitz, M., Franz, K.J., Maglathlin, R.L. et al. (2003). A powerful combinatorial screen to identify high-affinity terbium(III)-binding peptides. *Chembiochem* 4: 272–276.

Oliazadeh, M., Massinaie, M., Bagheri, A.S. et al. (2006). Recovery of copper from melting furnaces dust by microorganisms. *Minerals Engineering* 19 (2): 209–210.

Park, D.M., Reed, D.W., Yung, M.C. et al. (2016). Bioadsorption of rare earth elements through cell surface display of lanthanide binding tags. *Environmental Science & Technology* 50: 2735–2742.

Pat-Espadas, A.M., Razo-Flores, E., Rangel-Mendez, J.R. et al. (2013). Reduction of palladium and production of nano-catalyst by *Geobacter sulfurreducens*. *Applied Microbiology and Biotechnology* 97: 9553–9560.

Patra, P. and Natarajan, K.A. (2006). Surface chemical studies on selective separation of pyrite and galena in the presence of bacterial cells and metabolic products of *Paenibacillus polymyxa*. *Journal of Colloid and Interface Science* 298: 720–729.

Perez, J.A.M., Garcia-Ribera, R., Quesada, T. et al. (2008). Biosorption of heavy metals by the exopolysaccharide produced by *Paenibacillus jamilae*. *World Journal of Microbiology and Biotechnology* 24 (11): 2699–2704.

Pollmann, K. and Matys, S. (2007). Construction of an S-layer protein exhibiting modified self-assembling properties and enhanced metal binding capacities. *Applied Microbiology and Biotechnology* 75: 1079–1085.

Pollmann, K., Merroun, M., Raff, J. et al. (2006a). Manufacturing and characterization of Pd-nanoparticles formed on immobilized bacterial cells. *Letters in Applied Microbiology* 43: 39–45.

Pollmann, K., Raff, J., Merroun, M. et al. (2006b). Metal binding by bacteria from uranium mining waste piles and its potential applications. *Biotechnology Advances* 24: 58–68.

Pollmann, K., Kutschke, S., Matys, S. et al. (2018). Bio-recycling of metals: recycling of technical products using biological applications. *Biotechnology Advances* 36 (4): 1048–1062.

Pourhossein, F. and Mousavi, S.M. (2019). A novel step-wise indirect bioleaching using biogenic ferric agent for enhancement recovery of valuable metals from waste light emitting diode (WLED). *Journal of Hazardous Materials* 378: 120648, ISSN 0304-3894, https://doi.org/10.1016/j.jhazmat.2019.05.041.

Qu, Y. and Lian, B. (2013a). Bioleaching of rare earth and radioactive elements from red mud using Penicillium tricolor RM-10. *Bioresource Technology* 136: 16–23.

Qu, Y., Lian, B., Mo, B.B. et al. (2013b). Bioleaching of heavy metals from red mud using *Aspergillus niger*. *Hydrometallurgy* 136: 71–77.

Qu, Y., Li, H., Tian, W.J. et al. (2015). Leaching of valuable metals from red mud via batch and continuous processes by using fungi. *Minerals Engineering* 81: 1–4.

Quan, X.C., Sun, B., and Xu, H.D. (2015a). Anode decoration with biogenic Pd nanoparticles improved power generation in microbial fuel cells. *Electrochimica Acta* 182: 815–820.

Quan, X.C., Zhang, X., and Xu, H.D. (2015b). In-situ formation and immobilization of biogenic nanopalladium into anaerobic granular sludge enhances azo dyes degradation. *Water Research* 78: 74–83.

Raff, J., Soltmann, U., Matys, S. et al. (2003). Biosorption of uranium and copper by biocers. *Chemistry of Materials* 15: 240–244.

Ramanathan, T. and Ting, Y.P. (2016). Alkaline bioleaching of municipal solid waste incineration fly ash by autochthonous extremophiles. *Chemosphere* 160: 54–61.

Reed, D.W., Fujita, Y., Daubaras, A.L. et al. (2016). Bioleaching of rare earth elements from waste phosphors and cracking catalysts. *Hydrometallurgy* 166: 34–40.

Ritter, S.K. (2004). Biomass or bust. *Chemical and Engineering News* 82: 31–34.

Rizki, I.N., Tanaka, Y., and Okibe, N. (2019). Thiourea bioleaching for gold recycling from e-waste. *Waste Management* 84: 158–165.

Rotaru, A.E., Jiang, W., Finster, K. et al. (2012). Non-enzymatic palladium recovery on microbial and synthetic surfaces. *Biotechnology and Bioengineering* 109: 1889–1897.

Saha, P. and Rao, K.V.B. (2017). Biosurfactants – a current perspective on production and applications. *Nature, Environment and Pollution Technology* 16: 181–188.

Salamun, N., Triwahyono, S., Jalil, A.A. et al. (2015). Acid-vacuo heat treated low cost banana stems fiber for efficient biosorption of Hg(II). *RSC Advances* 5: 14129–14137.

Sarikaya, M., Tamerler, C., Jen, A.K.Y. et al. (2003). Molecular biomimetics: nanotechnology through biology. *Nature Materials* 2: 577–585.

Sawada, T., Asada, M., and Serizawa, T. (2016). Selective rare earth recovery employing filamentous viruses with chemically conjugated peptides. *ChemistrySelect* 1: 2712–2716.

Schönberger, N., Braun, R., Matys, S. et al. (2019a). Chromatopanning for the identification of gallium binding peptides. *Journal of Chromatography. A* 1600: 158–166.

Schönberger, N., Zeitler, C., Braun, R. et al. (2019b). Directed evolution and engineering of gallium-binding phage clones – a preliminary study. *Biomimetics* 4: 35.

Schrader, S., Kutschke, S., Rudolph, M. et al. (2017). Production of amphiphilic hydroxamate siderophores marinobactins by *Marinobacter* sp. DS40M6 for bioflotation process. *Solid State Phenomena* 262: 413–416.

Schwaminger, S., Blank-Shim, S.A., Borkowska-Panek, M. et al. (2018). Experimental characterization and simulation of amino acid and peptide interactions with inorganic materials. *Engineering in Life Sciences* 18: 84–100.

Seker, U.O.S. and Demir, H.V. (2011). Material binding peptides for nanotechnology. *Molecules* 16: 1426–1451.

Shen, L.F., Zhu, Y.Z., Zhang, P.F. et al. (2017). Capturing of nano-TiO_2 from complex mixtures by bisphosphonate-functionalized Fe_3O_4 nanoparticles. *ACS Sustainable Chemistry & Engineering* 5: 1704–1710.

Shin, D., Kim, J., Kim, B. et al. (2015a). Use of phosphate solubilizing bacteria to leach rare earth elements from monazite-bearing ore. *Minerals-Basel* 5 (2): 189–202.

Shin, D., Park, J., Jeong, J. et al. (2015b). A biological cyanide production and accumulation system and the recovery of platinum-group metals from spent automotive catalysts by biogenic cyanide. *Hydrometallurgy* 158: 10–18.

Sleytr, U.B., Schuster, B., Egelseer, E.M. et al. (2014). S-layers: principles and applications. *FEMS Microbiology Reviews* 38: 823–864.

Soltmann, U., Raff, J., Selenska-Pobell, S. et al. (2002). Biosorption of heavy metals by sol-gel immobilized *Bacillus sphaericus* cells, spores and S-layers. *Journal of Sol–Gel Science and Technology* 26: 1209–1212.

Sousa, C., Cebolla, A., and deLorenzo, V. (1996). Enhanced metalloadsorption of bacterial cells displaying poly-his peptides. *Nature Biotechnology* 14: 1017–1020.

Sousa, C., Kotrba, P., Ruml, T. et al. (1998). Metalloadsorption by *Escherichia coli* cells displaying yeast and mammalian metallothioneins anchored to the outer membrane protein LamB. *Journal of Bacteriology* 180: 2280–2284.

Stair, J.L. and Holcombe, J.A. (2005). Metal remediation and preconcentration using immobilized short-chain peptides composed of aspartic acid and cysteine. *Microchemical Journal* 81: 69–80.

Suhr, M., Unger, N., Viacava, K.E. et al. (2014). Investigation of metal sorption behavior of Slp1 from *Lysinibacillus sphaericus* JG-B53: a combined study using QCM-D, ICP-MS and AFM. *BioMetals* 27: 1337–1349.

Tan, L., Cui, H., Xiao, Y. et al. (2019). Enhancement of platinum biosorption by surface-displaying EC20 on *Escherichia coli*. *Ecotoxicology and Environmental Safety* 169: 103–111.

Terashima, M., Oka, N., Sei, T. et al. (2002). Adsorption of cadmium ion and gallium ion to immobilized metallothionein fusion protein. *Biotechnology Progress* 18: 1318–1323.

Ting, Y.P. and Sun, G. (2000). Comparative study on polyvinyl alcohol and alginate for cell immobilization in biosorption. *Water Science and Technology* 42: 85–90.

Toca-Herrera, J.L., Krastev, R., Bosio, V. et al. (2005). Recrystallization of bacterial S-layers on flat polyelectrolyte surfaces and hollow polyelectrolyte capsutes. *Small* 1: 339–348.

Tuo, Y., Liu, G.F., Zhou, J.T. et al. (2013). Microbial formation of palladium nanoparticles by *Geobacter sulfurreducens* for chromate reduction. *Bioresource Technology* 133: 606–611.

Valls, M., Gonzalez-Duarte, R., Atrian, S. et al. (1998). Bioaccumulation of heavy metals with protein fusions of metallothionein to bacterial OMPs. *Biochimie* 80: 855–861.

Valls, M., Atrian, S., de Lorenzo, V. et al. (2000a). Engineering a mouse metallothionein on the cell surface of Ralstonia eutropha CH_{34} for immobilization of heavy metals in soil. *Nature Biotechnology* 18: 661–665.

Valls, M., de Lorenzo, V., Gonzalez-Duarte, R. et al. (2000b). Engineering outer-membrane proteins in *Pseudomonas putida* for enhanced heavy-metal bioadsorption. *Journal of Inorganic Biochemistry* 79: 219–223.

Varshini, J.S.C., Das, D., and Das, N. (2015). Recovery of cerium (III) from electronic industry effluent using novel biohydrogel: batch and column studies. *Der Pharmacia Lettre* 7 (6): 166–179.

Volesky, B. (2007). Biosorption and me. *Water Research* 41: 4017–4029.

Volesky, B. and Holan, Z.R. (1995). Biosorption of heavy-metals. *Biotechnology Progress* 11: 235–250.

Vreuls, C., Genin, A., Zocchi, G. et al. (2011). Genetically engineered polypeptides as a new tool for inorganic nano-particles separation in water based media. *Journal of Materials Chemistry* 21: 13841–13846.

Vu, B., Chen, M., Crawford, R.J. et al. (2009). Bacterial extracellular polysaccharides involved in biofilm formation. *Molecules* 14: 2535–2554.

Wahl, R., Mertig, M., Raff, J. et al. (2001). Electron-beam induced formation of highly ordered palladium and platinum nanoparticle arrays on the S-layer of *Bacillus sphaericus* NCTC9602. *Advanced Materials* 13: 736–740.

Watling, H.R. (2015). Review of biohydrometallurgical metals extraction from polymetallic mineral resources. *Minerals* 5: 1–60.

Weinert, U., Vogel, M., Reinemann, C. et al. (2015). S-layer proteins as an immobilization matrix for aptamers on different sensor surfaces. *Engineering in Life Sciences* 15: 710–720.

Yang, T., Zhang, X.Y., Zhang, X.X. et al. (2015). Chromium(III) binding phage screening for the selective adsorption of Cr(III) and chromium speciation. *ACS Applied Materials & Interfaces* 7: 21287–21294.

Yang, T., Zhang, X.X., Yang, J.Y. et al. (2018). Screening arsenic(III)-bining peptide for colorimetric detection of arsenic(III) based on peptide induced aggregation of gold nanoparticles. *Talanta* 177: 212–216.

Yates, M.D., Cusick, R.D., and Logan, B.E. (2013). Extracellular palladium nanoparticle production using geobacter sulfurreducens. *ACS Sustainable Chemistry & Engineering* 1: 1165–1171.

Yong, P., Rowson, N.A., Farr, J.P. et al. (2002). Bioreduction and biocrystallization of palladium by *Desulfovibrio desulfuricans* NCIMB 8307. *Biotechnology and Bioengineering* 80: 369–379.

Yong, P., Rowson, N.A., Farr, J.P. et al. (2003). A novel electrobiotechnology for the recovery of precious metals from spent automotive catalysts. *Environmental Technology* 2003: 289–297.

Yong, P., Mikheenko, I.P., Deplanche, K. et al. (2009). Biorecovery of precious metals from wastes and conversion into fuel cell catalyst for electricity production. *Advanced Materials Research-Switzerland* 71–73: 729–732.

Yong, P., Mikheenko, I.P., Deplanche, K. et al. (2010). Biorefining of precious metals from wastes: an answer to manufacturing of cheap nanocatalysts for fuel cells and power generation via an integrated biorefinery? *Biotechnology Letters* 32 (12): 1821–1828.

Yunus, I.S. and Tsai, S.L. (2015). Designed biomolecule-cellulose complexes for palladium recovery and detoxification. *RSC Advances* 5: 20276–20282.

10

Processing of Nonmetal Fraction from Printed Circuit Boards and Reutilization

Amit Kumar and Maria E. Holuszko

University of British Columbia, NBK Institute of Mining Engineering, 6350, Stores Road, Vancouver, BC V6T 1Z4, Canada

10.1 Background

Printed circuit boards (PCBs) are the most important and valuable building block of electronic equipment. It is the platform that provides connections for all other electronic components such as resistors, capacitors, diodes, memory, and processors (Awasthi et al. 2017; Ghosh et al. 2015; Ning et al. 2017). With the development of newer technologies, the printed circuit board market is expected to grow at a compound annual growth rate of 4.3% from 2019 to 2014 and expected to reach an estimated US$ 89.7 billion by 2024 (Research and Markets 2019).

The percentage weight of PCB in an electronic device depends on the type of unit and can vary from 2% to 22% (Szałatkiewicz 2013). On average, PCB accounts for 3–6% (by weight) of the total e-waste (Das et al. 2009; Ghosh et al. 2015; Golev et al. 2019; Golev et al. 2016; Ning et al. 2017; Szałatkiewicz 2014a). With the metal concentration of 30–35% in PCB, it contains 40–80% of the total estimated value of waste PCB (Awasthi et al. 2017; Golev and Corder 2017; Golev et al. 2019; Park and Fray 2009). Although with a relatively smaller fraction by weight, waste PCB accounts for ~US$ 150 million per year of the total metal recovery value in e-waste that provides a major incentive for recycling organizations. Rocchetti et al. (2018) reported that until July 2017, over 200 patents were considered relevant to the field of PCB recycling, with more patents approved/filed toward the end of the decade. This shows the growing interest of various organizations in the field of PCB recycling.

As mentioned, average metal concentration in PCB is 30–35%, while the rest is the nonmetal fraction such as resin, glass fibers, cellulose, and flame retardants (Bizzo et al. 2014; Luda 2011; Szałatkiewicz 2014b; Veit et al. 2014; Zheng et al. 2009b). The separated nonmetal fraction (NMF) that accounts for approximately 70% of the total weight is usually sent to the landfills (EPA 2012; Ghosh et al. 2015; Hadi et al. 2015). Duan et al. (2016) reported that the total amount of NMF would reach 279 kT by 2020 in China without accounting for the waste NMF produced during printed circuit board manufacturing.

Electronic Waste: Recycling and Reprocessing for a Sustainable Future, First Edition.
Edited by Maria E. Holuszko, Amit Kumar, and Denise C. R. Espinosa.

10.2 Nonmetal Fraction Composition

The composition of the NMF would depend on the type of printed circuit boards. The PCB is classified based on the type of base material. More broadly, the National Electrical Manufacturers Association (NEMA) classification has been widely used for base materials. The base material for PCB consists of a reinforcement encapsulated in resin material. Phenolic, epoxy, and polyester resins are the most commonly used resins in PCB manufacturing, whereas cotton/cellulose paper and fiberglass are the major reinforcement. There are several types of PCB, such as FR-1 to FR-6, CEM-1 to CEM-8, and G-10 and 11, XXP, and XXXPC (Kelley and Trobough 2016; Weil and Levchik 2004). FR stands for flame retardant, and CEM stands for composite epoxy materials. The major difference in different types of PCB is the reinforcement and resin materials. These changes affect the physical properties and cost of the PCB, rendering it suitable for specific applications. Kelley and Trobough (2016) suggested that the most common types of PCB are FR-2, FR-4, CEM-1, and CEM-3. Table 10.1 lists the details of these four types of PCB. All four boards have flame-retardant properties.

CEM-1 provides improved electrical and physical properties compared to FR-2 but can only be used as a single-layered PCB. CEM-3 is more suitable for plating through holes that allow the circuitry to be printed on both sides and can be used as a complete replacement for FR-4 but has a higher cost than other PCB. FR-4 is the most common PCB type with outstanding electrical, mechanical, and thermal properties, rendering it useful for a wide range of applications. The price of alternatives to FR-4 is usually 2–4 times higher (Lassen and Løkke 1999).

In the field of consumer electronics, FR-2 and FR-4 are the most commonly used PCB. Lassen and Løkke (1999) suggested that 90% of the produced PCB is FR-4 based, and the remaining 10% includes FR-2 and other types. FR-2 is widely used in televisions and home electronics, whereas FR-4 is mostly used in high-value electrical and

Table 10.1 Description of the most common types of PCB.

Type	Resin	Reinforcement	Application
FR-2	Phenolic	Cotton paper	Low cost, radios, calculators and, toys
FR-4	Epoxy	Woven glass cloth	Computers, servers, data storage, telecommunications, industrial applications
CEM-1	Epoxy	Cotton paper core with woven glass cloth on surface	Consumer and industrial electronics
CEM-3	Epoxy	Nonwoven glass core with woven glass cloth on surface	Early home computers, automobiles, home entertainment products

Source: Kelley and Trobough (2016).

electronic equipment (EEE) such as computers and electronics for telecommunications (Guo et al. 2009; Hall and Williams 2007; Lassen and Løkke 1999; Marques et al. 2013). The metals present in the PCBs are extracted for their value, and the remaining NMFs containing the resins, reinforcing materials, and residue metals are often sent to landfills (EPA 2012; Ghosh et al. 2015; Hadi et al. 2015).

10.3 Benefits of NMF Recycling

Recycling is one of the most researched areas to deal with the ever-growing e-waste issue around the world. The materials present in the NMF could also provide economic benefits to the recyclers and reduce the environmental concerns associated with improper disposal. This section provides a detailed analysis of the benefits associated with NMF recycling. The benefits are categorized into two sections.

10.3.1 Economic Benefits

The recovery of residual metals before the NMF disposal could provide an economic incentive for NMF recycling. Several studies have suggested that the NMF consists of glass fibers (65% by wt.), epoxy resins (32% by wt.), and residual metals (copper and other metals). Zheng et al. (2009a) showed that the concentration of glass fibers is highest in the fine fraction, whereas the resins are mostly concentrated in coarse fractions. An X-ray fluorescence (XRF) analysis of the NMF obtained from the waste PCB showed a high concentration of silica, calcium oxide, and alumina, which is mostly from the fibers present in the circuit boards with a trace amount of bromine, copper oxide, magnesium oxide, and other metallic oxides (Muniyandi et al. 2014). Duan et al. (2016) used XRF analysis to show that, on average, NMF contains 2.95% copper, 0.35% lead, 0.41% tin, 0.07% nickel, 0.04% zinc, 0.02% manganese, and 0.02% chromium as residual metals after separating the copper. It should be noted that these studies were performed on waste PCB obtained from personal computers containing FR-4 boards. Studies performed by Kumar et al. (2018a) showed that metal content in NMF could range from 10% to 20% depending on the size fraction. The metal concentration vs. size analysis suggested that for most of the cases, the finest size fraction (−75 μm) has the highest concentration of metals. It was shown that the NMF contains 5.4% Ca, 3% Al, 1.3% Cu, 0.23% Sb, 0.16% Fe, 0.15% Fe, 0.04% Sn, 0.04% Sn, and 0.03% Pb. The test was conducted on NMF obtained from a PCB recycling plant representing a more realistic solution as the majority of waste PCB is processed in some fashion before disposal.

Apart from these metals, fiberglass present in the NMF also presents an economic benefit for recycling. Fiberglass is one of the most commonly used reinforcements in thermoplastic and, if recycled properly, could be used as fillers in different composites. A life cycle analysis of e-glass, the most common type of fiberglass used in the electronics industry by Dai et al. (2015), showed that the production of one ton of e-glass consumes 0.4 tonne of sand, 13.7 GJ of natural gas and electricity, and over 3600 l of water. Hence, recycling this fiberglass would also provide significant

Table 10.2 Amount of fiberglass locked in e-waste.

Description	2016	2021	2050
E-waste (million tonnes)	44.7	52.2	120
Average circuit boards in e-waste (%)	3–6	3–6	4.5
Average NMF in circuit boards (%)	65–70	65–70	65–70
Average fiberglass in NMF (%)	39	39	39
Fiberglass (million tonnes)	0.53 ± 0.19	0.62 ± 0.23	1.43 ± 0.52

resource savings. The fiberglass concentration was determined to provide an estimate of the amount of fiberglass locked in e-waste.

Kumar et al. (2018a) showed that the overall fiberglass content in the dried NMF is 39%. The concentration in the coarse fraction is as low as 26%, and the maximum concentration of 69% is reached for the finest size fraction. Zheng et al. (2009a) showed a similar trend with the highest concentration of over 70% in the NMF in −106 μm size fraction and concentration of 53% in the −710 + 180 μm (25–80 mesh) size fraction. The concentration in +710 μm size fraction was not reported.

Based on the amount of e-waste produced in the world, average circuit board concentration, average NMF and fiberglass concentration, the amount of fiberglass locked in e-waste was estimated and is shown in Table 10.2. It shows the presence of nearly half a million tonnes of fiberglass in e-waste in 2016, which would increase to 1.5 million tonnes by 2050.

10.3.2 Environmental Protection and Public Health

Printed circuit boards contain different heavy metals and toxic compounds in high concentrations that would be harmful to the environment and public safety. Literature has already reported the concentration and issue associated with these metals and compounds.

Research in the field of associated risks with its disposal is limited (Duan et al. 2016). A review of different literature has shown that the only study conducted on the NMF obtained from a waste printed circuit board recycling plant was from Anhui province, China. The processing plant uses a combination of eddy current and electrostatic separation to separate metals from nonmetal fraction (Wang et al. 2018; Zhang et al. 2017b; Zhang et al. 2018) and has only discussed the residual metal concentration in NMF. Most of the research is limited to the waste printed circuit boards from specific EEE units.

The issues associated with landfilling are also limited to the waste printed circuit boards. An analysis from milled waste PCB shows the presence of 133 mg/l lead and 22 mg/l cadmium (Bizzo et al. 2014), whereas leaching characteristics for the NMF have not been reported. Guanghan et al. (2016) suggested that the NMF doesn't decay for a long time, and it also reduces the air permeability of the soil and pollutes the groundwater due to the presence of heavy ions and curing agents.

Kumar et al. (2018a) showed that the lead leachate concentration for NMF obtained from the toxicity characteristic leaching procedure (TCLP) test exceeded the landfilling leachate limit in Canada. The lead leachate concentration was three times higher than the regulation limits. Lead is primarily present in the solder matrix and hard to completely recover through comminution and related physical processing methods due to its malleability. Compared to the TCLP test results for milled circuit boards obtained by Bizzo et al. (2014), the lead leachate concentration has reduced from 133 to 17.5 mg/l, whereas cadmium has decreased to almost negligible amount compared to 22 mg/l after processing. This suggests that the majority of lead and cadmium had been recovered during the metal extraction process.

Polybrominated diphenyl ethers (PBDEs) are one of the other toxic substances used in circuit boards and are added as flame retardants to the resins at the laminate product stage. These materials would not be recovered during the metal extraction stage and would report to the NMF. These compounds have been slowly phased out in different parts of the world; hence, the circuit boards manufactured in different parts of the world at different years would have highly varying concentrations of PBDEs. Researchers have shown the PBDEs concentration ranging from 1.33 to 14 567 mg/kg in some circuit boards before any processing in China (Cai et al. 2018; Chen et al. 2012; Guo et al. 2015; Yu et al. 2017; Zhou et al. 2013). Kumar et al. (2018a) showed that the total PBDEs concentration in the dried NMF was 294 mg/kg. The deca-BDE had a much higher contribution than other PDBE congeners, followed by tetra and penta-BDE, which is in line with the Environment Canada (2004) report showing that deca-BDE is widely used as flame retardants in computers, television casings, and electrical/electronic components.

Similar to PBDEs, polychlorinated biphenyls are another class of hazardous substances that are highly regulated in Canada and around the world. The production of these compounds was banned in 1979 in the USA, but the products manufactured before 1979 ban might contain PCBs. According to the most recent regulation in Canada, no person shall release a solid containing more than 50 mg/kg of polychlorinated biphenyls from equipment not in use (Minister of Justice 2015). The concentration of polychlorinated biphenyls is also limited to the soil or dust in the area around recycling centers. The polychlorinated biphenyls concentration in 22 tree bark samples from an e-waste recycling area in China showed a concentration of 0.112 mg/kg of dry sample (Wen et al. 2009). Kumar et al. (2018a) showed that the total polychlorinated biphenyls concentration in the dried NMF is 3.2 mg/kg, which is well below the acceptable concentration limit for the environmental release limit by a person. However, if the NMF containing these PCBs are incinerated, it would cause the release of dioxins and furans into the environment if ill-managed.

The presence of harmful flame retardants in the NMF poses the risk of the release of toxic dioxins and furans (polychlorinated dibenzo-p-dioxin (PCDD)/Fs) due to improper incineration. PCDD/Fs are classified under toxic substances under the Canadian Environmental Protection Act (CEPA). The toxic equivalent quotient (TEQ) is a tool to assess the cumulative toxicity of a mixture of PCDD/Fs. The toxicity levels (TEFs) for all compounds are assessed relative to the toxicity of 2,3,7,8-tetrachlorodibenzo-p-dioxin (tetra-CDD), which is the most toxic dioxin.

TEQ represents the product of the concentrations of individual congeners and their respective TEFs. The lower limit is calculated by substituting zero (0) for any nondetectable compound, whereas the upper bound represents the worst-case scenario where the detection limit is used for the nondetectable compounds (Johnston 2016). The concentration of PCDD/Fs in unheated printed circuit boards was up to 16 000 ng TEQ/kg (Duan et al. 2011). The PCDD/F concentration reported by Wen et al. (2009) in 22 tree bark samples was 1800 pg/g (19.2 pg WHO-TEQ/g). The PCDD/F concentration in the soil and snails near various e-waste recycling sites in China was 53.39 and 75.47 pg WHO-TEQ/g, respectively (Liu et al. 2008). The PCDD/F concentration from the soil and combusted residue from the e-waste recycling site in Guiyu (China) was 0.8–506 pg WHO TEQ/g in various sample sources (Leung et al. 2007). The results from the PCDD/Fs analysis by Kumar et al. (2018a) showed that the toxicity limit of PCDD/Fs ranged from 2880 to 3170 pg/kg TEQ.

Greene et al. (2003) suggested that maintaining a lifetime average daily bodily dosage below 1–10 pg/kg/day (TEQ) is required to eliminate the risk of cancer. In contrast, tolerable daily intake value used by Health Canada is 10 pg/kg/day TEQ (Canadian Council of Ministers of the Environment 2002). The amount of PCDD/Fs produced from a plant processing unit tonne of NMF per day would be much higher than the tolerable daily intake recommended by Health Canada, which emphasizes the requirement of proper disposal of the separated NMF from the processing of circuit boards.

10.4 Recycling of NMF

According to the Environmental Protection Agency (2012), the recycling of NMF from waste PCBs is not usually practiced, and up to 94% is discarded in landfills. Some researchers have suggested alternative use of NMF as secondary materials in various applications, and other researchers have shown the possibility of the physical and chemical recycling of NMF.

10.4.1 Physical Recycling

One of the challenges with the recycling of NMF is its heterogeneous nature. The presence of inorganics in the NMF decreases its usability as the bundles of glass fiber can form higher stress concentration points that would reduce the strength of composites (Zhang et al. 2017a, 2017b) and removal of inorganics from the NMF increased the flexural strength, flexural modulus, and impact strength of the high-density polyethylene composite and thus improved its usability. Wang et al. (2018) suggested that separating the components of NMF is necessary to increase its reuse. The organics can be used as a filler in composite or for energy recovery, fiberglass can be used in construction/building materials, and residual metal can be sent for metal purification. Various physical methods for NMF physical recycling are listed below.

10.4.1.1 Size Classification

Size classification relies on the difference in sizes of various particles. Researchers have shown that the concentration of glass fibers is highest in the fine fraction (−75 μm), whereas the resins are mostly concentrated in coarse fractions. Thus, the loss on ignition for the coarser fraction is higher than the finer fractions (Yang et al. 2019; Zhang et al. 2017b; Zheng et al. 2009a). Wang et al. (2018) have shown that glass fiber bundles were observed in hammer mill–crushed coarse particles, and single glass fibers were observed in fine fractions. The glass fibers in coarse particles were still embedded in epoxy resin that results in a higher loss on ignition. The surface of glass fibers present in the finer particles was abraded and rugged, thus liberating the glass fibers from epoxy and increasing the fiber/inorganic concentration in the finer size fraction and reducing the loss on ignition value for fines. However, the differences in the organic and inorganic content in different size fraction for nonmetal fraction without any additional grinding may or may not be significant enough for separation.

10.4.1.2 Gravity Separation

Gravity separation is one of the most common low-cost physical processing methods and is already being used for the separation of metals from nonmetals. It has been successfully used in the separation of different types of plastic wastes too. Processes such as air classifiers, jigs, hydrocyclones, and float–sink have been widely studied in the separation of different types of plastics (Bauer et al. 2018; Ruj et al. 2015; Serranti and Bonifazi 2019).

The density of various components of NMF is listed in Table 10.3. These differences in the densities provide a possibility of producing relatively cleaner organic and inorganic streams using gravity separation (Kumar et al. 2018b, 2018c)

The density-based float–sink separation showed promising results with organic resins being accumulated in the lighter density fraction, whereas inorganic fiberglass and residual metals reported to heavy density ranges. The gravity-based separation flowsheet with three product separations could potentially recover a cleaner organic fraction with 47% yield at 86% organic content and a reject stream with a 35% yield at 72% glass fiber content. The washability curves and various washability indices suggest that the material would be classified as difficult to clean. However, comparing

Table 10.3 Density of various components of the nonmetal fraction.

Components	Density (g/cm^3)	References
Epoxy fiberglass laminate (FR-4)	1.94	Li et al. (2014)
Epoxy resin	1.11–1.23	Wang et al. (2011)
Glass fiber	2.48–2.70	Shrivastava (2018)
Phenolic paper laminate (FR-2)	1.55	Li et al. (2014)
Phenolic resin	1.2–1.4	Dixit et al. (2016)
Cotton/paper fiber	1.5–1.6	Foulk et al. (2006) & Wakeham (1949)
Residual metals	>2.7	The Engineering ToolBox (2004)

the indices with the similar size ranges for fine coal suggested that the indices were relativity similar to that for the fine coal being processed or studied using a dense medium cyclone, spirals, or Reflux™ classifiers.

10.4.1.3 Magnetic Separation

Magnetic separation has been mostly used to recover ferrous materials from crushed waste printed circuit boards (Cui and Forssberg 2003; Guo et al. 2011; Silvas et al. 2015; Veit et al. 2006). Yoo et al. (2009) used magnetic separation to recover 83% of iron and nickel as magnetic concentrate from shredder printed circuit boards.

In the case of NMF recycling, magnetic separation was used to recover the residual magnetic material from the nonmetal fraction (Wang et al. 2018). It was shown that the magnetic concentrate accounted for nearly 6% of the total NMF feed and the recover concentrate had a low organic content (12%). It suggested that the magnetic separation could be used to recover residual magnetic materials from the NMF but not to separate organic resins and inorganic fiberglass as they both would report to the tailings due to their diamagnetic properties.

10.4.1.4 Electrical Separation

Similar to magnetic separation, the electrical properties (conductivity) have been widely used to separate metals from the nonmetals (Cui and Forssberg 2003; Guo et al. 2011; Veit et al. 2005; Zhang et al. 1998).

In the field of NMF recycling, researchers have shown that the triboelectric separation can be used to separate organic resins from the inorganic fiberglass. Zhang et al. (2017b) and Yang et al. (2019) used triboelectric separation to separate inorganics from the nonmetal fraction to improve the quality of the product. Overall, a concentrate yield of 47% was achieved with an organic content of 69%, recovery of 59%, and an upgradation ratio of 1.27 by Zhang et al. (2017b). In the case of Yang et al. (2019), an upgradation ratio of 1.17 was achieved.

It should be noted that the separation based on electrical properties is only feasible for completely dry materials and a plant using dry separation technology as the upstream process. A plant utilizing wet (water-based) separation would require drying the metal recycling rejects completely before the electrical separation of NMF.

10.4.1.5 Froth Flotation

Froth flotation uses the difference in surface properties of particles to separate different particles from each other. In the field of printed circuit boards, researchers have shown that the froth flotation process with or without collector could be used to recover the metals from waste circuit boards (He and Duan 2017; Mäkinen et al. 2015; Ogunniyi and Vermaak 2009; Zhu et al. 2019).

However, the applicability of the froth flotation process to separate the organic resins and inorganic fiberglass has not been studied in any literature. Since the fiberglass is mostly encapsulated in epoxy resins, the unliberated fiberglass would mostly be hydrophobic due to the presence of organic resins on the surface. Only the completely/partially liberated fiberglass would show any increase in its hydrophilic behavior.

10.4.2 Chemical Recycling

Guo et al. (2009) listed four ways to recycle the NMF by chemical/thermal methods. These are pyrolysis, gasification, supercritical fluid depolymerization, and hydrogenolytic degradation, with the primary goal of converting the polymers in the NMFs into chemical feedstocks or fuels.

The pyrolysis process converts the organic resins into oil and gases, and the glass fiber reported to the char residue. Major gases produced are carbon dioxide, carbon monoxide, and methane, whereas phenol and its derivatives are the major oils produced during the pyrolysis process. Zheng et al. (2009b) used a fluidized bed in combination with air cyclones at temperature range from 400 to 600 °C to recover high-value glass fiber from the NMF but acknowledged that the strength of fibers might have severely weakened. The gasification process converts the organic matter into synthesis gas (carbon monoxide and hydrogen) at reaction temperatures up to 1600 °C that can then be used as fuel. Supercritical water, methanol, or ammonia has been used to decompose resins, but the method requires high temperature and pressure (Guo et al. 2009). Yousef et al. (2017) extracted the woven fiberglass at a temperature of 50 °C with dimethylformamide chemical and ultrasonic treatment; however, the average strength of the recycled fiberglass decreased by 48%. The hydrogenolytic degradation process can solubilize epoxy resin with a hydrogen donor such as phthalic anhydride at 340 °C in two hours (Braun et al. 2001).

One of the most common disadvantages of these chemical/thermal processes is the higher operations cost and environmental risks associated with the chemicals. In comparison, physical recycling methods are relatively simple, environment friendly, and have lower energy and operating costs. However, the physical separation of individual components of NMF is more complicated than the chemical treatment due to the complex structure and encapsulation.

10.5 Potential Usage

Various researchers have shown the possible end use of NMF after metal recovery as secondary materials in various applications (Guo et al. 2009; Marques et al. 2013; Ning et al. 2017; Sohaili et al. 2012).

Zheng et al. (2009a) have suggested the use of NMF as reinforcing fillers in polypropylene composites that would enhance its tensile strength and improve its heat resistivity. Mou et al. (2005) suggested using NMF in the manufacturing of composite boards with enhanced strength due to its compatibility with epoxy resin that can be used in furniture, automobiles, and decorative materials. In contrast, Guo et al. (2010) have used NMF in producing wood–plastic composites. The NMF was also used with the recycled plasticized polyvinyl chloride from waste wires and cables to produce a composite with enhanced strength (Das et al. 2019). Hadi et al. (2013a, Hadi et al. 2015) have suggested that the presence of carbonaceous and siliceous materials in NMF provides an opportunity to be used as adsorbents and remove pollutants from aqueous solutions. Rajagopal et al. (2016) and Ke et al.

(2013) showed that the NMF from waste printed circuit boards could be used to prepare activated or porous carbon with high surface areas.

The most traditional approach for the use of NMF would be as an alternative fuel in cement manufacturing. The organic resins have a very high loss on ignition (LOI) suggesting that it would also have a relatively higher heating value. Alternative fuels such as biomass, household waste, and nonhazardous industrial and commercial waste have regularly been used in the cement industry to reduce fossil fuel consumption, provide cost savings, and significantly reduce the greenhouse gas emissions (Rahman et al. 2015).

One of the major challenges in using NMF as an alternative fuel in the cement industry is the composition of fuel ash as the fuel ash would combine with the raw materials and become a part of the clinker. A high level of alkali oxides (potassium and sodium), sulfur, and chlorine could also harm cement quality (Chinyama 2011). Myavagh (2013) showed the presence of 0.22% sulfur in the nonmetal fraction obtained from a plant in Hong Kong. In terms of chlorine, various researchers have shown the presence of chlorine in the circuit board resin matrix in the form of chlorine-based flame retardants and as a part of some plastic parts, cable insulators, coatings, rubber, and paper reinforcements (Nimpuno et al. 2009). The chlorine content varies widely depending on the type of board. Yao et al. (2015) have shown that the chlorine content in nonmetal fraction ranges from 0.02% to 1.67%.

Kumar et al. (2018b) showed that the heating value of NMF organic concentrate is 21.3 GJ/t, which is similar to or higher than some alternative solid fuels and would be classified as a medium-grade fuel. It is also higher than the heating value of peat coal and comparable to the lignite coals. The NMF ash residue would also be high in silica, alumina, and calcium oxide due to the presence of fiberglass, which could provide an additional source of raw materials for the clinker production.

Another use of the NMF is in the preparation of activated carbon due to its high carbon content. The organic materials in NMF would be pyrolyzed at a temperature ranging from 600 to 800 °C to remove the volatile matter. The remaining char residue can be mixed with solid KOH and then activated in nitrogen or carbon dioxide atmosphere at 600–900 °C for two hours (Kan et al. 2016; Ke et al. 2013; Rajagopal et al. 2016). Ke et al. (2013) used NMF with 72% initial carbon to produce porous carbon with a specific surface area up to 3112 m^2/g in a nitrogen atmosphere at 600 °C pyrolysis temperature and 900 °C activation temperature.

The NMF could also be directly used as a filler in building materials. Marques et al. (2013) have suggested that due to its lighter weight, finer grain size, and thick glass fibers, the NMF could improve the mechanical strength. Wang et al. (2012) have suggested that the NMF can be added in cement mortar that would increase its water retention property, decrease bulk density without significantly affecting the compressive and flexural strength.

The NMF could also be used to prepare high-purity porous silica. The glass fiber used in circuit boards is composed of 52–56% high-quality silica, which could be used to produce amorphous porous silica. Porous silica can be used in many applications ranging from life and space science to physics of high-energy particles due to its large surface area, optical properties, and very low thermal conductivity (Pajonk 2003).

Bazargan et al. (2014) have shown that the NMF with 51% silicon (from XRF) can be treated with 4 M nitric acid at 90 °C for six (6) hours followed by a heat treatment at 500 °C for three (3) hours and produce 99% pure silica with a specific surface area of over 300 m^2/g.

Researchers have also suggested the application of NMF as a toxic heavy metal adsorbent for wastewater treatment (Hadi et al. 2013b, Hadi et al. 2015). Hadi et al. (2013b) suggested to impregnate the NMF with the 1 M potassium hydroxide solution for three hours and then activating the slurry at 250 °C for three hours in a muffle furnace in an inert atmosphere. The activated sample is washed down several times with distilled water and then dried for later use. The activated material with a surface area of over 200 m^2/g had a similar elemental composition as the NMF without the carbon and bromine, which was burnt off during the activation process. The activated material could be used to adsorb copper, lead, cadmium, zinc, cobalt, and nickel ions from wastewater.

Some researchers have also shown the potential application of recycled fiberglass as a sound absorber. The recycled fiberglass has the capability of absorbing sound over a broad frequency range. In a study, the small residual organics were removed by a thermal breakdown at 500 °C, followed by calcination in a muffle furnace (Sun et al. 2015).

References

Awasthi, A.K., Zlamparet, G.I., Zeng, X., and Li, J. (2017). Evaluating waste printed circuit boards recycling: opportunities and challenges, a mini review. *Waste Management and Research* 35 (4): 346–356. https://doi.org/10.1177/0734242X16682607.

Bauer, M., Lehner, M., Schwabl, D. et al. (2018). Sink–float density separation of post-consumer plastics for feedstock recycling. *Journal of Material Cycles and Waste Management* 20 (3): 1781–1791. https://doi.org/10.1007/s10163-018-0748-z.

Bazargan, A., Bwegendaho, D., Barford, J., and McKay, G. (2014). Printed circuit board waste as a source for high purity porous silica. *Separation and Purification Technology* 136: 88–93. https://doi.org/10.1016/J.SEPPUR.2014.08.026.

Bizzo, W., Figueiredo, R., and de Andrade, V. (2014). Characterization of printed circuit boards for metal and energy recovery after milling and mechanical separation. *Materials* 7 (6): 4555–4566. https://doi.org/10.3390/ma7064555.

Braun, D., von Gentzkow, W., and Rudolf, A.P. (2001). Hydrogenolytic degradation of thermosets. *Polymer Degradation and Stability* 74 (1): 25–32. https://doi.org/10.1016/S0141-3910(01)00035-0.

Cai, C., Yu, S., Liu, Y. et al. (2018). PBDE emission from e-wastes during the pyrolytic process: emission factor, compositional profile, size distribution, and gas-particle partitioning. *Environmental Pollution* 235: 419–428.

Canadian Council of Ministers of the Environment (2002). Canadian soil quality guidelines for the protection of environmental and human health. http://ceqg-rcqe.ccme.ca/download/en/275 (retrieved 22 March 2021).

Chen, Y., Li, J., Chen, L. et al. (2012). Brominated flame retardants (BFRs) in waste electrical and electronic equipment (WEEE) plastics and printed circuit boards (PCBs). *Procedia Environmental Sciences* 16: 552–559.

Chinyama, M.P.M. (2011). Alternative fuels in cement manufacturing. In: *Alternative Fuel* (ed. M. Manzanera), 263–284. IntechOpen https://doi.org/10.5772/22319.

Cui, J. and Forssberg, E. (2003). Mechanical recycling of waste electric and electronic equipment: a review. *Journal of Hazardous Materials* 99 (3): 243–263. https://doi.org/10.1016/S0304-3894(03)00061-X.

Dai, Q., Kelly, J., Sullivan, J., and Elgowainy, A. (2015). *Life-Cycle Analysis Update of Glass and Glass Fiber for the GREET Model.* Argonne National Laboratory.

Das, A., Vidyadhar, A., and Mehrotra, S.P. (2009). A novel flowsheet for the recovery of metal values from waste printed circuit boards. *Resources, Conservation and Recycling* 53 (8): 464–469. https://doi.org/10.1016/j.resconrec.2009.03.008.

Das, R.K., Gohatre, O.K., Biswal, M. et al. (2019). Influence of non-metallic parts of waste printed circuit boards on the properties of plasticised polyvinyl chloride recycled from the waste wire. *Waste Management & Research* 37 (6): 569–577. https://doi.org/10.1177/0734242X19836725.

Dixit, D., Pal, R., Kapoor, G., and Stabenau, M. (2016). Lightweight composite materials processing. In: *Lightweight Ballistic Composites*, 2e (ed. A. Bhatnagar), 157–216. Woodhead Publishing https://doi.org/10.1016/B978-0-08-100406-7.00006-4.

Duan, H., Li, J., Liu, Y. et al. (2011). Characterization and inventory of PCDD/Fs and PBDD/Fs emissions from the incineration of waste printed circuit board. *Environmental Science & Technology* 45: 6322–6328.

Duan, H., Hu, J., Yuan, W. et al. (2016). Characterizing the environmental implications of the recycling of non-metallic fractions from waste printed circuit boards. *Journal of Cleaner Production* 137: 546–554.

Environment Canada (2004). Environmental screening assessment report on polybrominated diphenyl Ethers (PBDEs). https://www.ec.gc.ca/lcpe-cepa/documents/substances/pbde/pbde_draft-eng.pdf (retrieved 12 December 2019).

EPA (2012). Printed circuit board recycling methods. https://www.epa.gov/sites/production/files/2014-05/documents/handout-10-circuitboards.pdf (retrieved 8 November 2016).

Foulk, J.A., Chao, W.Y., Akin, D.E. et al. (2006). Analysis of flax and cotton fiber fabric blends and recycled polyethylene composites. *Journal of Polymers and the Environment* 14 (1): 15–25. https://doi.org/10.1007/s10924-005-8703-1.

Ghosh, B., Ghosh, M.K., Parhi, P. et al. (2015). Waste printed circuit boards recycling: an extensive assessment of current status. *Journal of Cleaner Production* 94: 5–19. https://doi.org/10.1016/J.JCLEPRO.2015.02.024.

Golev, A. and Corder, G.D. (2017). Quantifying metal values in e-waste in Australia: the value chain perspective. *Minerals Engineering* 107: 81–87. https://doi.org/10.1016/j.mineng.2016.10.021.

Golev, A., Schmeda-Lopez, D.R., Smart, S.K. et al. (2016). Where next on e-waste in Australia? *Waste Management* 58: 348–358.

Golev, A., Corder, G.D., and Rhamdhani, M.A. (2019). Estimating flows and metal recovery values of waste printed circuit boards in Australian e-waste. *Minerals Engineering* 137: 171–176. https://doi.org/10.1016/J.MINENG.2019.04.017.

Greene, J., Hays, S., and Paustenbach, D. (2003). Basis for a proposed reference dose (RfD) for dioxin of 1-10 PG/KG-day: a weight of evidence evaluation of the human and animal studies. *Journal of Toxicology and Environmental Health, Part B* 6 (2): 115–159. https://doi.org/10.1080/10937400306470.

Guanghan, S., Zhu, X., Wenyi, Y. et al. (2016). Recycling and disposal technology for non-metallic materials from waste printed circuit boards (WPCBs) in China. *Procedia Environmental Sciences* 31: 935–940.

Guo, J., Guo, J., and Xu, Z. (2009). Recycling of non-metallic fractions from waste printed circuit boards: a review. *Journal of Hazardous Materials* 168 (2–3): 567–590.

Guo, J., Tang, Y., and Xu, Z. (2010). Wood plastic composite produced by nonmetals from pulverized waste printed circuit boards. *Environmental Science & Technology* 44 (1): 463–468. https://doi.org/10.1021/es902889b.

Guo, C., Wang, H., Liang, W. et al. (2011). Liberation characteristic and physical separation of printed circuit board (PCB). *Waste Management* 31 (9–10): 2161–2166. https://doi.org/10.1016/J.WASMAN.2011.05.011.

Guo, J., Zhang, R., and Xu, Z. (2015). PBDEs emission from waste printed wiring boards during thermal process. *Environmental Science & Technology* 49 (5): 2716–2723.

Hadi, P., Barford, J., and McKay, G. (2013a). Toxic heavy metal capture using a novel electronic waste-based material – mechanism, modeling and comparison. *Environmental Science & Technology* 47 (15): 8248–8255, 130719140359006. doi:https://doi.org/10.1021/es4001664.

Hadi, P., Gao, P., Barford, J.P., and McKay, G. (2013b). Novel application of the nonmetallic fraction of the recycled printed circuit boards as a toxic heavy metal adsorbent. *Journal of Hazardous Materials* 252–253: 166–170. https://doi.org/10.1016/J.JHAZMAT.2013.02.037.

Hadi, P., Ning, C., Ouyang, W. et al. (2015). Toward environmentally-benign utilization of nonmetallic fraction of waste printed circuit boards as modifier and precursor. *Waste Management* 35: 236–246.

Hall, W.J. and Williams, P.T. (2007). Separation and recovery of materials from scrap printed circuit boards. *Resources, Conservation and Recycling* 51 (3): 691–709. https://doi.org/10.1016/J.RESCONREC.2006.11.010.

He, J. and Duan, C. (2017). Recovery of metallic concentrations from waste printed circuit boards via reverse floatation. *Waste Management* 60: 618–628. https://doi.org/10.1016/J.WASMAN.2016.11.019.

Johnston, J. (2016). Dioxins, furans, and dioxin-like PCB congeners: ecological risk calculation methodology for upland soil https://fortress.wa.gov/ecy/publications/documents/1609044.pdf (retrieved 22 March 2021).

Kan, Y., Yue, Q., Gao, B., and Li, Q. (2016). Comparison of activated carbons from epoxy resin of waste printed circuit boards with KOH activation by conventional and microwave heating methods. *Journal of the Taiwan Institute of Chemical Engineers* 68: 440–445. https://doi.org/10.1016/J.JTICE.2016.08.047.

Ke, Y., Yang, E., Liu, X. et al. (2013). Preparation of porous carbons from non-metallic fractions of waste printed circuit boards by chemical and physical activation. *New Carbon Materials* 28 (2): 108–113. https://doi.org/10.1016/S1872-5805(13)60069-4.

Kelley, E. and Trobough, D. (2016). Introduction to base materials. In: *Printed Circuits Handbook*, 7e (eds. C.F. Coombs and H.T. Holden), 171–195. McGraw-Hill Education https://www.mheducation.ca/printed-circuits-handbook-seventh-edition-9780071833950-can (retrieved 22 March 2021).

Kumar, A., Holuszko, M.E., and Janke, T. (2018a). Characterization of the non-metal fraction of the processed waste printed circuit boards. *Waste Management* 75: 94–102. https://doi.org/10.1016/J.WASMAN.2018.02.010.

Kumar, A., Holuszko, M., and Janke, T. (2018b). Separating inorganics from the non-metal fraction of the processed waste PCBS using heavy liquid separation. *International Journal of Waste Resources* 8 (4) https://doi.org/10.4172/2252-5211.1000355.

Kumar, A., Kuppusamy, V., Holuszko, M., and Janke, T. (2018c). Improving the energy concentration in waste printed circuit boards using gravity separation. *Recycling* 3 (2): 21. https://doi.org/10.3390/recycling3020021.

Lassen, C. and Løkke, S. (1999). Brominated flame retardants-substance flow analysis and assessment of alternatives. https://www2.mst.dk/Udgiv/publications/1999/87-7909-416-3/199987-7909-416-3.pdf (retrieved 22 March 2021).

Leung, A.O.W., Luksemburg, W.J., Wong, A.S., and Wong, M.H. (2007). Spatial distribution of polybrominated diphenyl ethers and polychlorinated dibenzo-p-dioxins and dibenzofurans in soil and combusted residue at Guiyu, an electronic waste recycling site in Southeast China. *Environmental Science & Technology* 41 (8): 2730–2737.

Li, J., Gao, B., and Xu, Z. (2014). New technology for separating resin powder and fiberglass powder from fiberglass–resin powder of waste printed circuit boards. *Environmental Science & Technology* 48 (9): 5171–5178.

Liu, H., Zhou, Q., Wang, Y. et al. (2008). E-waste recycling induced polybrominated diphenyl ethers, polychlorinated biphenyls, polychlorinated dibenzo-p-dioxins and dibenzo-furans pollution in the ambient environment. *Environment International* 34 (1): 67–72.

Luda, M.P. (2011). Recycling of printed circuit boards. In: *Integrated Waste Management – Volume II* (ed. M.S. Kumar). IntechOpen http://cdn.intechopen.com/pdfs/18491/InTech-Recycling_of_printed_circuit_boards.pdf (retrieved 22 March 2021).

Mäkinen, J., Bachér, J., Kaartinen, T. et al. (2015). The effect of flotation and parameters for bioleaching of printed circuit boards. *Minerals Engineering* 75: 26–31. https://doi.org/10.1016/J.MINENG.2015.01.009.

Marques, A.C., Cabrera Marrero, J.-M., and de Fraga Malfatti, C. (2013). A review of the recycling of non-metallic fractions of printed circuit boards. *SpringerPlus* 2: 521.

Minister of Justice (2015). *PCB Regulations*. Canada Gazette https://laws-lois.justice.gc.ca/PDF/SOR-2008-273.pdf (retrieved 22 March 2021).

Mou, P., Xiang, D., Pan, X., Wa, L., Gao, J., & Duan, G. (2005). New solutions for reusing nonmetals reclaimed from waste printed circuit boards. IEEE International

Symposium on Electronics and the Environment, New Orleans, LA, USA, 205–209. doi:https://doi.org/10.1109/isee.2005.1437026

Muniyandi, S.K., Sohaili, J., and Hassan, A. (2014). Encapsulation of nonmetallic fractions recovered from printed circuit boards waste with thermoplastic. *Journal of the Air & Waste Management Association* 64 (9): 1085–1092.

Myavagh, P.H. (2013). *Recovery of Sorption Products from Processing Non-metallic Waste PCBs*. The Hong Kong University of Science and Technology https://doi.org/10.14711/thesis-b1254392.

Nimpuno, N., McPherson, A., and Sadique, T. (2009). Greening consumer electronics-moving away from bromine and chlorine. https://www.issuelab.org/resources/9515/9515.pdf (retrieved 22 March 2021).

Ning, C., Lin, C.S.K., Hui, D.C.W., and McKay, G. (2017). Waste printed circuit board (PCB) recycling techniques. *Topics in Current Chemistry* 375 (2): 43.

Ogunniyi, I.O. and Vermaak, M.K.G. (2009). Investigation of froth flotation for beneficiation of printed circuit board comminution fines. *Minerals Engineering* 22 (4): 378–385.

Pajonk, G.M. (2003). Some applications of silica aerogels. *Colloid and Polymer Science* 281 (7): 637–651. https://doi.org/10.1007/s00396-002-0814-9.

Park, Y.J. and Fray, D.J. (2009). Recovery of high purity precious metals from printed circuit boards. *Journal of Hazardous Materials* 164 (2–3): 1152–1158. https://doi.org/10.1016/J.JHAZMAT.2008.09.043.

Rahman, A., Rasul, M.G., Khan, M.M.K., and Sharma, S. (2015). Recent development on the uses of alternative fuels in cement manufacturing process. *Fuel* 145: 84–99. https://doi.org/10.1016/J.FUEL.2014.12.029.

Rajagopal, R.R., Aravinda, L.S., Rajarao, R. et al. (2016). Activated carbon derived from non-metallic printed circuit board waste for supercapacitor application. *Electrochimica Acta* 211: 488–498. https://doi.org/10.1016/J.ELECTACTA.2016.06.077.

Research and Markets (2019). Printed circuit board (PCB) market report: trends, forecast and competitive analysis. https://www.researchandmarkets.com/reports/4791058/printed-circuit-board-pcb-market-report?utm_source=CI&utm_medium=PressRelease&utm_code=glpl3r&utm_campaign=1271813+-+Global+Printed+Circuit+Board+(PCB)+Market+Expected+to+Reach+an+Estimated+%2489.7+Billio (retrieved 22 March 2021).

Rocchetti, L., Amato, A., and Beolchini, F. (2018). Printed circuit board recycling: a patent review. *Journal of Cleaner Production* 178: 814–832. https://doi.org/10.1016/J.JCLEPRO.2018.01.076.

Ruj, B., Pandey, V., Jash, P., and Srivastava, V.K. (2015). Sorting of plastic waste for effective recycling. *Journal of Applied Sciences and Engineering Research* 4: 564–571. https://doi.org/10.6088.ijaser.04058.

Serranti, S. and Bonifazi, G. (2019). Techniques for separation of plastic wastes. In: *Use of Recycled Plastics in Eco-Efficient Concrete* (eds. F. Pacheco-Torgal, J. Khatib, F. Colangelo and R. Tuladhar), 9–37. Woodhead Publishing https://doi.org/10.1016/B978-0-08-102676-2.00002-5.

Shrivastava, A. (2018). Additives for plastics. In: *Introduction to Plastics Engineering* (ed. A. Shrivastava), 111–141. William Andrew Publishing https://doi.org/10.1016/B978-0-323-39500-7.00004-6.

Silvas, F.P.C., Jiménez Correa, M.M., Caldas, M.P.K. et al. (2015). Printed circuit board recycling: physical processing and copper extraction by selective leaching. *Waste Management* 46: 503–510. https://doi.org/10.1016/J.WASMAN.2015.08.030.

Sohaili, J., Muniyandi, S.K., and Mohamad, S.S. (2012). A review on printed circuit boards waste recycling technologies and reusing of recovered nonmetallic materials. *International Journal of Scientific & Engineering Research* 3 (2).

Sun, Z., Shen, Z., Ma, S., and Zhang, X. (2015). Sound absorption application of fiberglass recycled from waste printed circuit boards. *Materials and Structures/Materiaux et Constructions* 48 (1–2): 387–392. https://doi.org/10.1617/s11527-013-0190-6.

Szałatkiewicz, J. (2013). Metals recovery from waste of printed circuit boards processed in plasmatron plasma reactor. *IFAC Proceedings Volumes (IFAC-Papers Online)* 15 (PART 1): 478–483. https://doi.org/10.3182/20130825-4-US-2038.00109.

Szałatkiewicz, J. (2014a). Metals content in printed circuit board waste. *Polish Journal of Environmental Studies* 23 (6): 2365–2369.

Szałatkiewicz, J. (2014b). The amount of PCBs in WEEE. *Polish Journal of Environmental Studies* 23 (6): 2365–2369.

The Engineering ToolBox (2004). Metals and alloys – densities. https://www.engineeringtoolbox.com/metal-alloys-densities-d_50.html (retrieved 12 June 2012).

Veit, H.M., Diehl, T.R., Salami, A.P. et al. (2005). Utilization of magnetic and electrostatic separation in the recycling of printed circuit boards scrap. *Waste Management* 25 (1): 67–74. https://doi.org/10.1016/J.WASMAN.2004.09.009.

Veit, H.M., Bernardes, A.M., Ferreira, J.Z. et al. (2006). Recovery of copper from printed circuit boards scraps by mechanical processing and electrometallurgy. *Journal of Hazardous Materials* 137 (3): 1704–1709. https://doi.org/10.1016/J.JHAZMAT.2006.05.010.

Veit, H.M., Juchneski, N.C.d.F., and Scherer, J. (2014). Use of gravity separation in metals concentration from printed circuit board scraps. *Rem: Revista Escola de Minas* 67 (1): 73–79. https://doi.org/10.1590/S0370-44672014000100011.

Wakeham, H. (1949). The fine structure of cotton fibers from density measurements. *Textile Research Journal* 19 (10): 595–605. https://doi.org/10.1177/004051754901901001.

Wang, R.-M., Zheng, S.-R., Zheng, Y.-P. et al. (2011). Matrix materials. In: *Polymer Matrix Composites and Technology* (eds. R.-M. Wang, S.-R. Zheng and Y.-P. Zheng), 101–548. Woodhead Publishing https://doi.org/10.1533/9780857092229.1.101.

Wang, R., Zhang, T., and Wang, P. (2012). Waste printed circuit boards nonmetallic powder as admixture in cement mortar. *Materials and Structures/Materiaux et Constructions* 45 (10): 1439–1445. https://doi.org/10.1617/s11527-012-9843-0.

Wang, H., Zhang, G., Hao, J. et al. (2018). Morphology, mineralogy and separation characteristics of nonmetallic fractions from waste printed circuit boards. *Journal of Cleaner Production* 170: 1501–1507. https://doi.org/10.1016/J.JCLEPRO.2017.09.280.

Weil, E.D. and Levchik, S. (2004). A review of current flame retardant systems for epoxy resins. *Journal of Fire Sciences* 22: 25–40.

Wen, S., Yang, F., Li, J.G. et al. (2009). Polychlorinated dibenzo-p-dioxin and dibenzofurans (PCDD/Fs), polybrominated diphenyl ethers (PBDEs), and polychlorinated biphenyls (PCBs) monitored by tree bark in an E-waste recycling area. *Chemosphere* 74 (7): 981–987.

Yang, J., Wang, H., Zhang, G. et al. (2019). Recycling organics from non-metallic fraction of waste printed circuit boards by a novel conical surface triboelectric separator. *Resources, Conservation and Recycling* 146: 264–269. https://doi.org/10.1016/J.RESCONREC.2019.03.008.

Yao, Z.T., Yuan, W.Y., Xie, Z.M., and Tang, J.H. (2015). Non-metallic materials from waste printed circuit boards (PCBs): characteristics, environmental risk and recycling. *Applied Mechanics and Materials* 768: 576–587. https://doi.org/10.4028/www.scientific.net/AMM.768.576.

Yoo, J.-M., Jeong, J., Yoo, K. et al. (2009). Enrichment of the metallic components from waste printed circuit boards by a mechanical separation process using a stamp mill. *Waste Management* 29 (3): 1132–1137. https://doi.org/10.1016/J.WASMAN.2008.06.035.

Yousef, S., Tatariants, M., Bendikiene, R., and Denafas, G. (2017). Mechanical and thermal characterizations of non-metallic components recycled from waste printed circuit boards. *Journal of Cleaner Production* 167: 271–280.

Yu, D., Duan, H., Song, Q. et al. (2017). Characterization of brominated flame retardants from e-waste components in China. *Waste Management* 68: 498–507.

Zhang, S., Forssberg, E., Arvidson, B., and Moss, W. (1998). Aluminum recovery from electronic scrap by high-force® eddy-current separators. *Resources, Conservation and Recycling* 23 (4): 225–241. https://doi.org/10.1016/S0921-3449(98)00022-6.

Zhang, G., He, Y., Wang, H. et al. (2017a). Application of triboelectric separation to improve the usability of nonmetallic fractions of waste printed circuit boards: removing inorganics. *Journal of Cleaner Production* 142: 1911–1917.

Zhang, G., Wang, H., He, Y. et al. (2017b). Triboelectric separation technology for removing inorganics from non-metallic fraction of waste printed circuit boards: influence of size fraction and process optimization. *Waste Management* 60: 42–49. https://doi.org/10.1016/j.wasman.2016.08.010.

Zhang, G., Wang, H., Yang, J. et al. (2018). Application of electric field to a fluidized bed for recovering residual metals from fine particles of the non-metallic fraction of waste printed circuit boards. *Journal of Cleaner Production* 187: 1036–1042. https://doi.org/10.1016/J.JCLEPRO.2018.03.282.

Zheng, Y., Shen, Z., Cai, C. et al. (2009a). The reuse of nonmetals recycled from waste printed circuit boards as reinforcing fillers in the polypropylene composites. *Journal of Hazardous Materials* 163 (2–3): 600–606. https://doi.org/10.1016/j.jhazmat.2008.07.008.

Zheng, Y., Shen, Z., Ma, S. et al. (2009b). A novel approach to recycling of glass fibers from nonmetal materials of waste printed circuit boards. *Journal of Hazardous Materials* 170 (2–3): 978–982.

Zhou, X., Guo, J., Lin, K. et al. (2013). Leaching characteristics of heavy metals and brominated flame retardants from waste printed circuit boards. *Journal of Hazardous Materials* 246–247: 96–102.

Zhu, X., Nie, C., Zhang, H. et al. (2019). Recovery of metals in waste printed circuit boards by flotation technology with soap collector prepared by waste oil through saponification. *Waste Management* 89: 21–26. https://doi.org/10.1016/J.WASMAN.2019.03.061.

11

Life Cycle Assessment of e-Waste – Waste Cellphone Recycling

Pengwei He[1,2]*, Haibo Feng*[2]*, Gyan Chhipi-Shrestha*[2]*, Kasun Hewage*[2]*, and Rehan Sadiq*[2]

[1]*Central South University, School of Business, 932 Lushan S Rd, Yuelu District, Changsha, Hunan 410083, People's Republic of China*
[2]*University of British Columbia, School of Engineering, 3333 University Way, Okanagan, Kelowna, BC V1V 1V7, Canada*

11.1 Introduction

Life cycle assessment (LCA) is an international standardized methodological framework defined in the ISO 14040 series, mainly consisting of four phases: (i) the definition of goal and scope, (ii) life cycle inventory (LCI), (iii) life cycle impact assessment (LCIA), and (iv) life cycle interpretation (Ismail and Hanafiah 2019). Generally, LCA is developed to evaluate the environmental impact of products and the entire process in their life cycle. Typically, LCA is used to assess products' potential environmental impact, from the extraction of raw materials to production, the use phase of the products, and until their final disposal (Luo et al. 2018). However, LCA also has many other application aspects. One obvious example is that it has been utilized to evaluate a specific product's life cycle. Some LCA studies have focused on a specific product's production stage and during its end-of-life stage, while other LCA studies merely focused on waste management (Ismail and Hanafiah 2019).

LCA is an ideal tool to evaluate waste management's environmental impact and investigate various waste management strategies (Finnveden et al. 2007). Over 200 studies of LCA have been applied in waste management globally, including studies of LCA focused on waste electrical and electronic equipment (WEEE) management (Laurent et al. 2014a,b). In terms of LCA on waste management and certain products, obvious progress has been made in recent decades. According to a previous review by Laurent et al., only seven involved WEEE among the 222 studies published during 1995–2012 on solid waste management systems (Laurent et al. 2014a,b). LCA studies focused on consumer electronics are more common. For example, Andrae and Andersen indicated that by the end of 2010, there were five studies related to mobile phones, five studies related to television, five studies focused on laptop computers, and nine studies focused on desktop computers (Andrae and Andersen 2010). However, based on recent review research, Ismail et al. indicated that the number

Electronic Waste: Recycling and Reprocessing for a Sustainable Future, First Edition.
Edited by Maria E. Holuszko, Amit Kumar, and Denise C. R. Espinosa.

of WEEE management LCA studies had skyrocketed recently, among the 61 LCA studies in WEEE management. In conclusion, roughly three major research areas of LCA studies in WEEE management were identified, with various research scopes and WEEE types used as research subjects. Additionally, some of the recent studies combined LCA method with other methodologies to cope with evaluations in other aspects, such as economics and management (Ismail and Hanafiah 2019).

In this chapter, we shed light on the practical application of LCA for WEEE evaluation, and the context is organized as follows: Section 11.2 first presents some detailed background information related to the origin, development, and application of LCA and then introduces the literature review of LCA on WEEE. Section 11.3 presents an LCA case study of waste cellphones in China, since cellphones have a relatively short lifespan and are the most commonly used electronic products and are growing at an unprecedented rate among all WEEEs.

11.2 Background

11.2.1 Theory of Life Cycle Assessment

LCA is an instrumental methodology to calculate the environmental impacts of goods and services from "cradle to grave" (Hellweg and i Canals 2014). LCA is an internationally standardized methodology to systematically evaluate the environmental performance of a product or process from its origin to the final disposal. LCA can help decision makers better reach their environmental product or service goals through its holistic perspective in quantifying environmental impacts, which has been demonstrated to provide valuable recommendations to identify appropriate solutions for managing solid waste (Hu et al. 2020). To date, LCA has been a very popular analysis tool used in waste management to provide identifying strategies that minimize input–output burdens of products and services on ecosystems, human health, or natural resources.

The concept of LCA emerged in the 1960s, and Coca-Cola was its first user to investigate the influence by replacing all the glass bottles with plastic bottles in the 1970s (Bauman and Tillman 2004). Since then, global scientific awareness of business improvement has called for the development of LCA methodology, and its application in environmental science fields has received much more attention since the 1990s. The current ISO standards 14040 and 14044 describe a general methodology without giving the definite concept name to illustrate this environmental LCA application. Therefore, different names have been used to introduce this concept, such as resource and environment profile analysis (USA), eco-balancing (Germany, Switzerland, Austria, and Japan), environmental profiling and cradle-to-grave assessment (Roy et al. 2009). To spread out the understanding of the complex concept of LCA, The Society of Environmental Toxicology and Chemistry (SETAC) and the US Environmental Protection Agency (USEPA) sponsored a few projects to promote the development of LCI analysis and impact assessment in the 1990s. SETAC Europe and other international organizations, such as the International Organization for Standardization (ISO) and global LCA practitioners, also

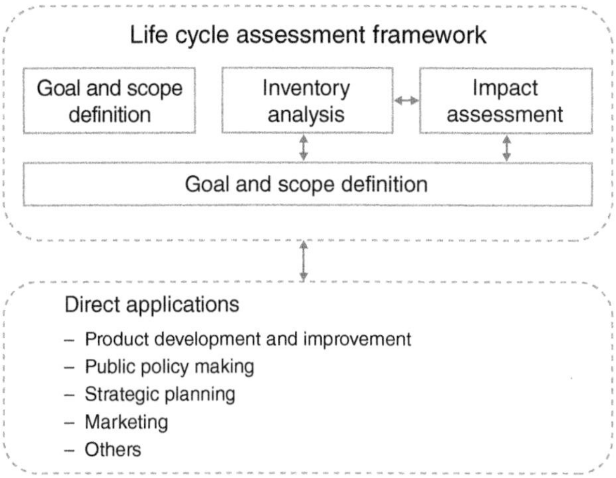

Figure 11.1 Stages of an LCA.

undertook similar efforts (Roy et al. 2009). Consensus was achieved for an overall LCA framework and a mature inventory development method, which was rapidly involved into a well-known tool for individuals, industries, and policymakers in the environmental science field (ISO 1997). Figure 11.1 shows the stages of an LCA (ISO 2006). The purpose of an LCA can be (i) comparison of alternative products, processes, or services; (ii) comparison of alternative life cycles for a certain product or service; and (iii) identification of promising parts of the life cycle to obtain the greatest improvements (environmental hotspots). There are four crucial steps in conducting a complete LCA assessment: (i) goal definition and scoping; (ii) LCI analysis; (iii) impact assessment; and (iv) interpretation (Roy et al. 2009).

The first step is to set up the goal and scope, which is probably the most important component. The whole LCA study is conducted based on the statement defined in this stage, including the purpose of this study, system boundaries, functional units (FUs) and assumptions, etc. A general input and output flow diagram is commonly used for the scheme illustration within the system boundaries. FUs, usually defined by the mass of the product under study, are used to provide a reference unit for the inventory data to be normalized (Roy et al. 2009; Feng et al. 2020).

LCI analysis is the most intensive step. Due to the complexity of data collection, LCI analysis is known as the most time-consuming and work-intensive step in an LCA. However, if the customers and suppliers are supportive and useful databases are available, the data collection process could be much easier. There are existing databases in LCA software, which usually contains environmental information, including transport, raw materials extraction, material processing, production of used products such as plastic and cardboard, material disposals, etc. For general data such as electricity, coal, or packaging production, the database can be used directly for processes. For product-specific data, site-specific data are required for different processes. In each process, the data should include all the inputs and outputs. Inputs include raw materials, water, energy (renewable and nonrenewable), etc., while outputs include the products and coproducts and emissions to air, water and soil, as well as solid waste generation (Roy et al. 2009; Feng and Hewage 2014).

Impact assessment is the transition step, which aims to understand and evaluate the environmental impacts served by the inventory analysis based on the study's goal and scope framework. In this phase, the inventory results are assigned to indicate the impact of various expected types on the environment. LCA impact assessment usually comprises four elements: classification, characterization, normalization, and valuation. Classification means assigning and initially aggregating LCI data into common impact groups. Characterization is the process of evaluating the magnitude of each inventory flow's potential impacts on its corresponding environmental impacts. For example, the potential impacts of methane and carbon dioxide on global warming. Normalization is the process to translate the potential impacts into a way that can be compared, and valuation is the process to measure the relevant importance of environmental impacts by assigning weights, which allows the results to be aggregated further compared with other products (Roy et al. 2009).

Regarding the interpretation, it is the final step of conducting an LCA. An LCA interpretation aims to draw conclusions that can support a decision based on the LCA results. In LCA interpretation, the LCI and impact assessment results are discussed based on the initial goal and scope setup, and the significant environmental impacts are highlighted for conclusions and recommendations. LCA interpretation is a systematic approach to identify, quantify, and evaluate the information based on the LCI and LCIA results, and communicated them effectively. The LCA interpretation might also lead to quantitative or qualitative improvement strategies, such as process and active design, consumer use and waste management, or changes in product. (Roy et al. 2009).

11.3 LCA Studies on WEEE

Regarding LCA studies on WEEE management, relatively little attention has been given to the management and recycling of WEEE from the LCA community. Laurent et al. (2014a,b) indicated that of the 222 studies published between 1995 and 2012 on solid waste management systems, only seven studies focused on WEEE management. The LCA on WEEE deserves more attention. In this study, we introduce three main research areas: LCA application to WEEE management strategies, LCA application to WEEE management systems, and LCA application to the potential of WEEE management and recycling.

11.3.1 Applications on WEEE Management Strategy

In China, Niu et al. (2012) compared the life cycle environmental impacts of three types of cathode-ray tubes (CRTs) treatment methods: incineration, mechanical dismantling, and manual dismantling. Lu et al. (2014) used sustainability-LCA to evaluate the environmental impacts of two recycling strategies for mobile phone components reuse and recovery. Wang et al. (2014) studied the liquid crystal recovery from LCD panel supercritical and distillation methods and used LCA to analyze this treatment method's environment impact. Song et al. (2018) analyzed the current CRT TV recycling practice based on the Chinese WEEE Directive and the proposed recycling strategy that includes extended treatment to recover lead from CRT TV, and compared the environmental impacts of two strategies through LCA studies. Yao et al. (2018) applied LCA method to predict the long-term environmental impacts from mobile phone recycling and recommend the optimize mobile phones.

Similar studies have been conducted in Italy. For example, Andreola et al. (2007) explored the CRT glass-based ceramic glaze production vs. standard ceramic glaze production, and assessed the environmental impact of CRT glass recycling strategy through LCA. Compagno et al. (2014) studied the current CRT recycling practices as well as the proposed recycling strategy with metallic lead recovery, and compared their environmental impact differences through standard LCA studies. Amato et al. (2017) also use LCA to measure the environmental impacts of four LCD monitoring treatments: landfill, incineration, traditional recycling, and innovative recycling with indium recovery.

In the UK, Zink et al. (2014) applied smart phones and traditional refurbish smartphones with battery power and solar power into parking meters and evaluated their environmental performance separately through LCA. Alston and Arnold (2011) also analyzed the environmental impacts of plastic mixtures from WEEE (i.e. WEEE plastic) under different recycling rates and calculated the environmental impacts through LCA. The environmental impacts of the treatment systems were assessed as well.

In Japan, Dodbiba et al. (2008) analyzed the recycling strategies of plastic residue from TV sets, and compared the environmental impacts between energy recovery strategy through thermal recycling and material recovery through mechanical recycling. Dodbiba et al. (2012) also explored two liberation methods in a pretreatment system to increase the recovery rate of indium from LCD monitoring and followed with an LCA study to understand the environmental performance.

11.3.2 Applications on WEEE Management System

In China, Song et al. (2013) examined the environmental impact of WEEE recycling systems for various WEEE products. Hong et al. (2015) analyzed formal and informal recycling systems for various WEEE products and used the LCA method to examine each system's environmental performance. Xue et al. (2015) analyzed treatment systems' environmental performance for wired printed boards (WPBs). Xiao et al. (2016) explored a variety of WEEE transportation scenarios (labeled S0, S1, S3, S4, and S5) for WEEE management and recycling systems of refrigerators (RFs) in China, and applied environmental assessment to understand if recycling refrigerators could balance with the reclamation processes in terms of emissions.

In Italy, Rocchetti et al. (2013) examined the environmental impact of treatment systems to recover various materials from four types of WEEE residues (i.e. residue from WEEE): fluorescent lamp, CRT, Li-ion accumulator, and PCB. Biganzoli et al. (2015) compared the environmental performance of five WEEE categories by Italian regulations (i.e. R1–R5) in terms of WEEE recycling system. Iannicelli-Zubiani et al. (2017) examined the environmental impact of a treatment system for printed circuit boards (PCBs).

In Belgium, Belboom et al. (2011) established a recycling system for refrigerators and freezers, and compared the environmental impact changes before and after the recycling system. Van Eygen et al. (2016) explored the recycling systems for laptops and desktop computers and calculated each system's environmental emissions. Tran et al. (2018) conducted an LCA study of a treatment system for batteries from mixed waste, and further merged the LCA method with criticality-based impact assessment (CIAM) method to evaluate the treatment system.

In Brazil, Foelster et al. (2016) compared the environmental impacts of refrigerators recycling systems under different approaches (i.e. informal recycling systems). Campolina et al. (2017) studied a WEEE treatment system for WEEE plastics to produce recycled high-impact polystyrenes (HIPS) pallets and acrylonitrile–butadiene–styrene (ABS) and compared their environmental impacts.

11.3.3 Applications on Hazardous Potential of WEEE Management and Recycling

In China, Song et al. (2015) examined the potential impact on PCB and CRT recycling's environmental and human health at recycling plants, where LCA methodology was utilized for the evaluation combined with noise assessment and heavy metal risk assessment. In the United States, Lim et al. (2011) combined the LCA method with chemical analysis and hazard assessment models to assess the potential environmental and human health of nine types of light-emitting diodes (LEDs) under two impact categories: resource depletion and toxicity. Hibbert and Ogunseitan (2014) also combined LCA method with chemical analysis and analyzed the potential human health and environmental impacts of ashes from incinerated mobile phones with ecotoxicity impact categories.

In conclusion, LCA is a very useful method and an important decision-support tool to assess a specific product or process's environmental impacts from its origin to the final disposal. LCA studies have been widely utilized for WEEE management, especially for the recycling of waste electronic products. In Section 11.4 of this chapter, we introduce a case study of cellphones in the Chinese scenario, focusing on LCAs of recycling different types of waste cellphones and common metals contained in them. The purpose of this case study is to illustrate how LCA methodology can elaborate a particular electronic product and assess the environmental impacts of the product from "cradle to grave."

11.4 Case Study

With rapid economic development and the provision of living standards, it is currently estimated that the quantity of hazardous electronic and electrical waste circulating in the world exceeds 6 kg, totaling 44.7 million tons in 2016 (Baldé et al. 2017; Awasthi et al. 2019). Meanwhile, an increasing number of natural resources have been produced, consumed, and accumulated in electronic and electrical products, generating an increasing volume of urban minerals. Among all WEEE, waste cellphones are the most commonly used and are considered to have the shortest lifespan. Therefore, as a specific small electronic and electrical product, cellphones will become an important focus in LCA studies. More importantly, regarding cellphones, the Chinese scenario cannot be neglected. Since 2004, with the acceleration of industrialization and urbanization, China has become the world's largest mobile phone producer and mobile phone consumer. China's mobile phone subscribers account for more than 1.3 billion of the world's 7.08 billion mobile phone users (ITU 2015).

In this case study, the recycling of waste mobile phones is considered. Mobile phones, the smallest electronic products, have many precious materials in addition to plastic, but they are in very small quantities. The use of recycled materials will definitely prevent the manufacturing of those materials, but the recycling starting from the collection, transportation, and recycling process of these small phones may have higher environmental impacts due to the consumption of large amounts of energy and materials, negating the environmental benefits of the recycled materials. Therefore, it is necessary to conduct LCA research on mobile phone use to evaluate recycling's overall benefits. Compliance with ISO 14040 standard is necessary to perform LCA on two types of used mobile phones: feature phones and smartphones. The environmental impacts from waste cellphone collection, transportation, and dismantling to waste cellphone component recycling/disposal were calculated. The negative environmental impacts due to the metals recycled from the waste cellphone circuit board were also estimated. The following Sections 11.4.1–11.4.3 introduce detailed LCA development procedures, such as target and scope definitions, LCI, LCIA methods, and result interpretation.

11.4.1 Goal and Scope Definition

This LCA aimed to quantify the potential environmental impacts of CO_2 emissions from the recycling process of two different types of waste cellphones and common metals in them. A wide range of models and brands of waste cellphones discarded by Chinese consumers were selected. This is because components or materials are varied based on the differentiation of cellphones. The quantified life cycle environmental impacts can provide industry managers with an overall view of the environmental impact of different types of waste mobile phone recycling methods to determine appropriate management methods that have little impact on the environment throughout the life cycle.

11.4.1.1 Functional Unit

The FU was defined as one unit of waste feature phone and one unit of waste smartphone. According to the literature reviews (Tan et al. 2017; Singh et al. 2018), the average composition of an average waste feature phone was as follows: average

Table 11.1 The average composition of feature phones and smartphones in China.

	Product categories	
	Feature phone (g/unit)	**Smartphone (g/unit)**
Average weight	37.13–67.44 g	51.86–112.9 g
Plastic	17.45–31.70 g	19.19–41.77 g
Screen	3.71–6.74 g	7.26–15.81 g
Battery	16–42 g	19–48 g
PCBs, including metallic wires	6.47–19.87 g	9.46–16.68 g

weight of 37.13–67.44 g; plastic, 47% of the average weight; screen, 10% of the average weight; battery, 16–42 g; PCBs, including metallic wires, 6.47–19.87 g. The average composition of an average waste smartphone was as follows: average weight of 51.86–112.9 g; plastic, 37% of the average weight; screen, 14% of the average weight; battery, 19–48 g; PCBs, including metallic wires, 9.46–16.68 g. Detailed information is shown in Table 11.1.

11.4.1.2 System Boundary

The second step of this LCA was to set up the system boundaries. The recycling process starts at the end of cellphone life. The whole process commences with a waste cellphone being sent to the mechanical dismantling plant after formal/informal collection and ends with the final disposal process. Formal collection means an officially certified e-waste recycling approach regulated by governments, informal collection means peddlers or scavengers recycling e-waste products without regulations; they are usually small and disadvantaged, with a high labor intensity (Chi et al. 2011; Gu et al. 2016). In most LCA studies on e-waste management, due to the relative availability of data or the relative impact on the entire process, the environmental impact of collection practices is not considered. In this study, however, the LCA of collection practices was taken into account based on the differentiation of waste cellphones. Additionally, LCA's time period was not considered because the purpose of this case study is to quantify the total environmental impacts caused by recycling two different types of waste cellphones rather than estimating the entire life cycle impacts within a specific temporal range.

11.4.2 Life Cycle Inventory

The data required for each type of cellphone as inputs mainly come from assembly plant and literature reviews. In addition, some inventory data for waste cellphone collection and waste cellphone dismantling were collected from recycling industry reports or recycling enterprises' official websites. The following sections from 11.4.2.1–11.4.2.7 describe the checklist established for all methods, and the detailed material flows of the two types of cellphone approaches can be found in the supporting file.

11.4.2.1 Formal Collection

The data for estimating the environmental impacts of the formal collection were obtained from the Ecoinvent 3.3 database and local industry reports. In the Chinese scenario, the formal collection process of waste cellphones has three main steps: (i) Customer to collection station; (ii) Collection station to secondhand market; (iii) Secondhand market to dismantling center. Because the collection station is near residential areas, the first step's collection distance can be negligible. The second step is the collection distance from various collection stations to the secondhand market. Taking Beijing city as an example, the major transportation method would be light commercial vehicles, and the estimated collection distance for the second step would be approximately 7–12 km (Yan 2018; Chen 2019). The third step is the collection distance from the secondhand market to the dismantling center. Taking Beijing city as an example, the majority of waste electronic products from all over the country would be transported to a secondhand market in Shenzhen (Chen 2019). Therefore, the major transportation method would be lorry, and the estimated collection distance for the third step would be approximately 2175–2239 km.

11.4.2.2 Informal Collection

The data for estimating the environmental impacts of informal collection were also obtained from the Ecoinvent 3.3 database and local industry reports. In the Chinese scenario, the informal collection process of waste cellphones has three main steps: (i) Customer to peddler; (ii) Peddler to secondhand market; (iii) Secondhand market to dismantling center. Because peddlers are usually near residential areas, the collection distance for the first step can be negligible. The second step is the collection distance from peddlers to the secondhand market. Taking Beijing city as an example, the major transportation method would be electric scooters, and the estimated collection distance for the second step would be approximately 1–30 km (Yan 2018; Chen 2019). The third step is the collection distance from the secondhand market to the dismantling center. Taking Beijing city as an example, the majority of waste electronic products from all over the country would be transported to a secondhand market in Shenzhen (Chen 2019). Therefore, the major transportation method would be a lorry, and the estimated collection distance for the third step would be approximately 2175–2239 km.

11.4.2.3 Mechanical Dismantling

The data for estimating the environmental impacts of waste cellphone mechanical dismantling were obtained from the Ecoinvent 3.3 database and literature reviews. In terms of the mechanical dismantling impact from waste feature phones, we utilized the unit weight information of one waste feature phone from the literature reviews (Tan et al. 2017; Singh et al. 2018) as well as the PCB shredder fraction information from the Ecoinvent 3.3 database. Regarding the mechanical dismantling impact from waste smartphones, we also utilized the unit weight information of one waste smartphone from the literature reviews (Tan et al. 2017; Singh et al. 2018) as well as the PCB shredder fraction information from the Ecoinvent 3.3 database. The LCI for estimating the environmental impacts of mechanical dismantling is related to machinery emissions. In this case, we did not consider the environmental impacts of labor.

11.4.2.4 Plastic Recycling

The LCI data for estimating plastic recycling's environmental impacts were obtained from the Ecoinvent 3.3 database and literature reviews. In terms of plastic recycling impact from waste feature phones, we utilized the unit weight information of one waste feature phone from the literature reviews (Tan et al. 2017; Singh et al. 2018) as well as the plastic manufacture information from the Ecoinvent 3.3 database. Regarding the plastic recycling impact from waste smartphones, we also utilized the unit weight information of one waste smartphone from the literature reviews (Tan et al. 2017; Singh et al. 2018) as well as the plastic manufacturing information from the Ecoinvent 3.3 database. Research shows that the percentage of plastic in one waste feature phone is approximately 47%, while the percentage of plastic in one waste smartphone is approximately 37% (Tan et al. 2017; Singh et al. 2018). The major emissions of plastic recycling are related to plastic manufacturing.

11.4.2.5 Screen Glass Recycling

The LCI data for estimating screen glass recycling's environmental impacts were obtained from the Ecoinvent 3.3 database and literature reviews. In terms of screen glass recycling impact from waste feature phones, we utilized the unit weight information of one waste feature phone from the literature reviews (Tan et al. 2017; Singh et al. 2018) as well as the glass manufacture information from the Ecoinvent 3.3 database. Regarding screen glass recycling impact from waste smartphones, we also utilized the unit weight information of one waste smartphone from the literature reviews (Tan et al. 2017; Singh et al. 2018) as well as the glass manufacturing information from the Ecoinvent 3.3 database. Research shows that the percentage of screen glass in one waste feature phone is approximately 10%, while the percentage of screen glass in one waste smartphone is approximately 14% (Tan et al. 2017; Singh et al. 2018). The major emissions of screen glass recycling are related to the emissions of glass manufacturing.

11.4.2.6 Battery Disposal

The LCI data for estimating the environmental impacts of battery disposal were obtained from the Ecoinvent 3.3 database and literature reviews. In terms of battery disposal impact from waste feature phones, we utilized the unit weight information of one waste feature phone from the literature reviews (Tan et al. 2017; Singh et al. 2018) as well as the battery treat in waste cell information from the Ecoinvent 3.3 database. Regarding battery disposal impact from waste smartphones, we also utilized the unit weight information of one waste smartphone from the literature reviews (Tan et al. 2017; Singh et al. 2018) as well as the battery treat in waste cell information from the Ecoinvent 3.3 database. Research shows that the weight range of batteries in one waste feature phone is 16–42 g, while the weight range of batteries in one waste smartphone is 19–48 g (Tan et al. 2017; Singh et al. 2018). The significant emissions of battery disposal are related to the emissions of battery treatment in waste cells.

11.4.2.7 Electronic Refining for Materials

The LCI data for estimating the environmental impacts of electronic refining for various materials were obtained from the Ecoinvent 3.3 database and literature reviews. Materials, only gold, silver, and cooper, were considered in this study. The unit weight information of waste cellphones came from the literature reviews (Tan et al. 2017; Singh et al. 2018), and material treatment in waste cell information of three metal materials came from the Ecoinvent 3.3 database. In addition, material production information for the three metal materials also came from the Ecoinvent 3.3 database. These are the necessary data for analyzing raw material substitution impacts. The significant emissions of battery disposal are related to the emissions of electronic refining and raw material substitution.

11.4.3 Life Cycle Impact Assessment

SimaPro 8.5.0.0™ software was used with the ReCiPe H midpoint method to calculate the global warming potential (GWP) impact. Due to the variety of cellphone data, the LCI analysis data were gathered as interval data. To reduce the data uncertainty, a Monte Carlo simulation (MCS) was applied using Excel. The LCIA results of each process gathered in LCI were analyzed using 10 000 MCS runs. The LCIA results from each MCS run were gathered and are presented in Section 11.4.4.

11.4.4 Results

We developed four distinctive scenarios to present our results. The life cycle environmental impacts generated from these scenarios are presented separately. Waste materials generated in the processing of phone parts or raw materials generated in the telephone recycling process can be reset to recycle the same type of product materials or the same amount of raw materials. Therefore, the influence of the recovery cause is calculated as negative.

11.4.4.1 Feature Phone Formal Collection Scenario

The life cycle environmental impacts of the feature phone (formal scenario) are shown in Figure 11.2. After 10 000 iterations of MCS , the results show that the total CO_2 emissions equal to or less than 0 kg CO_2e are approximately 76%. The cumulative probability of the total greenhouse gas (GHG) emissions in the feature phone (formal scenario) is presented in Figure 11.3. The corresponding impact comprises six parts: the impact from waste feature phone collection, the impact from waste feature phone dismantling, the impact from waste feature phone part recycling, the impact from battery disposal, the impact from metal electronic refining, and the substitution impact from waste feature phone metal recycling.

The results indicate that four processes, namely phone collection, phone dismantling, battery disposal, and metal electronic refining, have positive GHG emissions. In other words, these processes exert negative impacts on the environment. In particular, the metal electronic refining process has the highest impact, followed by the phone dismantling process, which has the second greatest impact. Impacts from

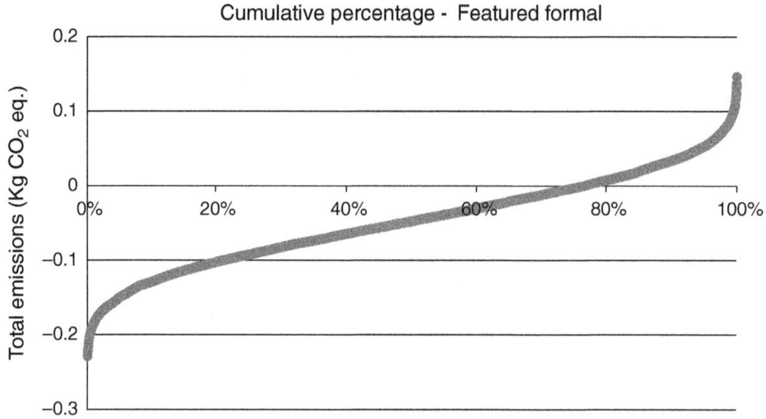

Figure 11.2 Life cycle environmental impacts of feature phones (formal scenario).

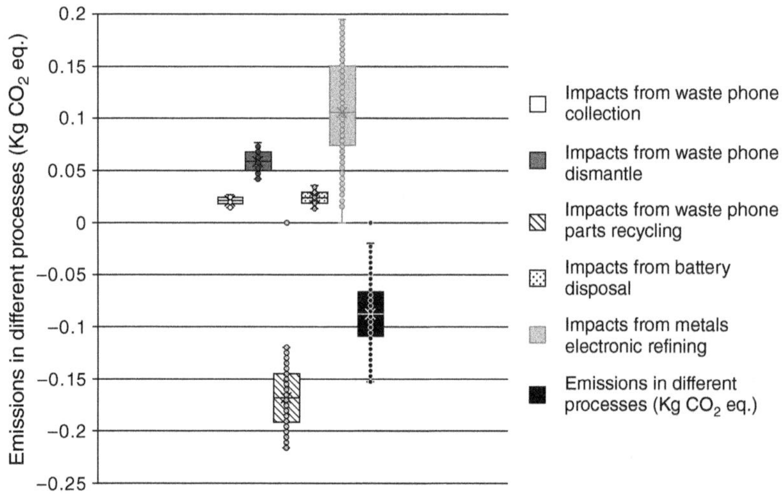

Figure 11.3 Cumulative probability of the total GHG emissions in the feature phone (formal scenario).

phone collection and battery disposal processes are minimal, between 0 and 0.05 kg CO_2e. The other two processes, namely, phone part recycling and phone metal recycling, have negative GHG emissions. In other words, they exert positive impacts on the environment. Particularly, by replacing waste plastic materials and waste screen glasses from waste feature phones, the phone part recycling process can reduce the large amount of GHG emissions, which is more than 0.2 kg CO_2e. Additionally, by substituting metals from waste feature phones, the phone metal recycling process can also reduce the amount of GHG emissions, which hits approximately 0.15 kg CO_2e. In this case, copper, gold, and silver were recycled and substituted for manufacturing.

In summary, regarding phone LCA analysis under the formal scenario, some processes show negative environmental impacts, while other processes indicate positive environmental impacts. Among these processes, the metal electronic refining process has the largest negative environmental impact, and the feature phone, part recycling process, has the highest positive environmental impact.

11.4.4.2 Feature Phone Informal Collection Scenario

The life cycle environmental impacts of the feature phone (informal scenario) are shown in Figure 11.4. After 10 000 iterations of MCS, the results show that the total CO_2 emissions equal to or less than 0 kg CO_2 are approximately 76%. The cumulative probability of the total GHG emissions in the feature phone (informal scenario) is presented in Figure 11.5. The corresponding impact also comprises six parts: the impact from waste feature phone collection, the impact from waste

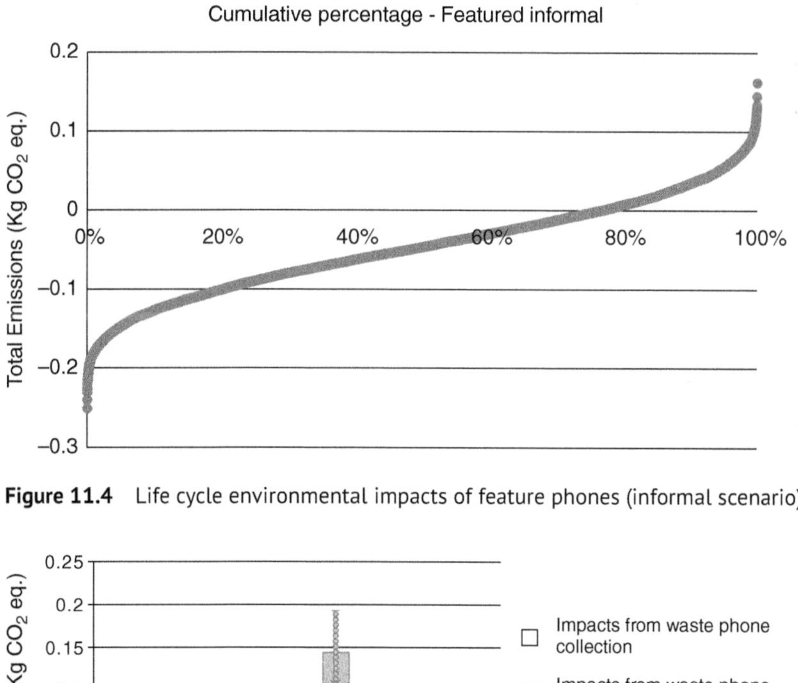

Figure 11.4 Life cycle environmental impacts of feature phones (informal scenario).

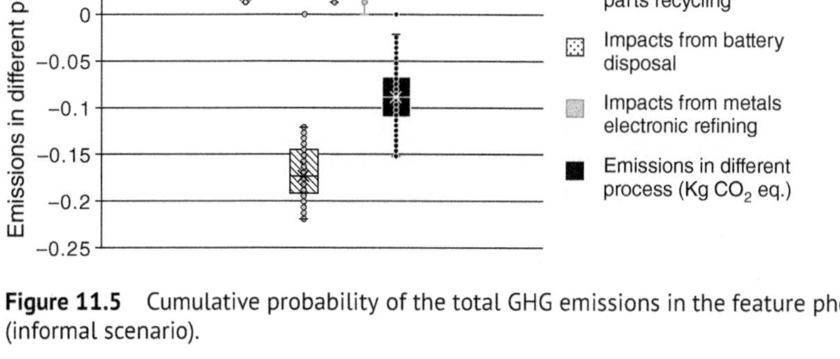

Figure 11.5 Cumulative probability of the total GHG emissions in the feature phone (informal scenario).

feature phone dismantling, the impact from waste feature phone part recycling, the impact from battery disposal, the impact from metal electronic refining, and the substitution impact from waste feature phone metal recycling.

The results indicate that four processes, namely phone collection, phone dismantling, battery disposal, and metal electronic refining, have positive GHG emissions. In other words, these processes exert negative impacts on the environment. In particular, the metal electronic refining process has the highest impact, followed by the phone dismantling process, which has the second greatest impact. Impacts from phone collection and battery disposal processes are minimal, between 0 and 0.05 kg CO_2e. The other two processes, namely phone part recycling and phone metal recycling, have negative GHG emissions. In other words, they exert positive impacts on the environment. Particularly, by replacing waste plastic materials and waste screen glasses from waste feature phones, the phone part recycling process can reduce the largest amount of GHG emissions, which is more than 0.2 kg CO_2e. Additionally, by substituting metals from waste feature phones, the phone metal recycling process can also reduce the amount of GHG emissions, which hits approximately 0.15 kg CO_2e. In this case, copper, gold, and silver are recycled and substituted.

In summary, LCA analysis of feature phones under informal scenario is very similar to that under formal scenario, except for the impact generated from the waste feature phone collection process, which shows a slight difference between the formal collection process and the informal collection process. It is also noted that some processes have shown negative environmental impacts, while other processes indicate positive environmental impacts. Among these processes, the metal electronic refining process has the highest negative environmental impact, and the feature phone part recycling process has the largest positive environmental impact.

11.4.4.3 Smartphone Formal Collection Scenario

The environmental impact of smartphones on the life cycle (formal scenario) is shown in Figure 11.6. After 10 000 iterations of MCS, the results show that the total CO_2 emissions equal to or less than 0 kg CO_2e are 70%. The cumulative probability of the total GHG emissions on smartphones (formal scenario) is presented in Figure 11.7. The corresponding impact comprises six parts: the impact from waste smartphone collection, the impact from waste smartphone dismantling, the impact from waste smartphone part recycling, the impact from battery disposal, the impact from metal electronic refining, and the substitution impact from waste smartphone metal recycling.

The results indicate that four processes, namely phone collection, phone dismantling, battery disposal and metal electronic refining, have positive GHG emissions. In other words, these mentioned processes exert negative impacts on the environment. In particular, the metal electronic refining process has the largest impact, followed by the phone dismantling process, which has the second greatest impact. The impacts from phone collection and battery disposal processes are similar, and both are very minimal, set between 0.03 and 0.04 kg CO_2e. The other two processes, namely phone part recycling and phone metal recycling, have negative GHG emissions. In other words, they exert positive impacts on the environment.

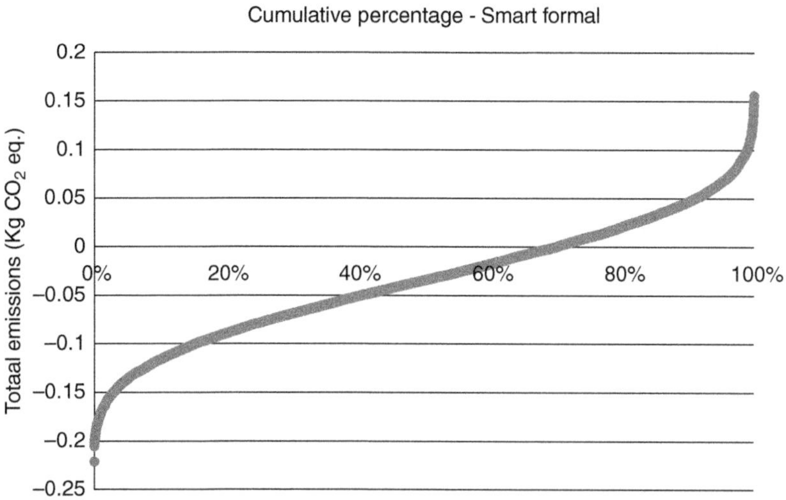

Figure 11.6 Life cycle environmental impacts of smartphones (formal scenario).

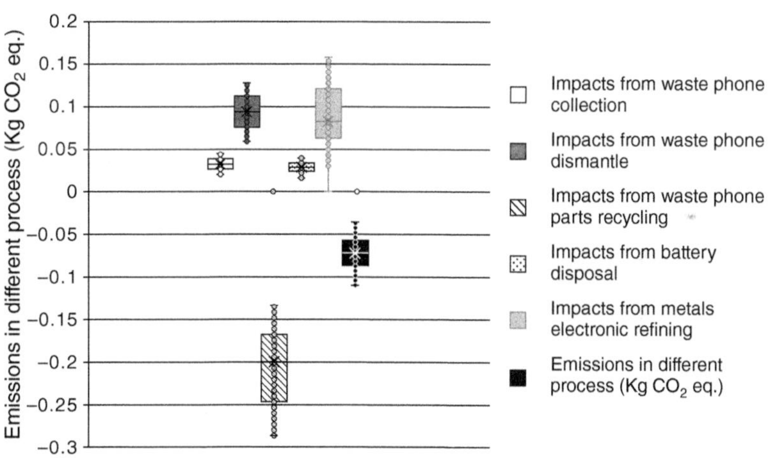

Figure 11.7 Cumulative probability of the total GHG emissions on smartphones (formal scenario).

Particularly, by replacing waste plastic materials and waste screen glasses from waste smartphones, the phone part recycling process can reduce the largest amount of GHG emissions, which is more than 0.2 kg CO_2e. Additionally, by substituting metals from waste smartphones, the phone metal recycling process can also reduce GHG emissions, which hits approximately 0.15 kg CO_2e. In this case, copper, gold, and silver were recycled and substituted.

In summary, regarding smartphone LCA analysis under formal scenarios, some processes have shown negative environmental impacts, while other processes indicate positive environmental impacts. Among these processes, the metal electronic refining process has the largest negative environmental impact, and the smartphone part recycling process has the largest positive environmental impact.

11.4.4.4 Smartphone Informal Collection Scenario

The impact of a smartphone's life cycle on the environment (informal situation) is shown in Figure 11.8. After 10 000 iterations of MCS, the results show that the probability of total CO_2 emissions equal to or less than 0 kg CO_2e is approximately 70%. The cumulative probability of the total GHG emissions in smartphones (informal scenario) is presented in Figure 11.9. The corresponding impact comprises six parts: the impact from waste smartphone collection, the impact from waste smartphone dismantling, the impact from waste smartphone part recycling, the impact from battery disposal, the impact from metal electronic refining, and the substitution impact from waste smartphone metal recycling.

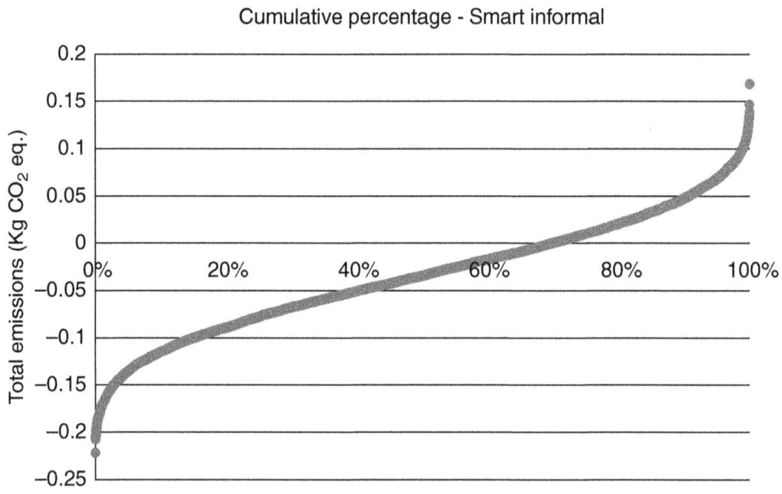

Figure 11.8 Life cycle environmental impacts of smartphones (informal scenario).

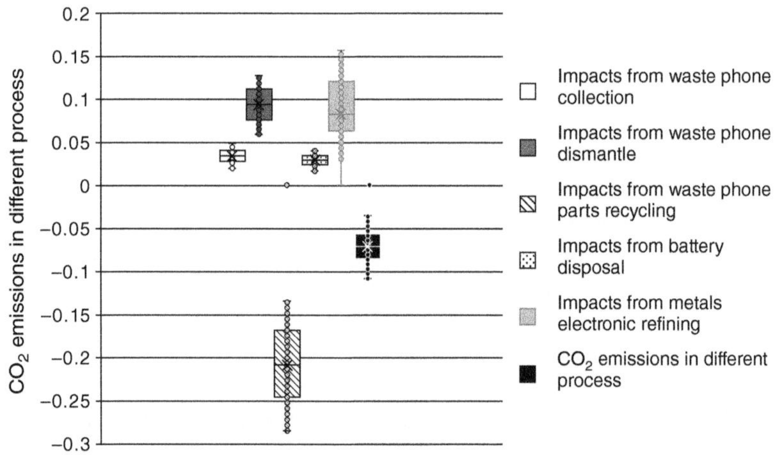

Figure 11.9 Cumulative probability of the total GHG emissions in smartphones (informal scenario).

The results indicate that four processes, namely phone collection, phone dismantling, battery disposal, and metal electronic refining, have positive GHG emissions. In other words, these mentioned processes exert negative impacts on the environment. In particular, the metal electronic refining process has the highest impact, followed by the phone dismantling process, which has the second greatest impact. The impacts from phone collection and battery disposal processes are similar, and both are very minimal, set between 0.03 and 0.04 kg CO_2e. The other two processes, namely phone part recycling and phone metal recycling, have negative GHG emissions. In other words, they exert positive impacts on the environment. Particularly, by replacing waste plastic materials and waste screen glasses from waste smartphones, the phone part recycling process can reduce the largest amount of GHG emissions, which is more than 0.2 kg CO_2e. Additionally, by substituting metals from waste smartphones, the phone metal recycling process can also reduce the amount of GHG emissions, which hits approximately 0.15 kg CO_2e. In this case, copper, gold, and silver were recycled and substituted.

In summary, LCA analysis of smartphones under informal scenario is very similar to that under formal scenarios, except for the impact generated from the waste smartphone collection process, which shows a slight difference between the formal collection process and the informal collection process. It is also noted that some processes have shown negative environmental impacts, while other processes indicate positive environmental impacts. Among these processes, the metal electronic refining process always has the largest negative environmental impact, and the smartphone part recycling process always has the highest positive environmental impact.

11.4.5 Discussion

Based on the LCA results, it can be repredicted that the metal electronic refining process always has the highest negative environmental impact in all four scenarios, and the part recycling process always has the greatest positive environmental impact. In addition, the phone dismantling process plays a more important role in smartphone scenarios than in feature phone scenarios since their life cycle environmental impacts are much higher. However, the phone metal recycling process shows a different context; it has a much higher impact in the feature phone scenario than in the smartphone scenario regarding the larger positive environmental impacts. Regarding phone collection and battery disposal processes, relatively similar impact results can be traced in all four scenarios: just between 0.1 and 0.5 kg CO_2e.

The reason for dividing the phone collection process into formal collection and informal collection is that the informal collection approach is actually widely accepted in the Chinese scenario. As the mobile phone weight, metal, and other material content information varies from feature phone and smartphone, we also intend to divide mobile phone types into feature phone and smartphone when we are conducting LCA analysis. Notably, there was only a very slight difference between the formal scenario and the informal scenario, and the impact from the phone collection process in all scenarios was relatively small among all processes. However, in regard to the impact from different mobile phone types, owing to the relatively higher metal content, the metal electronic refining process in waste feature phones actually poses a larger environmental impact than that in waste smartphones, which will significantly increase GHG emissions. Additionally, due to a higher metal content, the substitution impact from waste feature phone metal recycling is larger than that from waste smartphone metal recycling. However, because of the phone weight differences, the impact of waste smartphone part recycling is higher than that of waste feature phone part recycling, significantly reducing GHG emissions. Therefore, future studies should focus on those processes that have a greater impact on the environment. Examples include metal electronic recycling processes and mobile phone dismantling processes.

A great variety of valuable metals are stored in waste mobile phones, including some high-tech metals; in particular, the main high-tech minerals contained in waste feature phones are palladium and cobalt, while waste smartphones mainly contain cobalt, praseodymium, palladium, beryllium, neodymium, antimony, and platinum (Cucchiella et al. 2015; He et al. 2018; He et al. 2021). In this case study, we only considered three common metals, namely copper, gold, and silver. Copper, as an important common metal, has a very high content in waste mobile phones. That is, the impact of refining copper from waste mobile phones would be the greatest among all other metals. In other words, refining copper would cause the most significant environmental impact, which will dramatically increase GHG emissions. For precious metals such as gold or silver, the relative impact of refining them would be much less significant, but they still deserve attention. Future LCA studies should expand the category of metals in waste mobile phones. Expanding the category from precious metals to high-tech metals certainly comes with many challenges. For example, refining one particular type of high-tech metal may cause the loss of another type of high-tech metal. Therefore, it is important to develop a scientific standard on how to refine high-tech metals by sequence, with minimal influence on each other. Additionally, future studies should also combine life cycle cost (LCC) analysis with the LCA method in regard to various high-tech materials in waste mobile phones. From an economic perspective, our previous study indicated that the LCCs of extracting high-tech minerals (HTMs) from one waste feature phone are US$ 6.035 for 1 g of cobalt and US$ 0.014 for 1 g of palladium, while the LCCs of extracting 1 g of cobalt, palladium, antimony, beryllium, neodymium, praseodymium, and platinum from one waste smartphone are US$ 10.106, US$ 0.024, US$ 0.135, US$ 0.005, US$ 0.08, US$ 0.016, and US$ 0.006, respectively (He et al. 2020). Therefore, it is also crucial to investigate the LCAs of recycling high-tech materials from waste mobile phones from an environmental perspective. By combining the LCC and LCA methods, primary data would be available for decision makers or future recycling industry development. In China's case, the "Regulations on the Recycling of Waste Electrical and Electronic Products" (also known as the China WEEE Directive) was issued in 2009. However, only five main product categories were covered, namely freezer, washer, television, air conditioner, and computer. Until 2016, mobile phones were still included in the latest WEEE management directory (WEEE Catalog 2014). This was the first time that proper recycling management of mobile phones has been required and is subject to legal supervision. Even if the law stipulates that waste mobile phones can only be disposed of by recycling companies officially certified by the government, mobile phone manufacturers and importers must also pay fees to the central WEEE management fund. However, detailed guidelines on how to execute this mandated requirement are still ambiguous. As the results of this case study show that the metal electronic refining process always has the greatest negative environmental impact, it is indispensable to take necessary action to tackle this issue. For example, fund subsidy policies for some major electronic products can potentially be utilized in future mobile phone scenarios. Currently, formal government-certified recycling companies lack the motivation to collect and recycle waste mobile phones, both economically and efficiently. If a detailed mobile phone fund subsidy policy is released, such formal companies would be strongly motivated and begin to collect and recycle waste mobile phones. The metal electronic refining process's negative environmental impact is likely to be diminished, as these formal companies usually follow strict professional operational procedures.

11.5 Conclusion

The life cycle environmental impacts of the recycling process of two different types of waste cellphones and common metals contained in them were assessed by following the standardized ISO 14040 procedure. Four scenarios were possible: feature phone formal scenario, feature phone informal scenario, smartphone formal scenario, and smartphone informal scenario.

Specifically, in the feature phone formal scenario, impacts from phone collection and battery disposal processes are minimal, which is between 0 and 0.05 kg CO_2e. By replacing waste plastic materials and waste screen glasses from waste feature phones, the phone part recycling process can reduce the large amount of GHG emissions, which is more than 0.2 kg CO_2e. Additionally, by substituting metals from waste feature phones, the phone metal recycling process can also reduce the amount of GHG emissions, which hits approximately 0.15 kg CO_2e.

In the feature phone informal scenario, impacts from phone collection and battery disposal processes are minimal, which is between 0 and 0.05 kg CO_2e, by replacing waste plastic materials and waste screen glasses from waste feature phones, phone parts recycling process can reduce the biggest amount of GHG emissions, which is more than 0.2 kg CO_2e. Additionally, by substituting metals from waste feature phones, the phone metal recycling process can also reduce the amount of GHG emissions, which hits approximately 0.15 kg CO_2e.

In the smartphone formal scenario, impacts from phone collection and battery disposal processes are similar, both are very minimal, which set out between 0.03 and 0.04 kg CO_2e, by replacing waste plastic materials and waste screen glasses from waste smartphones, phone parts recycling process can reduce the biggest amount of GHG emissions, which is more than 0.2 kg CO_2e. Additionally, by substituting metals from waste smartphones, the phone metal recycling process can also reduce the amount of GHG emissions, which hits approximately 0.15 kg CO_2e.

In the smartphone informal scenario, impacts from phone collection and battery disposal processes are similar, both are very minimal, which set out between 0.03 and 0.04 kg CO_2e, by replacing waste plastic materials and waste screen glasses from waste smartphones, phone parts recycling process can reduce the largest amount of GHG emissions, which is more than 0.2 kg CO_2e. Additionally, by substituting metals from waste smartphones, the phone metal recycling process can also reduce the amount of GHG emissions, which hits approximately 0.15 kg CO_2e.

The results in all four scenarios show that there is over 70% chance that the recycling of waste cellphones has zero or negative emissions to the environment. The metal electronic refining process always has the largest negative environmental impact, while the part recycling process always has the largest positive environmental impact. The phone dismantling process plays a more important role in smartphone scenarios than in feature phone scenarios due to their much higher life cycle environmental impacts. Besides, phone metal recycling has a much higher impact in the feature phone scenario than in the smartphone scenario regarding the greater positive environmental impacts. Finally, phone collection and battery disposal processes and relatively similar impact results can be traced in all four scenarios. The results add new knowledge to LCA studies regarding the most commonly used phone products and common metals they contain.

References

Alston, S.M. and Arnold, J.C. (2011). Environmental impact of pyrolysis of mixed WEEE plastics part 2: life cycle assessment. *Environmental Science & Technology* 45 (21): 9386–9392.

Amato, A., Rocchetti, L., and Beolchini, F. (2017). Environmental impact assessment of different end-of-life LCD management strategies. *Waste Management* 59: 432–441.

Andrae, A.S. and Andersen, O. (2010). Life cycle assessments of consumer electronics – are they consistent? *The International Journal of Life Cycle Assessment* 15 (8): 827–836.

Andreola, F., Barbieri, L., Corradi, A. et al. (2007). Recycling of EOL CRT glass into ceramic glaze formulations and its environmental impact by LCA approach. *The International Journal of Life Cycle Assessment* 12 (6): 448–454.

Awasthi, A.K., Li, J., Koh, L., and Ogunseitan, O.A. (2019). Circular economy and electronic waste. *Nature Electronics* 2 (3): 86–89.

Baldé, C. P., Forti, V., Gray, V., Kuehr, R., & Stegmann, P. (2017). The Global E-waste Monitor–2017, United Nations University (UNU), International Telecommunication Union (ITU) & International Solid Waste Association (ISWA), Bonn/Geneva/Vienna. Electronic Version, 978–992.

Baumann, H. and Tillman, A.M. (2004). *The Hitchhiker's Guide to LCA*. Studentlitteratur.

Belboom, S., Renzoni, R., Deleu, X., Digneffe, J. M., & Leonard, A. (2011). Electrical waste management effects on environment using life cycle assessment methodology:

the fridge case study. In *SETAC EUROPE 17th LCA Case Study Symposium Sustainable Lifestyles*, Budapest, Hungary (Vol. 2).

Biganzoli, L., Falbo, A., Forte, F. et al. (2015). Mass balance and life cycle assessment of the waste electrical and electronic equipment management system implemented in Lombardia region (Italy). *Science of the Total Environment* 524: 361–375.

Campolina, J.M., Sigrist, C.S.L., de Paiva, J.M.F. et al. (2017). A study on the environmental aspects of WEEE plastic recycling in a Brazilian company. *The International Journal of Life Cycle Assessment* 22 (12): 1957–1968.

Chen Z (2019). Where do 400 million cellphones go? We filmed five months to restore the whole process (In Chinese). https://www.tmtpost.com/3803094.html (accessed 23 March 2021).

Chi, X., Streicher-Porte, M., Wang, M.Y., and Reuter, M.A. (2011). Informal electronic waste recycling: a sector review with special focus on China. *Waste Management* 31 (4): 731–742.

Compagno, L., Ingrao, C., Latora, A.G., and Trapani, N. (2014). Life cycle assessment of CRT lead recovery process. *International Journal of Product Lifecycle Management* 7 (2/3): 201.

Cucchiella, F., D'Adamo, I., Koh, S.L., and Rosa, P. (2015). Recycling of WEEEs: an economic assessment of present and future e-waste streams. *Renewable and Sustainable Energy Reviews* 51: 263–272.

Dodbiba, G., Takahashi, K., Sadaki, J., and Fujita, T. (2008). The recycling of plastic wastes from discarded TV sets: comparing energy recovery with mechanical recycling in the context of life cycle assessment. *Journal of Cleaner Production* 16 (4): 458–470.

Dodbiba, G., Nagai, H., Wang, L.P. et al. (2012). Leaching of indium from obsolete liquid crystal displays: comparing grinding with electrical disintegration in context of LCA. *Waste Management* 32 (10): 1937–1944.

Feng, H. and Hewage, K. (2014). Lifecycle assessment of living walls: air purification and energy performance. *Journal of Cleaner Production* 69: 91–99.

Feng, H., Liyannage, D.R., Karunathilake, H. et al. (2020). BIM-based life cycle environmental performance assessment of single-family houses: renovation and reconstruction strategies for aging building stock in British Columbia. *Journal of Cleaner Production* 250: 119543.

Finnveden, G., Björklund, A., Moberg, Å. et al. (2007). Environmental and economic assessment methods for waste management decision-support: possibilities and limitations. *Waste Management & Research* 25 (3): 263–269.

Foelster, A.S., Andrew, S., Kroeger, L. et al. (2016). Electronics recycling as an energy efficiency measure–a life cycle assessment (LCA) study on refrigerator recycling in Brazil. *Journal of Cleaner Production* 129: 30–42.

Gu, Y., Wu, Y., Xu, M. et al. (2016). The stability and profitability of the informal WEEE collector in developing countries: a case study of China. *Resources, Conservation and Recycling* 107: 18–26.

He, P., Wang, C., and Zuo, L. (2018). The present and future availability of high-tech minerals in waste mobile phones: evidence from China. *Journal of Cleaner Production* 192: 940–949.

He, P., Feng, H., Hu, G. et al. (2020). Life cycle cost analysis for recycling high-tech minerals from waste mobile phones in China. *Journal of Cleaner Production* 251: 119498.

He, P., Hu, G., Wang, C. et al. (2021). Analyzing present and future availability of critical high-tech minerals in waste cellphones: a case study of India. *Waste Management* 119: 275–284.

Hellweg, S. and I Canals, L. (2014). Emerging approaches, challenges and opportunities in life cycle assessment. *Science* 344 (6188): 1109–1113.

Hibbert, K. and Ogunseitan, O.A. (2014). Risks of toxic ash from artisanal mining of discarded cellphones. *Journal of Hazardous Materials* 278: 1–7.

Hong, J., Shi, W., Wang, Y. et al. (2015). Life cycle assessment of electronic waste treatment. *Waste management* 38: 357–365.

Hu, G., Feng, H., He, P. et al. (2020). Comparative life-cycle assessment of traditional and emerging oily sludge treatment approaches. *Journal of Cleaner Production* 251: 119594.

Iannicelli-Zubiani, E.M., Giani, M.I., Recanati, F. et al. (2017). Environmental impacts of a hydrometallurgical process for electronic waste treatment: a life cycle assessment case study. *Journal of Cleaner Production* 140: 1204–1216.

Ismail, H. and Hanafiah, M.M. (2019). An overview of LCA application in WEEE management: current practices, progress and challenges. *Journal of Cleaner Production.*

Laurent, A., Bakas, I., Clavreul, J. et al. (2014a). Review of LCA studies of solid waste management systems–part I: lessons learned and perspectives. *Waste Management* 34 (3): 573–588.

Laurent, A., Clavreul, J., Bernstad, A. et al. (2014b). Review of LCA studies of solid waste management systems–part II: methodological guidance for a better practice. *Waste Management* 34 (3): 589–606.

Lim, S.R., Kang, D., Ogunseitan, O.A., and Schoenung, J.M. (2011). Potential environmental impacts of light-emitting diodes (LEDs): metallic resources, toxicity, and hazardous waste classification. *Environmental science & technology* 45 (1): 320–327.

Lu, B., Li, B., Wang, L. et al. (2014). Reusability based on life cycle sustainability assessment: case study on WEEE. *Procedia Cirp* 15: 473–478.

Luo, L., Yang, L., and Hanafiah, M.M. (2018). Construction of renewable energy supply chain model based on LCA. *Open Physics* 16 (1): 1118–1126.

Niu, R., Wang, Z., Song, Q., and Li, J. (2012). LCA of scrap CRT display at various scenarios of treatment. *Procedia Environmental Sciences* 16: 576–584.

Rocchetti, L., Vegliò, F., Kopacek, B., and Beolchini, F. (2013). Environmental impact assessment of hydrometallurgical processes for metal recovery from WEEE residues using a portable prototype plant. *Environmental science & technology* 47 (3): 1581–1588.

Roy, P., Nei, D., Orikasa, T. et al. (2009). A review of life cycle assessment (LCA) on some food products. *Journal of Food Engineering* 90 (1): 1–10.

Singh, N., Duan, H., Yin, F. et al. (2018). Characterizing the materials composition and recovery potential from waste mobile phones: a comparative evaluation of cellular and smart phones. *ACS Sustainable Chemistry & Engineering* 6 (10): 13016–13024.

Song, Q., Wang, Z., Li, J., and Zeng, X. (2013). The life cycle assessment of an e-waste treatment enterprise in China. *Journal of Material Cycles and Waste Management* 15 (4): 469–475.

Song, Q., Zeng, X., Li, J. et al. (2015). Environmental risk assessment of CRT and PCB workshops in a mobile e-waste recycling plant. *Environmental Science and Pollution Research* 22 (16): 12366–12373.

Song, X., Zhang, C., Yuan, W., and Yang, D. (2018). Life-cycle energy use and GHG emissions of waste television treatment system in China. *Resources, Conservation and Recycling* 128: 470–478.

Tan, Q., Dong, Q., Liu, L. et al. (2017). Potential recycling availability and capacity assessment on typical metals in waste mobile phones: a current research study in China. *Journal of Cleaner Production* 148: 509–517.

Tran, H.P., Schaubroeck, T., Swart, P. et al. (2018). Recycling portable alkaline/ZnC batteries for a circular economy: an assessment of natural resource consumption from a life cycle and criticality perspective. *Resources, Conservation and Recycling* 135: 265–278.

Van Eygen, E., De Meester, S., Tran, H.P., and Dewulf, J. (2016). Resource savings by urban mining: the case of desktop and laptop computers in Belgium. *Resources, Conservation and Recycling* 107: 53–64.

Wang, Y.L., Hu, H.J., Qi, S., and Liu, G.F. (2014). Environmental impacts assessment of liquid crystal extraction from wasted LCD panels. *Applied Mechanics and Materials* 496: 55–62. Trans Tech Publications Ltd.

WEEE Catalog (2014 edition), www.ndrc.gov.cn/zcfb/zcfbgg/201502/t20150213_664456.html (accessed 23 March 2021).

Xiao, R., Zhang, Y., and Yuan, Z. (2016). Environmental impacts of reclamation and recycling processes of refrigerators using life cycle assessment (LCA) methods. *Journal of Cleaner Production* 131: 52–59.

Xue, M., Kendall, A., Xu, Z., and Schoenung, J.M. (2015). Waste management of printed wiring boards: a life cycle assessment of the metals recycling chain from liberation through refining. *Environmental Science & Technology* 49 (2): 940–947.

Yan L (2018). Beijing: Waste Electronic Products Collection Pilot Projects (In Chinese). https://www.sohu.com/a/227221601_114731 (accessed 23 March 2021).

Yao, L., Liu, T., Chen, X. et al. (2018). An integrated method of life-cycle assessment and system dynamics for waste mobile phone management and recycling in China. *Journal of Cleaner Production* 187: 852–862.

Zink, T., Maker, F., Geyer, R. et al. (2014). Comparative life cycle assessment of smartphone reuse: repurposing vs. refurbishment. *The International Journal of Life Cycle Assessment* 19 (5): 1099–1109.

12

Biodegradability and Compostability Aspects of Organic Electronic Materials and Devices

Abdelaziz Gouda[1], Manuel Reali[1], Alexandre Masson[1], Alexandra Zvezdin[1], Nia Byway[2], Denis Rho[2], and Clara Santato[1]

[1] *Polytechnique Montreal, Département de Génie Physique, C.P. 6079, Succ. Centre-ville, Montréal Québec, H3C 3A7, Canada*
[2] *National Research Council Canada, Aquatic and Crop Resource Development, 6100 Royalmount Avenue, Montréal Québec, H4P 2R2, Canada*

12.1 Introduction

12.1.1 Technological Innovation and Waste

Technological innovation, one of the mainstays of our society, aims at a better life for all. However, as most product life cycles become shorter, technological innovations are accompanied by an ever-increasing amount of waste. Electronics waste is the world's fastest-growing solid-waste stream. According to the Global E-Waste Monitor 2020, 53.6 Mt of e-waste have been produced worldwide in 2019, and this amount is expected to reach 74.7 Mt by 2030 (Forti et al. 2020).

Since waste management is one of the primary services that cities provide at a growing cost, the mayors of the world's great cities have recognized that "bold action on waste management is key to making our urban centres cleaner, healthier, more resilient and inclusive." That is the essence of the "C40 Cities: zero waste declaration" (C40 Group 2020). With this grand objective in mind, these cities have implemented a municipal recycling strategy partially detailed in Figure 12.1. Notably, **composting** of **organic waste** is one of those environmental technologies that the municipalities intend to implement. The achievement of the goal requires the participation of socially responsible citizens and the contribution of scientists and engineers to develop eco-friendly products and efficient environmental processes.

The current trend in the material design field is toward eco-design principles implemented in the first stages of product development (Morini et al. 2019). Quantifying resource utilization and environmental impacts of all steps into material production allows for a reduced ecological footprint without diminishing technical performance. Life cycle assessment (LCA) is the most recognized methodology for the evaluation of environmental burdens.

Electronic Waste: Recycling and Reprocessing for a Sustainable Future, First Edition.
Edited by Maria E. Holuszko, Amit Kumar, and Denise C. R. Espinosa.

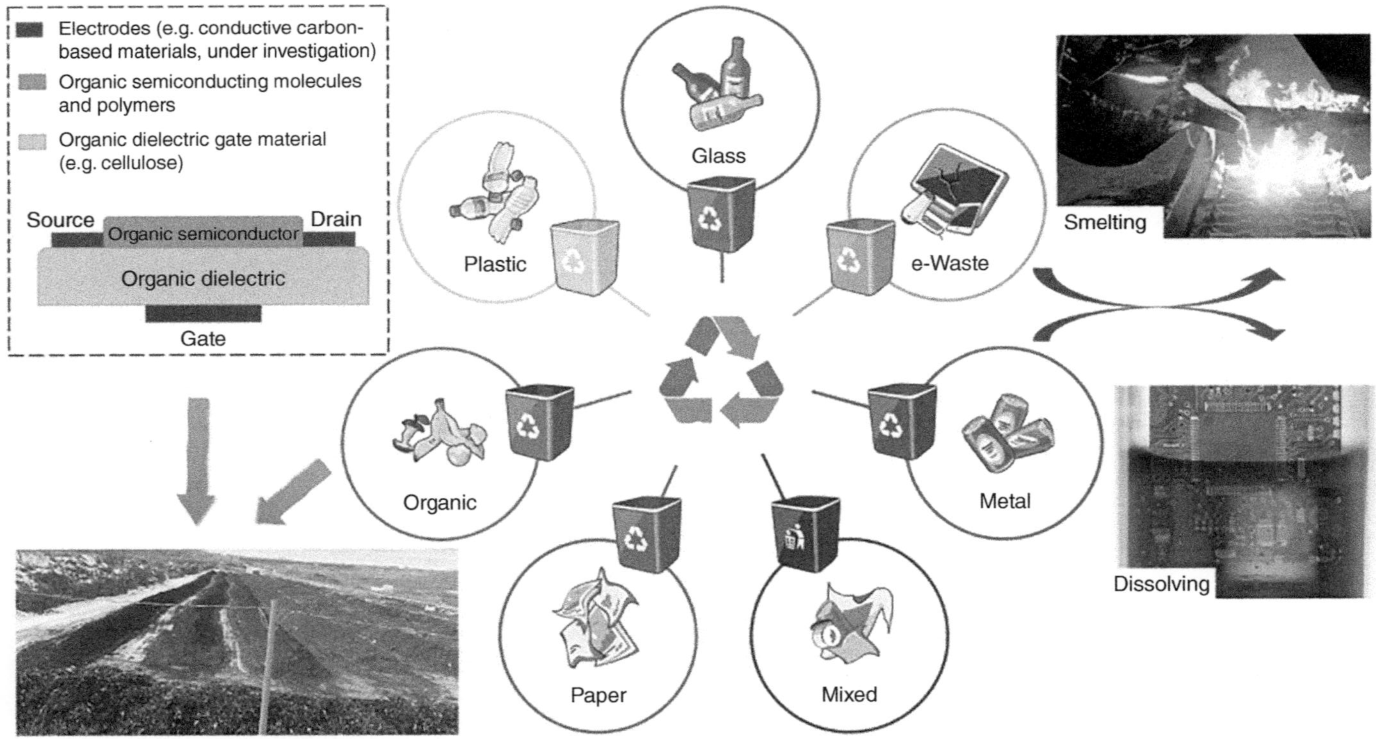

Figure 12.1 Illustration of currently adopted municipal recycling strategy and one potential route, which involves consumers being capable of dropping their compostable electronic waste into their compost bins. This strategy represents an environment-friendly alternative to current electronic waste red bins where the materials are brought to chemical treatments or smelting facilities.

12.1.2 Eco-friendliness

Eco-friendliness incorporates several aspects of materials' conception such as their design (e.g. the choice of materials and manufacturing processes) (International Organization for Standardization 2013, 2020; Lhopital and Bordignon 2018), their ingredients (e.g. ecotoxicity), and their origins (e.g. petrochemistry vs. biorefinery) (European Union 2006, 2015), and their end-of-life treatment (European Union 2012). To address the end-of-life issues of organic materials, we need to focus on the ability of biological ecosystems to efficiently induce changes in the microstructure and polymerization degree of the materials. A biosourced organic material does not automatically qualify as eco-friendly or biodegradable. In the near future, electronic devices biodegradable or compostable, based on biosourced materials, could alleviate the environmental of the electronics sector (Irimia-Vladu 2014; Gouda et al. 2019).

12.1.3 Organic Electronics

Organic electronics are based on organic molecules and polymers featuring electronic conjugation (alternation of single and double carbon–carbon bonds) (Heeger 2001). Organic conjugated molecules (e.g. anthracene, tetracene, pentacene) and polymers (e.g. polythiophene, polyacetylene, polypyrrole [PPy], polyaniline, poly(*p*-phenylene), poly(*p*-phenylene vinylene)) feature electrical conductivity from $\sim 10^{-10}$ S/m (insulating materials) to 10^5 S/m (metallic materials). The electrical conductivity of organic semiconductors generally goes from within 10^{-4} to 10^4 S/m. The conductivity of organic semiconductors can be modified by electrostatic (field-effect), chemical, or electrochemical doping approaches. Chemical organic synthesis, through modification of the molecular structures, opens the opportunity to control the functional properties of the semiconductors. A few examples of molecular structures of organic semiconductors are reported in Figure 12.2.

Weak van der Waals intermolecular interactions confer on organic electronic materials mechanical softness for applications in flexible, stretchable, conformable, and wearable electronics and enable mixed ionic/electronic conduction for applications in imperceptible electronics, bioelectronics (i.e. the field interfacing biology and electronics), smart packaging, and environmental sensing (Root et al. 2017; Berggren and Malliaras 2019).

Organic semiconducting thin films, typically ~50 nm thick, are critical components of organic photovoltaic cells, organic light-emitting diodes, and field-effect transistors (FETs) (Baumgartner et al. 2017a; Ragni et al. 2017).

In the context of e-waste, reusing and recycling electronic devices is not only environmentally desirable but also economically viable and increasingly socially acceptable. In parallel, novel approaches are being investigated in the field of sustainable (green) electronics adopting the idea of circular design, or eco-design, applied to the electronics sector. **Electronic devices based on organic biosourced or synthetic molecules have the potential to be biodegradable in controlled environmental conditions, thereby "integrating" into the environmental cycles.**

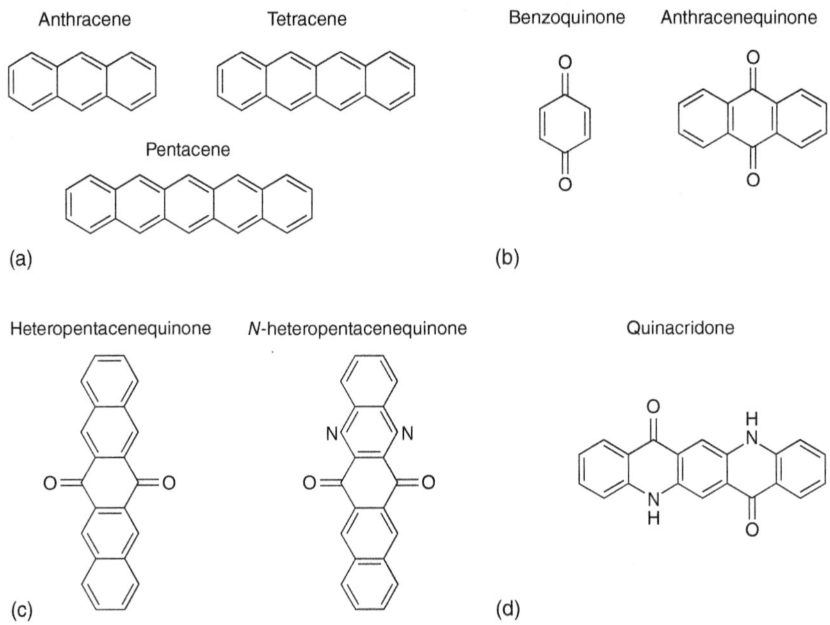

Figure 12.2 Structures of small molecules of interest in organic electronics (a) anthracene, tetracene, and pentacene Source: (a) Wang et al. (2014), (b) benzoquinone and anthracenequinone Source: (b) Mamada et al. (2010), (c) heteropentacenequinone and *N*-heteropentacenequinone Source: (c) Liang et al. (2010), and (d) quinacridone. Source: (d) Yanagisawa et al. (2008).

12.1.4 Opportunities for Green Organic Electronics

Nature could be a source of organic electronic materials extracted from biomass feedstock. The synthesis and processing of organic electronics materials and corresponding thin films through green chemistry principles amenable to devices is also under investigation. Solution-based processing (e.g. through printing) of organic electronic materials is expected to lead to devices featuring lower embodied energy than their inorganic counterparts, usually fabricated using high-vacuum and high-temperature techniques (Irimia-Vladu et al. 2019; Baumgartner et al. 2017b). Some organic electronic materials and devices offer the opportunity to be biodegraded in natural ecosystems or composted at their end of life (Baumgartner et al. 2017b). **In this chapter, we discuss the possibility to compost organic (carbon-based) electronic devices and related organic powering elements at their end of life, with the ultimate goal of limiting the environmental footprint of the electronics sector.**

12.2 State of the Art in Biodegradable Electronics

Biodegradable electronics have produced *transient* devices degrading in aqueous solutions, mainly for biomedical applications (Hwang et al. 2012, 2014a; Fu et al. 2016; Luo et al. 2013; Feig et al. 2018; Tran et al. 2019). The combination of organic and inorganic materials enables high-performance devices. Although the degradation of inorganic materials has been studied for years in the context of corrosion science, most studies have focused on bulk materials submerged in corrosive environments such as strong acidic or basic solutions. The study of degradation in biological solutions of electronic materials and devices in *thin-film* form is the object of increasing attention by the research community. Literature has shown that mono-silicon-nanomembranes (Si-NMs) (30–300 nm thick), polycrystalline silicon (poly-Si), amorphous silicon (a-Si), germanium (Ge), silicon–germanium alloy (SiGe), indium–gallium–zinc oxide (a-IGZO), and zinc oxide (ZnO) are dissolvable in physiological aqueous solutions (Hwang et al. 2012, 2014a,b; Kang et al. 2015; Dagdeviren et al. 2013; Yin et al. 2015b). Together with degradable inorganic dielectrics, metals, and polymer substrates, dissolvable Si-NMs enable fully biodegradable electronics with superior operation characteristics that can also be compatible with semiconductor foundry processes (Yin et al. 2015a; Chang et al. 2017). Demonstrated devices include arrays of basic building blocks for integrated circuits (resistors, inductors, capacitors, diodes, and transistors) (Hwang et al. 2012), temperature sensors (Hwang et al. 2012; Salvatore et al. 2017; Kang et al. 2016), dissolvable stretchable electrophysiological and pH sensors (Hwang et al. 2015), degradable power devices (radio frequency electronics (Hwang et al. 2013), ZnO-based piezoelectric energy harvesters (Dagdeviren et al. 2013), supercapacitors (Lee et al. 2017), and batteries (Yin et al. 2014).

The only class of materials for organic electronics whose biodegradation, mainly in aqueous environments, has been studied is represented by H-bonded pigments (Irimia-Vladu et al. 2019). H-bonded organic semiconductors (mostly pigments and dyes) are among the most important classes of organic electronic materials. Their core functional groups are diketo pyrrolo pyrrole (DPP), anthraquinone, acridone, indigo, and naphthoquinone, among others (Irimia-Vladu et al. 2019). Such materials feature air stability, processability into thin films, and, despite the limited intramolecular π–π conjugation, long-range order, thanks to extensive van der Waals interactions and intermolecular π–π conjugated networks. Bao's group incorporated conjugation breakers with different H-bonding chemistries and linker flexibilities in DPP and studied the mechanical and electrical properties of the corresponding polymer films in view of their application in stretchable electronics: chemical moieties causing a stronger H-bonding interaction result in a higher modulus and higher crack on-set strain (Zheng et al. 2020).

Most of H-bonded organic pigments are highly resistant to microbial attack. Toxicity to microbes, plants, and animals has been reported for indigo and indigo carmine dyes (Irimia-Vladu et al. 2019). Interestingly, the white-rot fungus *Phanerochaete chrysosporium* can degrade indigo and its derivatives owing to the production of two families of extracellular lignin peroxidase (LiP) and manganese-dependent peroxidase (MnP) and a H_2O_2-generating system (Irimia-Vladu et al. 2019). The metabolic pathways and biological mechanisms of this process are being elucidated, but toxicity studies of the intermediates and final degradation products have yet to be done (Fabara and Fraaije 2020; Ma et al. 2018; Montazer and Sadeghian Maryan 2010).

Bao's group (Feig et al. 2018) recently proposed to categorize biodegradable electronic systems into two groups: type I, where only the inert host matrix is biodegraded; and type II, where both the electroactive polymers and the host matrix are biodegraded. Examples of type I include biodegradable electronic systems based on cellulose nanofibril paper for edible and eco-friendly applications (Jung et al. 2015). In type II, the electroactive polymer is reverted to its monomer components allowing for monomer recycling, representing a great opportunity toward reducing electronic waste. Current strategy for achieving fully biodegradable electric components can be separated into two main approaches, namely the use of naturally occurring conjugated biopolymers and the fabrication of synthetic polymers using conjugation breakage and hydrolysable bonds.

Naturally occurring conjugated polymers such as indigo (Irimia-Vladu et al. 2012), melanins (Gouda et al. 2019), or β-carotene (Ramachandran et al. 2003) spark the interest of researchers as they are often nontoxic and undergo biodegradation processes. Nonconjugated linkers and hydrolysable bonds, like ester or imine functions, hinder the electrical conductivity of synthetic type II polymers but allow to break the backbone into reusable monomer molecules. Using imine-based monomer building blocks, Bao's group formed a conductive polymer that could then be reverted to its original components by exposition to slightly acidic aqueous solutions. By combining this type of polymer, a cellulose substrate and iron current collector, the group demonstrated a fully degradable transistor (Lei et al. 2017). Finally, by phase segregating a semiconducting polymer (built by stepwise polymerization of dialdehyde-functionalized (DPP) and *p*-phenyldiamine (PPD): [*p*(DPP-PPD)]) in a biodegradable elastomer (an urethane-based polymer: E-PCL), Bao's group was able to create a fully biodegradable and stretchable conductive material that could be used for the design of medical implanted self-degrading sensors (Tran et al. 2019).

12.3 Organic Field-Effect Transistors (OFETs)

12.3.1 Fundamentals

Transistors are three-terminal semiconductor devices used as current switches and amplifiers. In modern electronics, billions of amplifier and switch transistor chips are embedded in a single integrated circuit (IC) to assemble microprocessors for computer memories. In this process, known as chip bonding, chips are connected to the lead frame of chip carriers by metals such as Au, Al, Cu, and Pb soldered wires and, afterward, covered with molded plastic-based packages to protect them from damage. Nowadays, transistor technology focuses mainly on crystalline silicon (c-Si) and germanium (c-Ge) as the semiconductors and Au, Al, Cu, and Ni as metals.

Organic field-effect transistors (OFETs) are assembled in thin-film architectures: source and drain metal electrodes are usually directly deposited onto the organic thin film channel, separated from the metallic gate contact by a dielectric layer. To limit power consumption and operate at low voltage, the dielectric layer should display high specific capacitance. For superior field effect performance, the Schottky energy barrier for injection of electrons and holes at the metal electrode/organic semiconductor interface should be minimized. This mismatch, known as *energy band bending*, causes a nonohmic injection at the interface and stems from the offset between the work function of the metal electrodes and the highest occupied molecular orbital (HOMO) and/or lowest unoccupied molecular orbital (LUMO) levels of the organic semiconductor (Kymissis 2009) (HOMO and LUMO can be seen as the valence band edge and the conduction band edge of conventional semiconductors). Holes (electrons) are injected from the source (drain) electrode into the LUMO (HOMO) level of the organic semiconductor. In an optimized OFET architecture, high work function metals, e.g. Au, Al, and Cu, inject holes and low work function metals, e.g. Ca, Mg, inject electrons. However, Au is the most commonly used metal electrode for both source and drain because of its air stability (i.e. it does not form oxides) and quite good electron injection. The band-bending issue can be solved by using carbon nanotube (CNT)-based electrodes (Valitova et al. 2013). CNTs are hollow low-dimensional, high-aspect-ratio rolled-up graphene sheets featuring nanometer scale diameter and conductivity approaching that of metals ($\sim 10^5$ S/cm).

Most of natural and synthetic organic semiconductors suffer from high structural disorder because of weak intermolecular van der Waals bonds. This causes short-range π-electron delocalization, a limiting factor to promote efficient charge carrier transport (Malliaras and Friend 2005). In the quest of high-performance OFETs, attention has been mainly devoted to fused planar structures with high degree of intramolecular π-conjugation and close molecular packing such as single crystal acenes (Wang et al. 2014; Aleshin et al. 2004; De Boer et al. 2003; Butko et al. 2003; Schön et al. 2000; Kim et al. 2007) (i.e. anthracene, tetracene, and pentacene, Figure 12.2a). Single crystals' acene-based field-effect mobility outperforms that of amorphous silicon by about 2 orders of magnitude. However, their air stability is rather poor. Recently, air-stable hydrogen-bonded organic semiconductors (i.e. indigo [Irimia-Vladu et al. 2012] and tyrian purple [Głowacki et al. 2012]) were explored as novel materials for OFETs, reaching mobility close to $2\,cm^2/V/s$ (Irimia-Vladu 2014). In these systems, the lack of an effective intramolecular conjugation is balanced by the occurrence of inter- and intramolecular H-bonds. Such bonds improve charge carrier delocalization by enhancing the electronic coupling between neighbor molecules and confer high lattice energy to H-bonded semiconductors. Thus, H-bonded organic semiconductors feature, in principle, superior thermal stability with respect to conventional, single-crystal van der Waals–bonded semiconductors.

Along with indigo and tyrian purple, quinones are small organic molecules that contain two adjacent (or separated) carbonyl groups in an unsaturated six-carbon ring structure (Allen and Martin 2007; Blankenship 2014). In nature, there are over 2400 quinone species (Wu et al. 2017). Quinones and their derivatives (i.e. anthraquinones (AQ) (Zhao et al. 2020; Mamada et al. 2009), benzoquinones (BQ) (Mamada et al. 2010), *N*-heteropentacenequinones (HAQ) (Liang et al. 2010; Shi et al. 2014), quinacridones (QA) (Yanagisawa et al. 2008; Berg et al. 2009; Głowacki et al. 2013), Figure 12.2) are among the most interesting H-bonded semiconductors (Irimia-Vladu 2014). In particular, their molecular packing and HOMO–LUMO gap can be tuned by functionalization (Zhao et al. 2019). These features make quinones attractive for the development of greener electronic devices. In Table 12.1, we report a few functionalized quinone thin film-based organic semiconductor channels for OFETs.

12.3.2 Anthraquinone, Benzoquinone, and Acenequinone

The excellent electron-acceptor properties of AQ, BQ, and HAQ make them ideal candidates for conceiving n-type channel OFETs (Liang et al. 2010; Mamada et al. 2009; Shi et al. 2014; Zhao et al. 2019; Zhang et al. 2019; Tang et al. 2010). The good field-effect characteristics in quinone FETs stem from the H-bonded molecular structure that is expected to offer good electron delocalization. Yet, for achieving very high performance and stability in air, quinone derivatives with large electron affinity and deep LUMO levels are needed. To meet these requirements, the molecular structure of quinones can be properly modified by functionalization (Liang et al. 2010; Mamada et al. 2009; Shi et al. 2014; Zhao et al. 2019; Zhang et al. 2019). For example, an electron mobility as high as 0.28 $cm^2/V/s$ is reported for OFETs based on bis(trifluoromethylethynyl)-9,10-anthraquinones (Zhao et al. 2019). The 9,10-anthraquinones (9,10-AQ) are functionalized by attaching trifluoromethylethynyl groups to their molecular backbone.

Similarly, the air-stable films made of 2,6-bis[2-(4-trifluoromethylphenyl) thiazol-5-yl]benzo[1,2-b:4,5-b′]dithiophene-4,8-dione feature electron mobility of about 0.1 $cm^2/V/s$, and can be fabricated by attaching thiophene groups to the sidechains of BQ. This functionalization is followed by trifluoromethyethylation (Mamada et al. 2010). Air-stable fluorinated pentacenequinone (FPCQ) and *N*-heteropentacenequinones (NHPCQ) can be synthesized by attaching F and N atoms at different sites of pentacenequinones (Ragni et al. 2017). Both types of functionalization translate into chemically stable, high-electronic quality quinone films, with field-effect mobility between 0.1 and 0.2 $cm^2/V/s$ (Liang et al. 2010; Shi et al. 2014).

12.3.3 Quinacridones

Quinacridones (QAs) are synthetic derivatives of acridones in natural dyes. They are planar molecules forming different H-bonded, π-stacked crystals (Lincke 2002). The significant interest around QAs is originally derived from their extensive use as colorants for automobile and industrial coatings. It is only recently that attention was paid to QAs as organic thin-film semiconductor channels for OFETs. QA thin films exhibiting limited air degradation with respect to conventional acene films have been demonstrated (Głowacki et al. 2013). In OFET configuration, these films show both p-type and n-type behavior (ambipolarity), and feature hole mobility as high as 0.2 $cm^2/V/s$ (Table 12.1).

Table 12.1 State of the art of high-performance quinone-based OFETs.

References	Active material	–	Transistor channel type	Electron/hole mobility ($cm^2/V/s$)
Zhang et al. (2019)	Tetraphenyl diphenoquinone	TPDPQ	p-type	$\mu_h = 8.7 \times 10^{-7a}$
Zhao et al. (2019)	Bis (trifluoromethylethynyl) – 9,10-anthraquinones	—	n-type	$\mu_e = 0.28$
Liang et al. (2010)	2,6-Bis [2-(4-trifluoromethylphenyl) thiazol-5-yl] benzo [1,2-b:4,5-b′] dithiophene-4,8-dione	—	n-type	$\mu_e = 0.15$
Tang et al. (2010)	5,7,12,14-tetraaza-6,13-pentacenequinone	TAPCQ	n-type	$\mu_e = 9.0 \times 10^{-2}$
Tang et al. (2010)	6,11-diaza-5,12-tetracenequinone	DATCQ	n-type	$\mu_e = 2.0 \times 10^{-4}$
Liang et al. (2010)	Fluorinated pentacenequinones	FPCQ	n-type	$\mu_e = 0.15$
Root et al. (2017)	*N*-heteropentacenequinones	NHPCQ	n-type	$\mu_e = 0.15$
Shi et al. (2014)	Pentaceno [2,3-c:9,10-c′] bis ([1,2,5] thiadiazole)-6,14-dione	PCBTDQ	n-type	$\mu_e = 0.11$[a)]
				$\mu_e = 1.4 \times 10^{-3}$
Yanagisawa et al. (2008)	Quinacridone	QA	p-type	$\mu_h = 8.23 \times 10^{-6b}$
				$\mu_h = 1.08 \times 10^{-5c}$
			p-type	$\mu_h = 1.5 \times 10^{-3}$
			Ambipolar	$\mu_h = 0.2$
				μ_e from 0.01 to 0.2

a) Single crystal.
b) Latent pigments.
c) Commercial.

12.4 Electrochemical Energy Storage

12.4.1 Quinones

Quinones are biosourced, low-cost, and nontoxic alternatives to inorganic energy-storage materials (Wu et al. 2017; Han et al. 2019; Liang et al. 2017). Quinones act as biological proton/electron transfer agents in natural systems. The quinones' role as electron mediators has inspired researchers to improve the efficiency of human-made energy conversion systems, such as solar cells and photochemical water-splitting devices, in which water decomposes into oxygen and hydrogen (Cheng et al. 2012, 2013; Kato et al. 2012; Kim et al. 2014; Rodenberg et al. 2016). In addition, quinones are involved in the production of energetic molecules such as ATP carriers, through a mechanism analogous to reversible electron/ion transport in energy-storage devices (Osyczka et al. 2004). Quinone derivatives (e.g. BQ, AQ) (Figure 12.2b) are investigated as alternatives to inorganic redox couples in redox-flow batteries (Lin et al. 2015; Huskinson et al. 2014; Perry and Weber 2016). They also have been used in the development of sustainable electrodes for rechargeable batteries (Liang et al. 2017; Park et al. 2015; Ding and Yu 2016; Ma et al. 2016; Kim et al. 2016). The 1,4-benzoquinone (1,4-BQ) can deliver a theoretical capacity of ~496 mA h/g at 2.8 V vs. Li/Li^+ (Yokoji et al. 2016), which can be compared to conventional inorganic cathode materials such as $LiCo_2O_4$ (~140 mA h/g at 3.9 V vs. Li/Li^+) (Dong et al. 2018) and $LiFePO_4$ (~170 mA h/g at 3.45 V vs. Li/Li^+) (Figure 12.3) (Xu et al. 2015).

Researchers are currently conceiving cost-effective bio-inspired systems for energy harvesting and conversion that may provide superior performance without

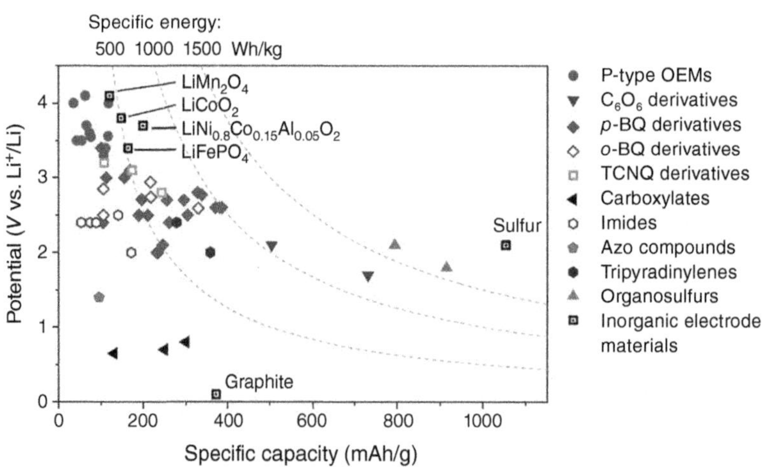

Figure 12.3 Comparison between organic electrode materials (OEMs) and inorganic ones for Li batteries in discharge potentials and specific capacities. Source: Poizot et al. (2020).

Besides the ability to bind monovalent (e.g. H^+, Na^+, Li^+, and K^+) and multivalent cations (e.g. Mg^{2+}, Zn^{+2}, Ca^{+2}, and Al^{+3}), quinones feature high-power density, long cycle life, and structure stability (Liang et al. 2017; Mauger et al. 2019; Han et al. 2019).

posing a threat to the environment. In recent years, a large class of quinone-based redox-active biomolecules and biopolymers (Figure 12.4) have been studied as energy-storage materials in various electrochemical technologies such as aqueous batteries (Liang et al. 2017; Ma et al. 2016; Song et al. 2014, 2015; Vonlanthen et al. 2014; Zhou et al. 2015; Huskinson et al. 2014), supercapacitors (Kumar et al. 2016; Mukhopadhyay et al. 2017; Nagaraju et al. 2014), and light-assisted supercapacitors (Xu et al. 2019; Zhu et al. 2016a).

12.4.2 Dopamine

Dopamine (DA) is a biochemical molecule that mimics some key chemical structures of the adhesive proteins (l-3,4-dihydroxyphenylalanine [DOPA]) in mussels. It self-polymerizes into polydopamine (PDA) under alkaline aqueous conditions in which the catechol functional group oxidizes to quinone (d'Ischia et al. 2014).

Polydopamine-coated few-walled carbon nanotubes (FWCNTs) were reported as free-standing and flexible cathode material for organic rechargeable Li- and Na-ion batteries in organic electrolytes (e.g. $LiPF_6$ 1 M in a mixture of ethylene carbonate [EC] and dimethyl carbonate [DMC] [3 : 7; v : v] and $NaPF_6$ 1 M in a mixture of EC/DMC [3 : 7; v:v]). Delivering gravimetric capacities of 133 mA h/g in Li-cells and 109 mA h/g in Na-cells, exploiting both the double-layer capacitance of FWCNTs and multiple redox reactions of polydopamine, and featured outstanding cycling stability (Liu et al. 2017). A biohybrid electrode material from dopamine, copolymerized with pyrrole in nanofibrillar PPy/DA structure, was reported as cathode material for rechargeable Li-ion batteries. This hybrid electrode material reversibly stored 160 and 90 mA h/g at discharging rates of 100 and 800 mA/g, respectively, with a Coulombic efficiency of almost 100% in 1 M $LiPF_6$ in EC/DMC (1 : 1; v:v) (Liedel et al. 2018). Polydopamine on CNTs was not only used in rechargeable Li-ion and Na-ion batteries but was also as cathode material for aqueous Zn-ion batteries. This biohybrid electrode exhibited a specific capacity of 126 mA h/g and capacity retention of 96% after 500 cycles in 3.3 M $ZnSO_4$ aqueous solution (Yue et al. 2019).

12.4.3 Melanins

Melanins are disordered macromolecules poorly soluble in most organic solvents. Eumelanin, pheomelanin, neuromelanin, allomelanin, and pyomelanin are subgroups of melanins (d'Ischia et al. 2020). Neuromelanin is present in two areas of the human nervous system: the locus coeruleus and the substantia nigra of the brain; it plays an important role in brain aging and Parkinson's disease (Segura-Aguilar et al. 2014). Eumelanin has fascinated the scientific community because of its widespread occurrence in the biosphere and unique physicochemical properties such as broadband optical absorption (Meredith and Sarna 2006), metal-ion chelation (binding affinity) (Lydén et al. 1984; Hong and Simon 2007), hydration-dependent electrical and photoelectrical response (Mostert et al. 2012; Mostert et al. 2018).

Due to its electronic conjugation, nontoxicity, biocompatibility, redox functionality, and potential for biodegradability (Bettinger et al. 2009; Solano 2017),

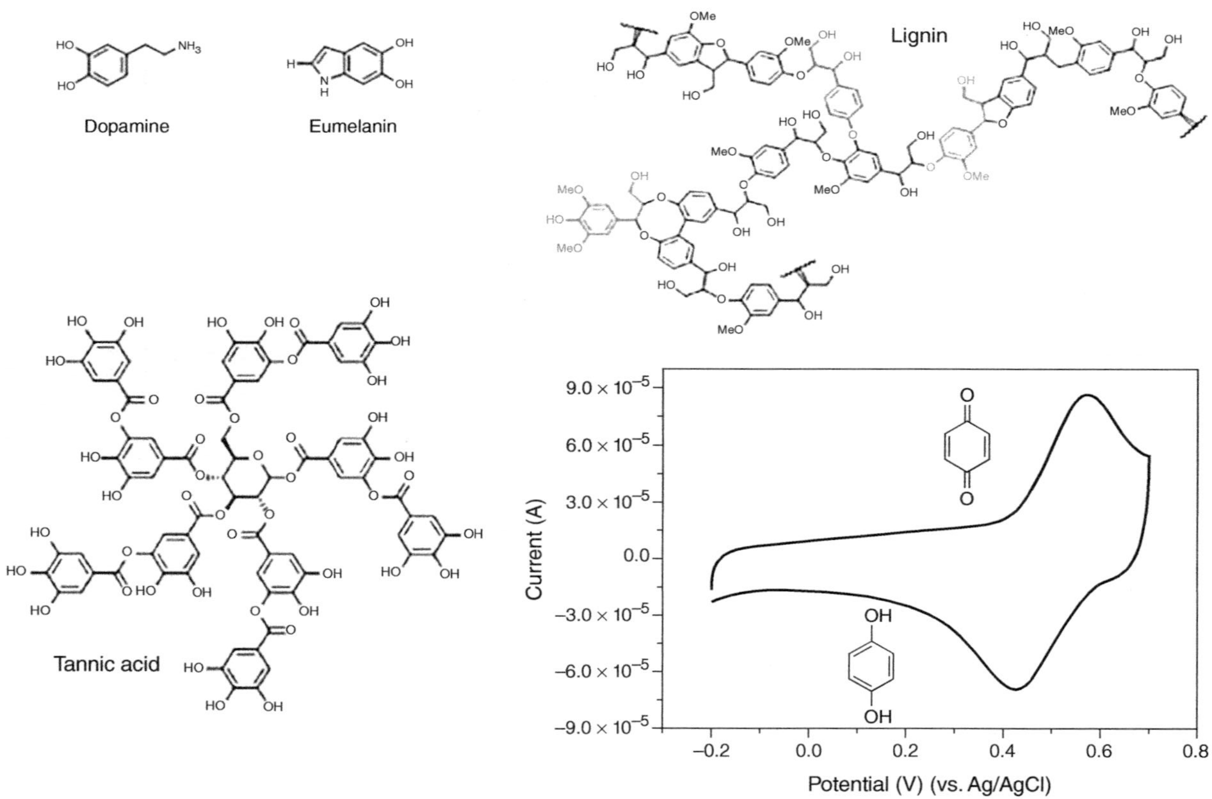

Figure 12.4 Chemical structures of dopamine, building blocks of eumelanin where R is −H in DHI and −COOH in DHICA, lignin with p-hydroxyphenyl (H), guaiacyl (G), and syringyl (S) monolignol units, and tannic acid. Cyclic voltammogram relative to the hydroquinone and quinone redox couple. Source: Ajjan et al. (2019).

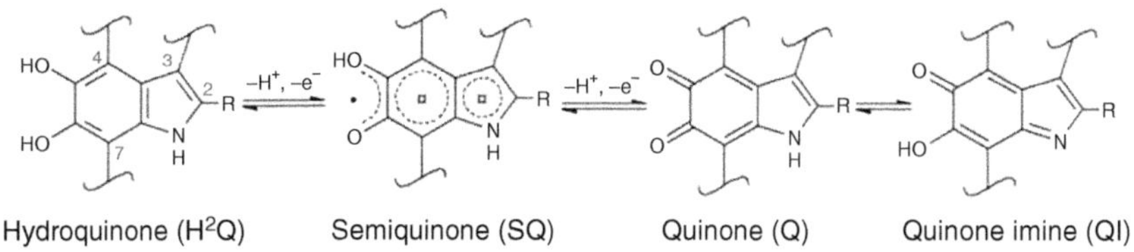

Figure 12.5 Hydroquinone (H2Q), semiquinone (SQ), and quinone (Q) redox forms of the building blocks of eumelanin: 5,6-dihydroxyindole (DHI) and 5,6-dihydroxyindole-2-carboxylic acid (DHICA). R is −H in DHI, whereas R is the −COOH group in DHICA. The quinone imine form (QI) is the tautomer of Q. Source: Xu et al. (2019).

eumelanin has emerged as a promising material for a wide range of applications in bioelectronics.

Eumelanin originates from the oxidative polymerization of tyrosine via two building blocks: 5,6-dyhidroxindole (DHI) and 5,6-dyhidroxindole-2-carboxylic acid (DHICA) (Figure 12.5).

The development of synthetic eumelanin based only on one of the two building blocks, namely DHICA-melanin and DHI-melanin, has provided a tremendous tool in the exploration of the poorly understood physicochemical properties of eumelanin for different applications such as electrochemical energy storage, memory devices, and sensors (Xu et al. 2019). The capability of melanin to reversibly bind mono- and multivalent cations constitutes the foundation for its use in energy-storage systems. Bettinger's group demonstrated the use of eumelanin electrodes in Mg^{2+} batteries in which the redox behavior is governed by two-electron, two-proton processes that bring catechols to ortho-quinones, coupled with Mg^{2+} insertion/removal (Jo et al. 2014). Melanin cathodes showed charge-storage capacity of 60 mA h/g at 0.1 A/g for more than 500 cycles, approximately twice the capacity of melanin anodes (30 mA h/g) in Na^+ batteries (Kim et al. 2013). Santato's group reported on the use of eumelanin grafted on carbon paper electrodes as pseudocapacitive material with a gravimetric specific capacity of 24 mA h/g in slightly acidic aqueous electrolyte, with delivered maximum power density of 20 mW/cm^2. Once in a symmetric supercapacitor configuration, melanin/carbon paper electrodes showed a Coulombic efficiency of 99.7%, proving good reversibility in the charge/discharge process (Kumar et al. 2016). Additionally, flexible micro-supercapacitor (miniaturized supercapacitors with a footprint close to several square millimeters) on polyethylene terephthalate (PET) was reported with 5 mW/cm^2 power density, energy density of 0.44 mJ/cm^2, and a specific capacitance of ~11 F/g. Recently, the same research group reported on the use of chemically controlled melanins (DHI- and DHI/DHICA-melanins) formed by solid-state polymerization at room temperature as sustainable organic redox materials for light-assisted pseudosupercapacitors where the storage performance of the device was improved under solar light (Xu et al. 2019). Under solar light conditions, the capacity and capacitance of DHI-melanin electrodes significantly increased, from 2.3 to 2.8 mC/cm^2 (+22%) and from 3.8 to 5.3 mF/cm^2 (+39%). Once in supercapacitor configuration, besides featuring a Coulombic efficiency close to 100% after 5000 cycles, the capacitance and capacity retention of the DHI-melanin electrodes improved after prolonged illumination, as did the energy from 44 to 52 mJ/g (i.e. 0.0122 to 0.0144 mW h/g) and power density from 3.8 to 5.9 W/g.

12.4.4 Tannins

Tannins are secondary plant metabolites with molecular weights between 300 and 3000 Da used since antiquity for leather production (Sieniawska and Baj 2017). They regroup over a thousand molecules with several different properties (Khanbabaee and van Ree 2001). If tannins have been used for various applications since c. 1500 BC, most recent studies focus on medicinal applications (Wang et al. 2018) for their antioxidant (Yilmaz and Toledo 2004), antitumor (Yang et al. 1999; Yoshizawa et al. 1987), and antiviral (Akiyama et al. 2001) properties. The molecular structure of tannins presents the highest phenol content among natural phenolic biopolymers with 5.6 mol/g, which is 5000 more than lignin (Melone et al. 2013). Tannins regroup a large number of different molecules with different electrochemical signatures making their application to electrochemical storage more complex and less explored with respect to previously observed phenolic biopolymers such as lignin (Mukhopadhyay et al. 2017). However, owing to their reversible redox activity, they are studies for applications in energy-storage devices (Temmink et al. 1989).

Tannins can be categorized into two groups: hydrolyzable tannins comprising the gallotannins, ellagitannins, and complex tannins subgroups; and condensed nonhydrolyzable tannins featuring high molecular weight, up to 20 000 Da (Khanbabaee and van Ree 2001). After cellulose and lignin, they are the most common plant metabolites and compose most of forest industry waste (12–16% of bark by weight [Hathway 1958]). Tannins can be extracted at low cost with environment-friendly aqueous media (Vázquez et al. 2008). The extraction process consists of immersing dry bark in water and shaking at room temperature for 8–10 hours or boiling for 10–15 minutes. Afterward, the extracted solutions are filtered and concentrated in a rotary evaporator and dried in a vacuum chamber to obtain a concentrated tannin powder (Temmink et al. 1989). Rainwater has been shown to extract tannins from discarded barks resulting in water contamination and toxicity, giving a further incentive in the use of tannins for other applications (Temmink et al. 1989).

The association of tannins and conductive polymers such as PPy can yield impressive capacitances. Mukhopadhyay et al. report an increase from 94 F/g for PPy alone to 370 F/g at 0.5 A/g (+293%) and from 7 to 196 F/g at 25 A/g (+2700%) using gold as current collector (Mukhopadhyay et al. 2017). Furthermore, tannins have been associated with metallic elements such as lithium to serve as natural anodic material and enhance battery properties. Rechargeable batteries using lithiated ellagic acid as anode material showed high reversible capacities of 450 and 200 mA h/g at C/10 and C/2.5 discharge rate, respectively (Goriparti et al. 2013). The C-rate is derived from the current that is required to charge the cell within one hour.

12.4.5 Lignin

Lignin is a three-dimensional amorphous and highly branched polyphenolic polymer (Figure 12.4) obtained through radical oxidative polymerization of three monolignol units (p-coumaryl, coniferyl, and sinapyl alcohol) (Zhu et al. 2016b; Thakur et al. 2014; Laurichesse and Avérous 2014). Lignin is the second most abundant organic polymer on Earth as it comprises c. 20–30% of the solid weight of plants. Moreover, lignin can be found in large amounts in papermaking and cellulosic biofuel industries waste. There are three main monolignols, called *p*-hydroxyphenyl (H), guaiacyl (G), and syringyl (S), that constitute its network and vary based on their plant source. Lignin is extracted using a sulfate process that leads to the formation of Kraft lignins (alkaline lignin) and lignosulfonates, which are soluble in water and acidic solutions (Van den Bosch et al. 2015). Milczarek and Inganäs were the first to report the use of lignin/PPy as hybrid electrode material for supercapacitors with specific capacitance range from 350 F/g for the thick electrode to 1000 F/g for the thin one (Milczarek and Inganäs 2012). The same research group proposed the use of PEDOT with lignin in a hybrid electrode material that was synthesized using chemical polymerization and electrochemical polymerization, showing a specific capacitance of 170 F/g and capacitance retention (83% after 2000 cycles) (Ajjan et al. 2016). Suarez et al. assembled lignin/PEDOT in asymmetric and symmetric supercapacitor configuration. The former configuration, designed with the lignin/PEDOT as the positive electrode and with partially reduced graphite oxide (p-rGrO) as the negative electrode, exhibited specific capacitance c. 35 F/g (Navarro-Suárez et al. 2017). The latter configuration, where lignin/PEDOT and p-rGrO were combined into a homogeneous composite material for both positive and negative electrodes, showed specific capacitance of 36 F/g with capacitance retention up to 79% after 1000 cycles. Recently, the same research group reported the use of lignin/PEDOT composite as cathode material in an all-organic aqueous polymer battery (Hernández et al. 2018). The full-cell configuration exhibited an output voltage of 1.0 V and delivered 40 mA/hg discharge capacity at 100 °C with 85% capacity retention and c. 93% Coulombic efficiency for 800 cycles. Lignin has been used as cathode electrode material and also as natural binder in lithium-ion batteries (Gnedenkov et al. 2015; Maleki et al. 2000). Moreover, lignin has been investigated as an additive material for lead–acid batteries to extend cycling life by enhancing the mechanical properties (Hirai et al. 2006).

12.5 Biodegradation in Natural and Industrial Ecosystems

To realize our vision of eco-friendly green organic electronics, it is first important to have an understanding of the abiotic and biotic processes associated with degradation and biodegradation phenomena, respectively. Billions of years of evolution have provided microorganisms with the ability to benefit from the energy stored in the chemical bonds of materials via enzyme production. Interestingly, this knowledge led to the development of technologies like composting and many others – in essence, our vision is inspired by nature.

12.5.1 Degradation and Biodegradation

Firstly, degradation relates to numerous abiotic phenomena taking place under different bio/geochemical environments for varying periods of time. These physical and chemical processes result in some degree of deleterious change in the chemical structure, physical properties, or appearance of a material. However, as is the case for most petroleum-based plastics, degradation is more a question of size reduction rather than of true chemical transformation into harmless products. This reduction plays an important role in the migration of chemicals within or across terrestrial and aquatic ecosystems, with countless harmful effects on environmental homeostasis and human health.

Interestingly, cellulose, one of the most abundant biopolymers on Earth, is a very stable compound under abiotic conditions with an estimated half-life of 5–8 Ma at 25 °C (Wolfenden and Snider 2001). In contrast, under biotic conditions, wood cellulosic fibers decay in soil within weeks or months in the presence of cellulolytic microorganisms. Even more interesting is nature's ability to degrade lignin, the second most abundant biopolymer on Earth. For millions of years, nothing could break down lignin, the strongest substance in plant cell walls. When the fossil record started showing trees being decomposed, around 300 Ma ago, Robinson (1990), a geographer, proposed a new theory: something played a major role, something evolved the ability to "break down" lignin. Then, Morgenstern et al. (2008) and Floudas and numerous collaborators (2012) discovered a group of fungi known as "white rot fungi," which had evolved the ability to decompose lignin; that was at approximately the same geological time that coal formation drastically decreased. As we will see in the Microbiome section, tropical termites and ruminants are two other beautiful examples of coevolution between the host's microflora and its diet, the lignocellulosic biomasses. Such research activity depicts the concept of evolution of one of the most important biogeological cycles, the Carbon cycle. Finally, evolution has also been observed in a man-made ecosystem, i.e. a landfill site filled with PET (polyethylene terephthalate) plastic waste for over 60 years. Such petroleum-based material was thought to be nonbiodegradable until a Japanese research team found and isolated the bacterium *Ideonella sakaiensis* 201-F6, which proved to be capable of enzymatically breaking down the PET backbone (Yoshida et al. 2016). These cases greatly illustrate the fundamental role of research activities in bringing to light the extraordinary potential of microorganisms and their enzymes to utilize substrates with diverse chemical structures.

Secondly, the word "biodegradation" is employed specifically to describe biotic processes occurring in different terrestrial and aquatic ecosystems. These ecosystems may be natural or man-made, and the process of biodegradation can occur over varying lengths of time. As both physicochemical (abiotic) and biological (biotic) phenomena are interconnected, they participate concurrently in the biotransformation of natural organic matter as well as anthropogenic organic waste and pollutants. In essence, that is the definition of a dynamic ecosystem participating in the biogeological cycles of C, N, K, P, and some metal ions. Although the latter are essential to sustain biological life, at sufficiently high concentrations, they become toxic to most biota and disrupt key biological processes.

Finally, composting is regarded in the field of bioengineering as a solid-state fermentation process. The bioreactor design, process optimization strategy, and microbiome characteristics are cornerstones in the development of an efficient environmental biotechnology. The bioreactor design aims at incorporating a microbiome into a reactor adequately equipped for complete mineralization of organic waste, in a short time. The total duration of a full composting cycle, which can span over one to two years, is dictated by parameters such as pH, temperature, and humidity, at the composting site. However, as it will be discussed below, inoculating a compost heap with selected microbes may increase the rate of biodegradation, thereby decreasing the total time of the process or allowing a recalcitrant material to biodegrade. The optimization of the process is achieved through variables such as carbon-to-nitrogen ratio, pH, water content, water activity (a_w), O_2, and particle size.

12.5.2 Composting Process

The following Sections 12.5.3–12.5.4 are, first, a summary of the knowledge and a description of the methods for the successful composting of organic waste, i.e. the main substrate, and secondly, a discussion of the advantages and challenges of green organic electronics to be biodegraded as a cosubstrate. This approach, presented by Alves et al. (2019), was envisaged for the co-composting of truly recalcitrant organic matter, namely petroleum hydrocarbon residues blended with easily degradable substrates, i.e. "regular" organic waste. Similarly, Arias et al. (2019) studied the co-composting of forest and industrial wastes, which constitutes a feasible management technique.

Although numerous videos and review papers present in endless detail the science of composting (USDA Natural Resources Conservation Service 2010), engineering principles relevant to the modus operandi (MDDELCC 2018; Josée 2006; Zhang et al. 2012), and microbiology of composting, it is worth reviewing such bioprocess on the basis of the fundamentals mentioned above. As shown in Figure 12.6, the composting process is essentially the same process as natural decomposition with the exception of taking place in a designated facility. Composting is the aerobic biological decomposition of a blend of shredded solid organic matter of various origins (e.g. from wood debris, leaves, garden waste, and grass clippings to animal manure, fish waste, and crustacean shells, as well as papers, fabrics, and polylactic acid [PLA]-based or starch-based cutlery) into mature compost with the concomitant production of simple molecules (CO_2 and H_2O) and heat. Its relative simplicity of operation, low-labor requirement, and, more importantly, its efficiency, make composting an environmental technology favored by municipalities to manage household organic waste. The uniqueness of this process relies essentially upon its microbiome structure and diversity.

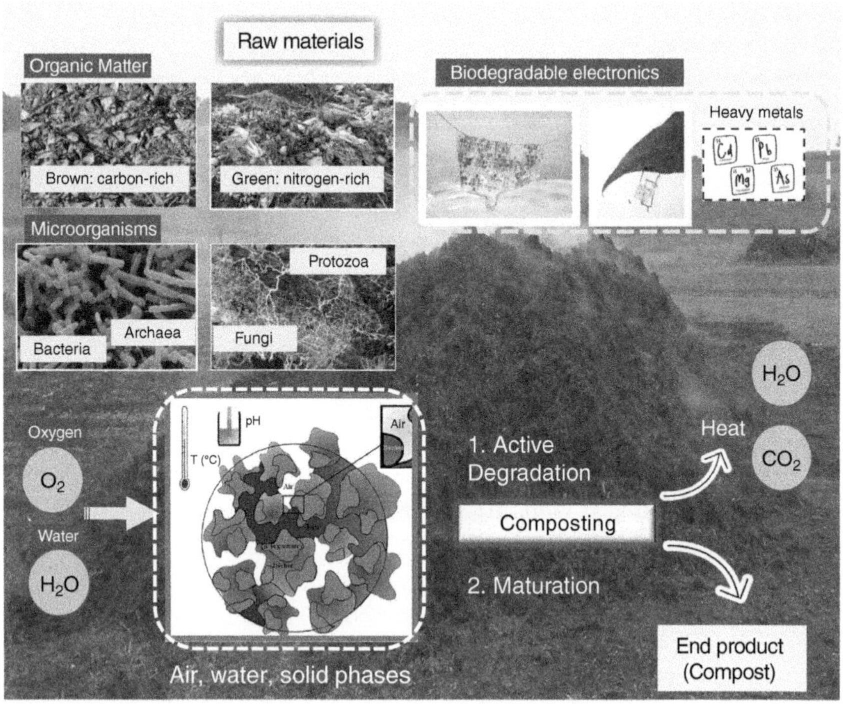

Figure 12.6 Illustration of the end of life of hypothetical biodegradable electronics. Composting is a two-step process. The first step is the thermophilic degradation, that is, when temperature profile can attain 70 °C for a maximal rate of biodegradation. The second step refers to the mesophilic maturation phase or curing phase. Here, mesophilic organisms, mainly bacteria and fungi, recolonize the compost pile, the rate of biodegradation gradually decreases, and the core reaches ambient temperature. pH, water, and oxygen are essential parameters to the microbial community (archaea, bacteria, fungi, and protozoa) for transformation of organic substances into H_2O, CO_2 and humus-like materials. The microbial activity occurs at the microscopic level within a porous air/water/solid structure, highlighted in the white square. Source: Lei et al. (2017) and Pujol (2012) Licensed under CC BY 4.0.

Although the microbiome is the cornerstone of the composting process, it is only one piece of the multiparameter puzzle illustrated in Figure 12.7. The design of the bioprocess is the first step required to ensure its effectiveness. To date, numerous composting strategies have been designed; however, the windrow is the classical approach for composting large quantities of organic waste in open-air facilities. Windrows (Figure 12.7a) of 250 m^3 typically have dimensions of 2 m × 5 m × 50^+ m to allow for optimal surface area to volume ratio; one of the parameters reviewed in great detail by Mason (2007). According to his findings, the range of values of full-scale systems were estimated to be from 0.4 : 1 to 3.8 : 1 m^2/m^3. To ensure proper ventilation of the compost pile, either a passive or forced air system should be used (Figure 12.7a).

A cross-section of a compost pile reveals schematically the interconnection between the biological activity and key physical phenomena, such as compaction,

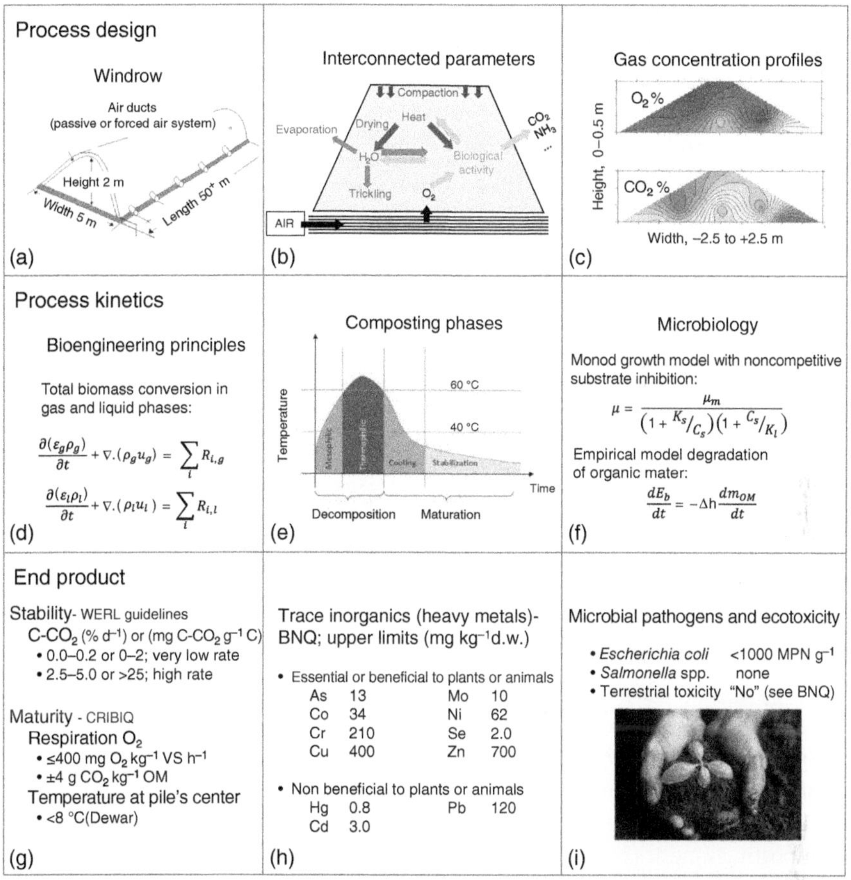

Figure 12.7 Process kinetics overview. (a) process design, (b) interconnected parameters, (c) gas concentration profiles, (d) process kinetics, (e) composting phases, (f) microbiology, (g) end product, (h) trace inorganics, (i) microbial pathogens and ecotoxicity. Source: (a) Abu Qdais and Al-Widyan (2016), (b, d–h) Pujol (2012), (c) Stegenta-Dąbrowska et al. (2019).

heat generation, drying, and aeration (Figure 12.7b). Notably, the material compaction occurs as it is being degraded, a process accompanied by the liberation of heat. This heat leads to the evaporation of water from the solid matter that results in greater compaction of the organic matter. To prevent an unfavorable outcome, it is advantageous to add some structural ingredients to the organic matter as well as to install adequate control systems for maintaining O_2 and CO_2 levels and water content within the acceptable range. Mason (2007) and Oudart (2014) modeled the O_2 and CO_2 profiles for a better understanding of the composting process efficiency (Figure 12.7c). The second set of considerations that bioengineers must evaluate to achieve successful composting is process kinetics (Hamelers 2001). Mason (2007) provides excellent explanations of mathematical models of the composting process that have been derived from both energy and mass balance considerations, with solutions typically derived in time, and, in some cases, spatially (Figure 12.7d,e). Microbiologists, on the other hand, regard growth and organic matter utilization from the microorganisms' perspective. Empirical degradation models or the classic Monod growth model are typically employed by microbiologists, while also taking into consideration inhibition by foreign matters like heavy metals (Figure 12.7f). The last step in the evaluation of the composting process is the monitoring of specific endpoint parameters. They provide clear indications that the composting process is reaching its steady state and that the product can be used and safely transferred from the facility to other ecosystems such as farmland, municipal parks, and residential areas. For example, as established in the scientific literature, the endpoint of the degradation parallels the microbiome's respiration activity, which can be measured by the monitoring of CO_2 production (Brinton and Evans 1993) or O_2 consumption (S.C.O. Canada 2016) (Figure 12.7g). Another indicator of process completion can be obtained by measuring the temperature at the center of the compost pile. Finally, to complete the compost's quality and eco-safety assessment, one needs to measure the levels of trace inorganics (heavy metals) (CCME 2005) (Figure 12.7h), as well as to confirm no microbial pathogens are present, as per regulatory guidelines (CCME 2005; Hébert 2012) (Figure 12.7i).

12.5.3 Materials Half-Life Under Composting Conditions

Composting kinetics modeling is necessary to design and operate composting facilities that comply with strict market demands and stringent environmental regulations (Hamelers 2001). One of the most commonly used approaches in modeling the composting kinetics is the substrate degradation models, where the independent variable is either the quantity or concentration of the organic materials remaining. The process kinetics are of great importance because they allow for quantification of the recalcitrant substrate or contaminant at any time; they can also formulate predictions on concentrations likely to be present at any point in time as prescribed by either a test method or by other considerations (e.g. site-specific climate). In complex environments such as a compost pile or contaminated soil, the biodegradation rate constant (k) of organic compounds by microorganisms is often described by mathematical equations typical of first-order environmental fate models. In other words, since the degradation of organic matter as a function of time (t) follows first-order kinetics, the rate of degradation can be expressed by the by Eq. (12.1):

$$\frac{d[OM]}{dt} = -k[OM] \tag{12.1}$$

where [OM] is the concentration of biodegradable organic matter at any time, in grams; t is the time, in days; and k is the first-order reaction rate constant, in day^{-1}. Integrating Eq. (12.1) by letting $[OM] = [OM]_0$ at $t = 0$, the concentration of organic matter at any time in the compost mass can be expressed as Eq. (12.2):

$$\ln\left(\frac{[OM]}{[OM]_0}\right) = -kt \tag{12.2}$$

Once k can be obtained, the material's half-life can be calculated using Eq. (12.3):

$$t_{1/2} = \frac{\ln(2)}{k} \tag{12.3}$$

The biological half-life is the time required for a substance to lose half of its mass. Biodegradation half-lives are needed for many applications, such as chemical screening, environmental fate modeling, and describing the transformation of pollutants.

The following are a few examples of estimated materials or pollutants half-lives once in contact with a soil or a compost. A soil burial biodegradation experiment of three cellulosic fabrics made of rayon, cotton, and Lyocell®, incubated up to 112 days, showed $t_{1/2}$ of 22, 40, and 94 days, respectively (Warnock et al. 2011). Under similar experimental conditions, Nam et al. (2015) demonstrated that rayon and raw cotton fibers showed $t_{1/2}$ of 7.6 and 12.6 d, respectively. Whereas PP fibers and PLA fibers are not biodegradable as no weight loss was observed. Likewise, Pradhan et al. (2010) evaluated the extent of degradation of different green PLA-based composites, in simulated composting bioreactor, as per ASTM D5338. All materials showed 85–100% degradation following 55–100 d incubation. Finally, according to our own laboratory composting tests in accordance with ASTM D5338, we determined that a 100 g sample of cutlery, bags, or plates (1 cm^2 pieces) buried in 1.2 kg of wet compost undergoes complete degradation in less than 60 d at 58 °C (unpublished results); while a fine microcrystalline cellulose powder (20 μm particle size) was used as a positive reference material. Our results indicate that cellulose has a half-life ($t_{1/2}$) of 5.0 days and a biodegradation rate constant (k) of 0.140 per day. As established by ASTM D5338 and ISO 14855, the two important threshold values, i.e. set endpoints that relate specifically to biodegradation kinetics are the maximal processing time and temperature (180 days at 58 °C) and the level of mineralization to be attained; total C-CO_2 measured must represent a minimum of 90% of the theoretical C content. This test method measures the biodegradation that leads to mineralization of a test material, i.e. the respired CO_2 by the microorganisms. This method does not consider the amount of organic C converted into the residual humus-like material obtained at the end of the long composting process, which lasted between two and six months. Over such a long incubation period, from an ecological point of view, the composting of organic waste is similar to the process of humification of the organic matter (leaves, etc.) observed in the forest. In addition, t_0, the initial concentration of the test material $[OM]_0$ is the third parameter to be used in the environmental fate model, which makes it possible to estimate the maximum theoretical $t_{1/2}$ value of the test material. For some reasons, ASTM D5338 set the initial concentration at a wet weight ratio of 8.5%, i.e. 100 g of test material buried in 1200 g of wet weight compost. From these three constraints, it is possible to estimate a maximum theoretical $t_{1/2}$ of 43 days, beyond which a newly designed eco-material could not be qualified, or even certified, as compostable under conditions of aerobic thermophilic composting.

12.5.4 Biodegradation in the Environment

In this section, we describe the journey of a green electronics that would end up in a composting installation through municipal compost bins (Figure 12.1).

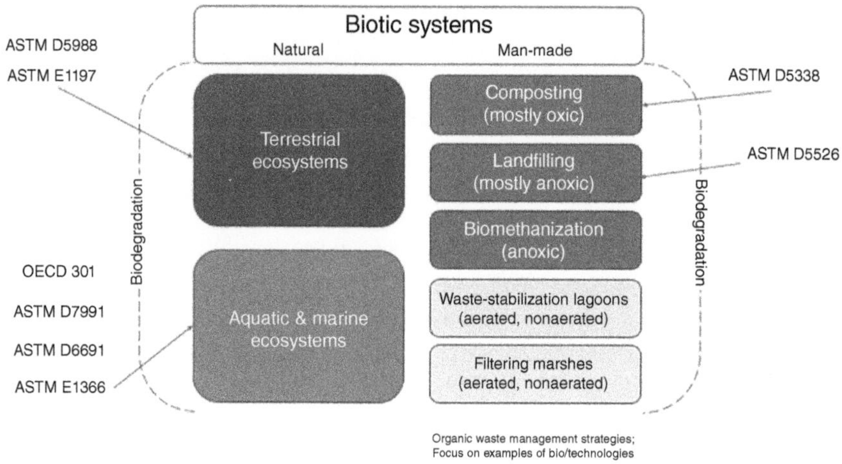

Figure 12.8 Organic waste treatment facilities can be considered as ecosystems on their own. Technical standards and guidelines created by ASTM, ISO, OECD, CAN/BNQ and other regulation bodies aim to indicate methodology, testing conditions, expected endpoint results, and, if applicable, eco-safety tests.

Indeed, after having extensively explored the end of life of organic electronics in a composting facility, it seems justified and very relevant to say a few words about other plausible end-of-life scenarios. Among other possibilities, we believe that green electronics could be intended for short-term uses such as monitoring or spotting in the natural environment (e.g. forest, lakes, marine ecosystems). Consequently, other test methods are to be envisaged to qualify a material as "biodegradable." Figure 12.8 gives a brief overview of the complexity of the situation regarding the definition of a "biodegradable product." For an extensive discussion on the subject of standard methods and guidelines, the readers are referred to these excellent references (Haider et al. 2019; Tian and Bilal 2020).

12.6 Microbiome in Natural and Industrial Ecosystems

Conceptually, in microbial ecology, microbiome refers to an interdependent microbial community that can be described by its functional microbial groups specialized in the utilization of particular biochemical substrates as their energy source. In essence, a microbiome emerges as a "product" of coevolution between a host and its environment resulting in a uniquely shaped community structure.

Their composition is most commonly studied using DNA metagenomic techniques (de Gannes et al. 2013), although classical microbiology techniques can also be applied. Whereas DNA metagenomic techniques allow for identification and quantification of microbes present in an environment (Figure 12.9b), classical microbiology has its own unique advantages. Culture methods allow for morphological observations as well as the detection of specific substrate usage and enzyme activity (Figure 12.9a). The application of both techniques has allowed

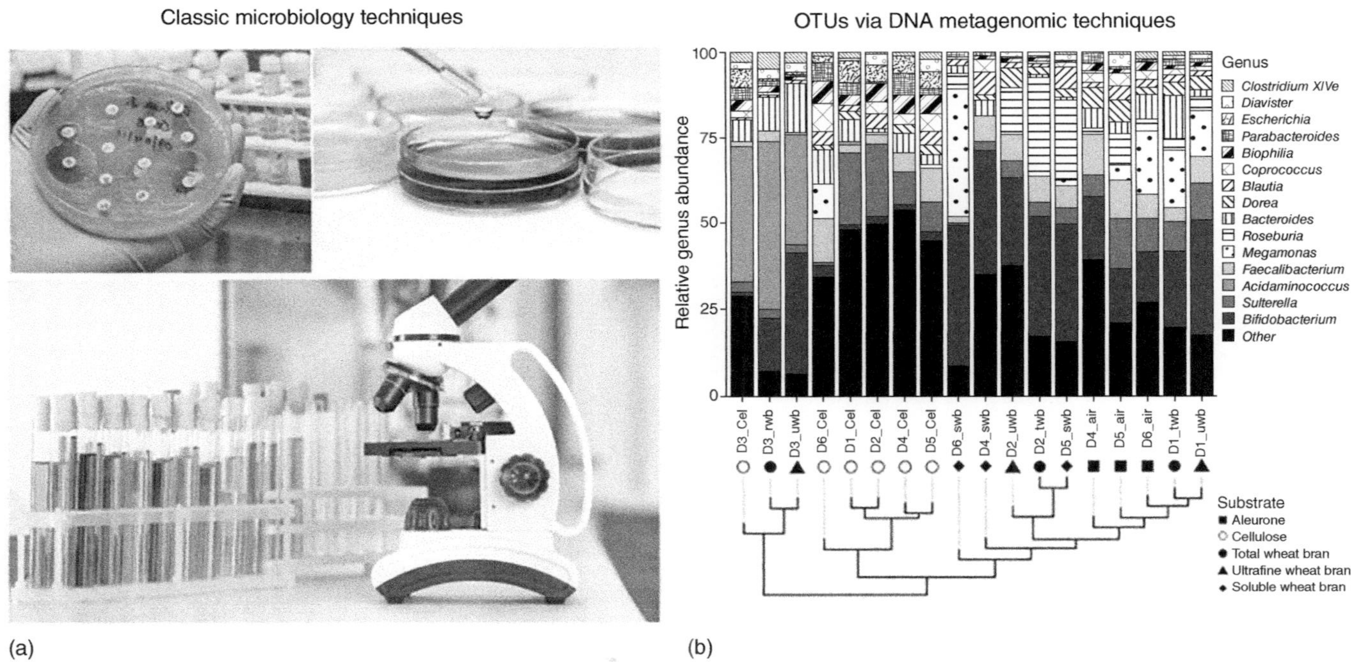

Figure 12.9 Microbiology techniques used in the study of microbiomes. (a) Classic microbiology techniques allow observations about morphology, function, isolation, and quantification of microbes. (b) DNA metagenomic techniques measure microbiome diversity and precisely describe composition, phylogeny and taxonomy of microbial communities. Source: (b) D'hoe et al. (2018). Licensed under CC BY 4.0.

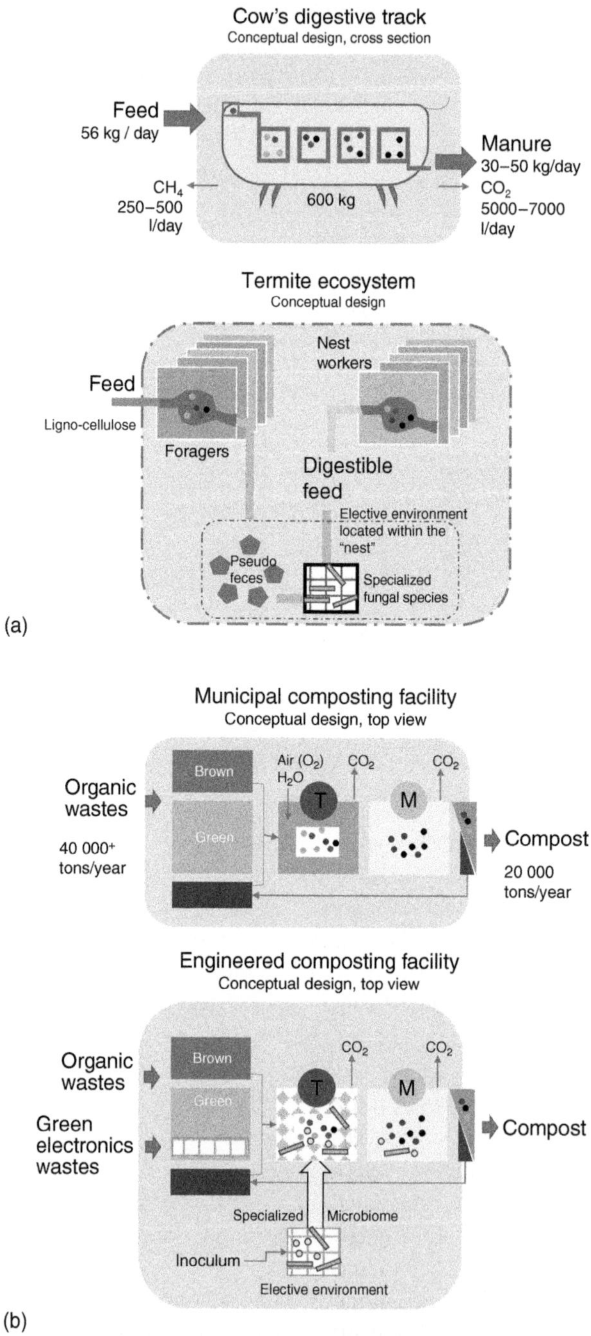

Figure 12.10 Schematic representation of (a) two natural ecosystems, i.e. cow and termites, and (b) two composting facilities, i.e. municipal and engineered.

microbiologists to make significant advancements in the characterization of microbiomes. Three specific microbiomes are described below (Figure 12.10) to provide the reader with a better understanding of each of their counterparts, namely three artificial ecosystems intended for the management of waste or pollutants. Interestingly, the analysis of these man-made ecosystems reveals that they are designed and operated as their "natural" counterparts. To position biodegradable green electronics in the main recycling strategy (Figure 12.1), engineered composting technology will be proposed as well as temporary options will be discussed to deal with the presence of heavy metals.

Details of two natural ecosystems will be presented to provide a basic understanding for the presentation of two composting strategies: firstly, the composting of (regular) organic waste inspired by nature, i.e. the rumen–hay ecosystem; then secondly, a biological engineering composting of organic waste with green electronics also inspired by nature, i.e. the termite–wood ecosystem.

12.6.1 The Ruminant–Hay Natural Ecosystem

An example (Figure 12.10) of a unique host–microbiome relationship is the rumen. The primary diet of dairy cows consists of lignocellulose biomasses or hay (43%, dry weight) and forage corn silage (47%). The biochemical composition of such substrate, which includes numerous complex and highly organized macromolecules such as lignin, cellulose microfibrils, hemicellulose, pectin, starch granules, fructan, organic acids, and protein, has led to the selection of a rumen microbiome adapted for efficient digestion. Together with mastication, an array of depolymerizing enzymes degrades the lignocellulosic biomass ingested by the ruminants. It is interesting to note that among the general microbial population, those identified as being largely responsible for the biodegradation process of food constituents can be grouped into a few functional microbial groups by means of DNA technology. With that, Söllinger et al. (2018) revealed the presence of bacteria, archaea, fungi, and ciliates in the rumen. The anaerobic bacteria *Prevotella*, *Butyrivibrio*, *Ruminococcus*, *Lachnospiraceae*, *Ruminococcaceae*, *Bacteroidales*, and *Clostridiales* constitute the predominant functional group in the rumen, representing over half of the microbiome. The other three functional groups are less predominant than the first group but are nevertheless crucial for the digestion of the lignocellulosic biomass. Their species belong to the ciliated protozoa, which make up over a third of the microbial population, and to *Methanobrevibacter archaea* and fungi, which represent only a small fraction of the microbiome (Matthews et al. 2019; Henderson et al. 2015). Briefly, the key role of anaerobic bacteria is to initiate the enzymatic depolymerization of most non-water-soluble macromolecules mentioned above, by means of the production of a range of extracellular cellulases (endo-β-1,4-glucanase, cellobiohydrolase, and β-glucosidase), hemicellulases (endoxylanase, β-xylosidase, and glucuronoxylan hydrolase), amylases, and other glucanases (Söllinger et al. 2018). Essentially, this enzymatic process leads to the production of water-soluble oligomers and monomers, which will then participate in a cascade of beneficial events for the host and the rumen microbiome, which is the true definition of a symbiotic relationship. So far, we have covered the microbiology of the C-cycle, but the N-cycle is just as important to the health of the host. For the cows, a functional group has evolved to accomplish the latter cycle, but other smaller ruminants (e.g. rabbits, hares, and pikas) have overcome their limited digestive capacity due to their body size with a special kind of digestion called hindgut fermentation: to assimilate the N-source nutrients they need, they reingest their *cecotropes (autocoprophagy)*.

In terms of process efficiency, here are some interesting facts that should give the reader an idea of the magnitude of a rumen microbiome to accomplish its role. In addition to utilizing lignocellulose as their sources of C, N, and energy, rumen microbes must also be adapted to temperatures averaging 39 °C and a neutral pH range. These microbes allow a 500 kg dairy cow feeding on an average 50 kg feed per day to achieve a digestive passive rate in the rumen of approximately 60 hours. As shown in Figure 12.10, the cow's digestive system can be represented schematically by few interdependent compartments. Apart from the fact that rumen microbiome operates in an anaerobic environment, we foresee a conceptual similitude with the aerobic composting strategy (i.e. shredding the incoming ingredients, depolymerization during the thermophilic phase, and full mineralization during the mesophilic phase) explained hereafter.

12.6.2 The Termite–Wood Natural Ecosystem

As a second example, described in Figure 12.10, termites have become one of the most efficient natural "bioreactor" specialized in the degradation of lignocellulose. Termites have adopted two contrasting microbiomes as a result of having evolved into two orders: higher and lower. Their differing diets are what have driven the composition of their microbiomes in separate directions; lower termites feed principally on wood, whereas higher termites have a much broader diet. Lower termites possess complex symbiotic relationships with eukaryotic flagellates and bacteria, whereas higher termites have external symbiotic relationships with Basidiomycetes fungi and/or have a microbiome solely populated by prokaryotes. Flagellates of the groups *Tritrichomonadea*, *Hypotrichomonadea*, and *Parabasalid* are found in all lower termites and can actively maintain their position in the gut to prevent washout. Contrarily, *Macrotermes* higher termites possess a gut microbiota dominated by bacteria that belong to the genus *Treponema*. To compensate for their lack of cellulolytic protozoa, higher termites have adopted different diets, developed more elaborate anatomy, and increased alkalinity in their gut, which is believed to play an important role in lignin degradation. All of this is simply a remarkable and extraordinary symbiotic relationship with fungi of the genus *Termitomyces*, which allows the termites to use the fungi as an *extracorporeal* digestive system (Figure 12.9). To digest the lignocellulosic components of the woody material collected by forager termites, the nest worker termites construct fungus combs from the pseudofeces excreted by the foragers. This undigested macerated woody material is shaped into fungus combs by the nest workers, which provide a structure for the *Termitomyces* fungus to grow in. Both the foragers and the nest worker termites are responsible for the inoculation of fungi spores in the combs. These spores will eventually germinate, grow, and propagate into the comb by means of their hyphae and their extracellular lingocellulolytic enzymes, which provides the termites with easily digestible substrates (Vesala et al. 2017).

Since the substrate consumed by termites is quite similar to the diet of ruminants, there are many overlapping enzymes produced by the termite gut microbiome, namely those of the glucanase family (amylases, cellulases, β-glucosidases, and hemicellulases). Previous reports suggested that within 24 hours, termites can convert 74–99% of cellulose, 65–87% of hemicellulose, and up to 83% of lignin into smaller and easily digestible polysaccharides such as acetate and glucose (Wong et al. 2014). Compared to their lignocellulose-degrading counterpart, ruminants, which only decompose 35% of polysaccharides in their forage, it becomes clear why termites are considered as some of the most efficient organisms on Earth at degrading lignocellulose (Wong et al. 2014).

12.6.3 The Industrial Composter–Biowaste Ecosystem

12.6.3.1 Municipal Composting Facility

As the practice of composting has been around for some time, the microbial community dynamics and composition of the process have been extensively studied. As shown in Figure 12.5, composting is a two-step process; biodegradation commences with the active degradation phase, followed by a maturation phase. The high temperatures of the active degradation phase are only suitable for microbes well adapted to those extreme conditions, namely thermophiles. The majority of the degradation occurs during the thermophilic phase owing to the fact that the increased temperature accelerates the breakdown of the organic matter in addition to the cellulases, hemicellulases, lipases, and lignocellulolytic enzymes primarily produced by Actinomycetes, Bacilli, and Thermus bacterial thermophiles (Trautmann and Olynciw 2020). Once the carbon substrate availability becomes insufficient for the thermophiles, the temperature of the compost gradually decreases leading to mineralization process which is accomplished over a much longer time period. This maturation phase is when the mesophilic microbes begin to recolonize the residual organic matter. The predominant organisms that take over during this phase include acid-producing bacteria like *Lactobacillus* spp., *Acetobacter* spp., as well as mesophilic fungi, and *Acinetobacter johnsonii*, *Mesorhizobium thiogangeticum*, and *Flavobacterium ceti*. Other predominant mesophilic species include *Staphylococcus sciuri*, *Bacillus flexus*, *B. subtilis*, and *B. pumilus*, with the latter two bacterial species also being prevalent in the thermophilic phase of composting. During the maturation phase, fungi are one of the most crucial microbes in the process of degradation as they can produce extracellular enzymes capable of breaking down lignin, which many of the thermophiles are incapable of producing (Palaniveloo et al. 2020). Altogether, microbial community dynamics is one of numerous components that must be well understood for successful composting results.

Mass balance is a significant factor to be assessed to measure the actual performance of organic waste treatment plants, as highlighted by Pognani et al. (2012). Their mass balance results (calculated and expressed in terms of t/y of wet [total] matter, carbon, nitrogen, and phosphorus) show valuable information. Among others, here are three relevant facts: (i) the compost final weight represents only 27.6% of the initial wet matter; (ii) it could be found in 49.7% of the initial N_{tot} and 86.4% of P_2O_5; (iii) but only 21.6% of the total carbon treated in the facility. The final compost also achieves a high level of stabilization with a dynamic respiration index of 0.3 ± 0.1 g O_2 per kg of total solids per hour, which represents a reduction of 93% from that of the raw organic waste. The work of Zhang and Matsuto (2011) is also valuable. They investigated the performance of several composting facilities that treat different types of organic waste by questionnaire survey and via chemical analysis of material collected at the facilities.

Although microbial process, community, and efficiency vary greatly between the two, the composting biodegradation process can be seen as an example of a man-made version of the cow's digestive system (Figure 12.9). The composting biodegradation process can also be seen as an example of a man-made version of the cow's digestive system. Thus, just as the cow's digestive system can be seen as a superior bioprocessing strategy, which has evolved the appropriate abiotic and biotic factors (listed above), the composting facility offers an "adequate" environment for the thermophilic and mesophilic microbiomes to degrade cellulosic substrates (organic waste) in a two-step process. Although the degradation process of both ecosystems differs in many aspects, particularly in regard to the oxic condition (presence/absence of O_2), the microbial activity, and diversity, the starting material and end products of either process are quite similar. The resulting product of the ruminant digestion process is manure, whereas that of the compost process is humus; both are stable and rich in nutrients, including nitrogen, phosphorus, and potassium, which makes them both excellent organic alternatives to commercial fertilizers.

12.6.3.2 Engineered Composting Facility

As we begin to better understand environment–microbe and microbe–microbe relationships and functions, we are starting to develop biotechnology methods to benefit from these microbiomes and to enhance biological processes. Dairy production enhanced through rumen transfaunation is a prime example of the role of the microbiome (Henderson et al. 2015). Similar to the consumption of probiotic foods among humans, the cow rumen is inoculated to improve digestion of the feed. With the conceptual strategy of modifying the microbiome's structure in mind, the following includes some interesting strategies to promote a higher rate of composting of organic matter by adding inoculant and also address the issue of heavy metals.

12.6.4 Specialized Inoculant Adapted to Organic Matter

In the initial phase of a composting process characterized by a temperature range of 40 to 70 °C, Wei et al. (2019) used the thermophilic actinomycetes to inoculate cellulose-, hemicellulose-, and lignin-rich substrates. When they compared results with those from the noninoculated compost pile, they found that the inoculated bacteria had led to an increase of over 20% in the degradation rate for each of the three substrates. On the other hand, Bolta et al. (2003) conducted an experiment in which the inoculant used was obtained directly from the active phase of a compost. At t_0, the inoculated material contained six times more viable microbial biomass than the noninoculated one. It was clear that the inoculated reactor had entered immediately into the thermophilic phase rather than after six days for the noninoculated reactor. The degradation also increased in the inoculated compost, where 37.8% of organic matter was mineralized after 198 days, as opposed to 9.6% in the noninoculated compost. Similarly, Karnchanawong and Nissaikla (2014) investigated the acceleration of the composting process by either using a mature compost's microbiome or a commercial inoculant. Their results demonstrated clearly that adding a mature compost's microbiome was significantly beneficial as the degradation of the organic matter was found to be improved by 20%. Finally, Manu et al. (2017) also investigated the effect of inoculating a compost pile with an active microbial consortium. Both the inoculated and noninoculated composts had a starting weight of 80 kg, their final mass after 60 days was 19 and 48 kg, respectively. Interestingly, lignin reduction in the inoculated compost was 78%, whereas it was 50% in the noninoculated compost. In short, across all of the above-mentioned studies, the inoculation strategy apparently improved the degradation rate of organic matter by over 20%.

Accordingly, the inoculation of compost is of great interest to scientists developing biodegradable electronics. As discussed, the vision for the disposal of green electronics is to have consumers to dispose some of their electronics in their household compost bins. Hence, it is key that we keep exploring a variety of microbial consortia that demonstrate high biodegradation efficiency. Although the functions (i.e. metabolic capabilities) of most microbes in compost are largely understudied, it has been widely accepted that "efficient microbes" can significantly increase the degradation rate (Rastogi et al. 2020). These microbes can consist of a single bacterial strain grown in a lab, or a consortium of bacteria which have been recovered from mature compost (Rastogi et al. 2020).

12.6.5 Specialized Inoculant Adapted to Heavy Metals

As environmental engineers become increasingly aware of the fundamental role of microbiomes, it has become evident that the path to successful composting of green electronics will depend on, firstly, the formulation of the chemical species and heavy metals used, and secondly but as importantly, their inhibitory effect on the microbial metabolism of the compost microbial community. Therefore, the formulation of chemical species is a direct determinant of recalcitrance to biodegradation, while heavy metals continue to pose a significant problem for the composting process (Wuana and Okieimen 2011; Petruzzelli 1996). This being said, ongoing research activities reveal promising results concerning heavy metals in compost (Zennaro et al. 2005; Singh and Kalamdhad 2013). Recent studies have shown that most microbes have minimum inhibitory concentrations of heavy metals at which their metabolic activity remains unaffected. For example, the soil bacterium *Alcaligenes eutrophus* is capable of metabolizing 2,4-dichlorophenoxyacetic acid, namely in the presence of Cd^{2+} at a concentration of 60 μg/g or lower (Sandrin and Maier 2003). Moreover, an indigenous soil microbial community is capable of tolerating 1.0 mg/l of Cd^{2+} without the rate of phenanthrene degradation being affected (Sandrin and Maier 2003). Similarly, Damasceno et al. (2015) studied the oxidation state of heavy metals (Sn, Pb, Ni, Zn, and Cu) from electronic waste as chopped printed circuit boards (PCBs) in compost. They buried 2 kg of PCBs in 58 kg of compost and were thus able to confirm that microbes actively oxidize the metals present in their environment by means of powerful oxidants such as 2-ketogluconic acid. Indeed, the "danger" of the solubilization of metal ions is that they migrate toward the organic fraction of the compost, and lead to an inhibition of the microbial activity involved in the composting of organic matter. Equally important is that the end product is classified as not eco-safe. Another avenue that presents a possible solution to the inhibitory effects of heavy metals is to inoculate a compost with microbes that possess heavy metal resistance genes. For example, Liu et al. (2019) were capable of identifying a novel heavy metal–tolerance gene in *Burkholderia fungorum*. This bacterium, isolated from an oil-contaminated soil in Xinjiang, was capable of utilizing phenanthrene (300 mg/l) as a sole C source in the presence of 40, 400, and 1200 mg/l of Pb(II), Cd(II), and Zn(II), respectively. Another study (Gupta et al. 2012) revealed that some Firmicutes Gram+ bacteria isolated from the rhizosphere of plants, can tolerate a high concentration of heavy metals such as Co, Ni, Pb, Cr, and Fe. Further microbiological analyses show that the alkaliphilic *Bacillus weihenstephanensis* and *Exiguobacterium aestuarii*, showing high Ni_2^+ and Cr_6^+ removal capabilities, respectively, can be prospective candidates for bioremediation. The studies mentioned above provide evidence that introducing green electronics into a compost with a specialized inoculum should successfully restructure the existing microbiome while maintaining the same rate of degradation. At first glance, this composting strategy (Figure 12.10) appears a little sophisticated or even complex, it is nonetheless realistic and feasible.

12.7 Concluding Remarks and Perspectives

The challenge is to define environmental requirements and characteristics already in the conceptual design and initial phases, where the selection of the materials to be used in the product development is made (Morini et al. 2019; Allione et al. 2012). The tailoring of green electronics to these environmental setpoints, established by organizations such as ASTM, ISO, and OECD, remains a challenge as they are ecosystem specific.

Contrary to the environmental fate of plastic waste, our understanding of green electronics biodegradation in different ecosystems is very limited. The biodegradation principle, i.e. that the material's structure and chemical composition (e.g. degree of substitution on OH groups of polymer backbone structure, benzene rings and double bonds, cross-linkages between macromolecules, crystallinity, etc.) dictates the overall degradation rate of a material, should guide the research efforts. The astonishing work of Chamas et al. (2020) on scrutinizing existing literature to obtain environmental degradation rates and pathways for the major types of thermoplastic polymers as well on a metric based on specific surface degradation rate to extrapolate half-lives, should lead the way for future work in the field of end-of-life scenario for green organic electronics.

Process efficiency of most environmental biotechnologies, such as composting or anaerobic digestion, can be improved either by changing the reactor design (optimal volume-to-surface ratio) and its processing strategy (better aeration and mixing) or by selecting, cultivating, and reinoculating the compost pile with a specialized microbiome.

With the knowledge that has been accumulated on microbiomes as well as our continuous efforts to know more about them, they could become key players in demonstrating compostable organic electronics. Interestingly, the proposed advanced industrial composting process is truly *Inspired by Nature* as it mimics the termite microbiome, one of the most advanced ecosystems, perfected by millions of years of coevolution. Moreover, the characterization of microbiomes found within man-made environments is a promising start to overcome some of the hurdles that come with compostable organic electronics, such as chemical heterogeneity and heavy metal contamination. Certain microbes can adapt to evolving environmental conditions, a promising start for answering some of the questions regarding the end of life of biodegradable electronics.

Acknowledgment

We thank Barbara Ann Jardin, MScA, Bioengineering, McGill University, for her valuable comments and suggestions regarding process kinetics and modelization.

References

Abu Qdais, H. and Al-Widyan, M. (2016). Evaluating composting and co-composting kinetics of various agro-industrial wastes. *International Journal of Recycling of Organic Waste in Agriculture* 5 (3): 273–280.

Ajjan, F., Casado, N., Rębiś, T. et al. (2016). High performance PEDOT/lignin biopolymer composites for electrochemical supercapacitors. *Journal of Materials Chemistry A* 4 (5): 1838–1847.

Ajjan, F.N., Mecerreyes, D., and Inganäs, O. (2019). Enhancing energy storage devices with biomacromolecules in hybrid electrodes. *Biotechnology Journal* 14 (12): 1900062.

Akiyama, H., Fujii, K., Yamasaki, O. et al. (2001). Antibacterial action of several tannins against *Staphylococcus aureus*. *Journal of Antimicrobial Chemotherapy* 48 (4): 487–491.

Aleshin, A., Lee, J., Chu, S. et al. (2004). Mobility studies of field-effect transistor structures basedon anthracene single crystals. *Applied Physics Letters* 84 (26): 5383–5385.

Allen, J.F. and Martin, W. (2007). Evolutionary biology: out of thin air. *Nature* 445 (7128): 610.

Allione, C., De Giorgi, C., Lerma, B., and Petruccelli, L. (2012). From ecodesign products guidelines to materials guidelines for a sustainable product. Qualitative and quantitative multicriteria environmental profile of a material. *Energy* 39 (1): 90–99.

Alves, D., Villar, I., and Mato, S. (2019). Thermophilic composting of hydrocarbon residue with sewage sludge and fish sludge as cosubstrates: microbial changes and TPH reduction. *Journal of Environmental Management* 239: 30–37.

Arias, O., Viña, S., and Soto, M. (2019). Co-composting of forest and industrial wastes watered with pig manure. *Environmental Technology* 42 (5): 705–716.

Baumgartner, M., Coppola, M.E., Sariciftci, N.S. et al. (2017a). Emerging "green" materials and technologies for electronics. In: *Green Materials for Electronics* (eds. M. Baumgartner, M.E. Coppola, N.S. Sariciftci and E.D. Glowacki), 1–53. Weinheim: Germany: Wiley-VCH.

Baumgartner, M., Coppola, M.E., Sariciftci, N.S. et al. (2017b). *Emerging "green" materials and technologies for electronics*. In: *Green Materials for Electronics* (ed. M. Irimia-Vladu), 101. Weinheim, DE: Wiley-VCH.

Berg, D., Nielinger, C., Mader, W., and Sokolowski, M. (2009). Quinacridone organic field effect transistors with significant stability by vacuum sublimation. *Synthetic Metals* 159 (23–24): 2599–2602.

Berggren, M. and Malliaras, G.G. (2019). How conducting polymer electrodes operate. *Science* 364 (6437): 233–234.

Bettinger, C.J., Bruggeman, J.P., Misra, A. et al. (2009). Biocompatibility of biodegradable semiconducting melanin films for nerve tissue engineering. *Biomaterials* 30 (17): 3050–3057.

Blankenship, R.E. (2014). *Molecular Mechanisms of Photosynthesis*, 2e. Wiley-Blackwell.

Bolta, S.V., Mihelic, R., Lobnik, F., and Lestan, D. (2003). Microbial community structure during composting with and without mass inocula. *Compost Science & Utilization* 11 (1): 6–15.

Brinton, W.F. Jr., and Evans, E. (1993). *On-Farm Composting: Guidelines for Use of Dairy and Poultry Manures in Composting Formulations*. Mount Vernon, ME: Woods End Research Laboratory.

Butko, V., Chi, X., and Ramirez, A. (2003). Free-standing tetracene single crystal field effect transistor. *Solid State Communications* 128 (11): 431–434.

CCME (2005). *Guidelines for Compost Quality*. Canada.

C40 Group (2020). C40: Zero waste declaration. https://www.c40.org/other/zero-waste-declaration (accessed 23 March 2021).

Chamas, A., Moon, H., Zheng, J. et al. (2020). Degradation rates of plastics in the environment. *ACS Sustainable Chemistry & Engineering* 8 (9): 3494–3511.

Chang, J.-K., Fang, H., Bower, C.A. et al. (2017). Materials and processing approaches for foundry-compatible transient electronics. *Proceedings of the National Academy of Sciences of the United States of America* 114 (28): E5522–E5529.

Cheng, M., Yang, X., Zhang, F. et al. (2012). Efficient dye-sensitized solar cells based on hydroquinone/benzoquinone as a bioinspired redox couple. *Angewandte Chemie International Edition* 51 (39): 9896–9899.

Cheng, M., Yang, X., Chen, C. et al. (2013). Dye-sensitized solar cells based on hydroquinone/benzoquinone as bio-inspired redox couple with different counter electrodes. *Physical Chemistry Chemical Physics* 15 (36): 15146–15152.

Dagdeviren, C., Hwang, S.-W., Su, Y. et al. (2013). Transient, biocompatible electronics and energy harvesters based on ZnO. *Small* 9 (20): 3398–3404.

Damasceno, O.I.d.C., Reis, C., Reis, E.L. et al. (2015). Assessment of bioavailability of heavy metals after vermicomposting in the presence of electronic waste. *Revista Brasileira de Ciência do Solo* 39: 1786–1795.

De Boer, R., Klapwijk, T., and Morpurgo, A. (2003). Field-effect transistors on tetracene single crystals. *Applied Physics Letters* 83 (21): 4345–4347.

D'hoe, K., Conterno, L., Fava, F. et al. (2018). Prebiotic wheat bran fractions induce specific microbiota changes. *Frontiers in Microbiology* 9: 31.

Ding, Y. and Yu, G. (2016). A bio-inspired, heavy-metal-free, dual-electrolyte liquid battery towards sustainable energy storage. *Angewandte Chemie International Edition* 55 (15): 4772–4776.

d'Ischia, M., Napolitano, A., Ball, V. et al. (2014). Polydopamine and eumelanin: from structure–property relationships to a unified tailoring strategy. *Accounts of Chemical Research* 47 (12): 3541–3550.

d'Ischia, M., Napolitano, A., Pezzella, A. et al. (2020). Melanin biopolymers: tailoring chemical complexity for materials design. *Angewandte Chemie International Edition* 59 (28): 11196–11205.

Dong, T., Zhang, J., Xu, G. et al. (2018). A multifunctional polymer electrolyte enables ultra-long cycle-life in a high-voltage lithium metal battery. *Energy & Environmental Science* 11 (5): 1197–1203.

European Union (2006). Directive 1907, regulation on registration, evaluation, authorisation and restriction of chemicals (REACh).

European Union (2012). Directive 19, regulation on WEEE management.

European Union (2015). Directive 863, restriction of hazardous substances directive (RoHS).

Fabara, A.N. and Fraaije, M.W. (2020). An overview of microbial indigo-forming enzymes. *Applied Microbiology and Biotechnology* 104 (3): 925–933.

Feig, V.R., Tran, H., and Bao, Z. (2018). Biodegradable polymeric materials in degradable electronic devices. *ACS Central Science* 4 (3): 337–348.

Floudas, D. and Numerous Collaborators (2012). The paleozoic origin of enzymatic lignin decomposition reconstructed from 31 fungal genomes. *Science* 336 (6089): 1715–1719.

Forti, V., Balde, C.P., Kuehr, R., and Bel, G. (2020). *The Global E-waste Monitor 2020: Quantities, Flows and the Circular Economy Potential*. Bonn, Geneva, Rotterdam: United Nations University/United Nations Institute for Training and Research, International Telecommunication Union, and International Solid Waste Association.

Fu, K.K., Wang, Z., Dai, J. et al. (2016). Transient electronics: materials and devices. *Chemistry of Materials* 28 (11): 3527–3539.

de Gannes, V., Eudoxie, G., and Hickey, W.J. (2013). Prokaryotic successions and diversity in composts as revealed by 454-pyrosequencing. *Bioresource Technology* 133: 573–580.

Głowacki, E.D., Voss, G., Leonat, L. et al. (2012). Indigo and tyrian purple–from ancient natural dyes to modern organic semiconductors. *Israel Journal of Chemistry* 52 (6): 540–551.

Głowacki, E.D., Irimia-Vladu, M., Kaltenbrunner, M. et al. (2013). Hydrogen-bonded semiconducting pigments for air-stable field-effect transistors. *Advanced Materials* 25 (11): 1563–1569.

Gnedenkov, S.V., Opra, D.P., Zemnukhova, L.A. et al. (2015). Electrochemical performance of Klason lignin as a low-cost cathode-active material for primary lithium battery. *Journal of Energy Chemistry* 24 (3): 346–352.

Goriparti, S., Harish, M., and Sampath, S. (2013). Ellagic acid – a novel organic electrode material for high capacity lithium ion batteries. *Chemical Communications* 49 (65): 7234–7236.

Gouda, A., Bobbara, S., Reali, M., and Santato, C. (2019). Light-assisted melanin-based electrochemical energy storage: physicochemical aspects. *Journal of Physics D: Applied Physics* 53: 043003–043019.

Gupta, K., Chatterjee, C., and Gupta, B. (2012). Isolation and characterization of heavy metal tolerant gram-positive bacteria with bioremedial properties from municipal waste rich soil of Kestopur canal (Kolkata), West Bengal, India. *Biologia* 67 (5): 827–836.

Haider, T.P., Völker, C., Kramm, J. et al. (2019). Plastics of the future? The impact of biodegradable polymers on the environment and on society. *Angewandte Chemie International Edition* 58 (1): 50–62.

Hamelers, H.V.M. (2001). *A Mathematical Model for Composting Kinetics*, 295. Wageningen, Netherlands: University of Wageningen.

Han, C., Li, H., Shi, R. et al. (2019). Organic quinones towards advanced electrochemical energy storage: recent advances and challenges. *Journal of Materials Chemistry A* 7: 23378–23415.

Hathway, D. (1958). Oak-bark tannins. *Biochemical Journal* 70 (1): 34–42.

Hébert, M. (2012). Les critères de qualité des composts du BNQ et du CCME.

Heeger, A.J. (2001). *Semiconducting and Metallic Polymers: The 4th Generation of Polymeric Materials.* ACS Publications.

Henderson, G., Cox, F., Ganesh, S. et al. (2015). Rumen microbial community composition varies with diet and host, but a core microbiome is found across a wide geographical range. *Scientific Reports* 5 (1): 14567.

Hernández, G., Casado, N., Zamarayeva, A.M. et al. (2018). Perylene polyimide-polyether anodes for aqueous all-organic polymer batteries. *ACS Applied Energy Materials* 1 (12): 7199–7205.

Hirai, N., Tanaka, T., Kubo, S. et al. (2006). Density and hardness of negative pastes of lead–acid batteries containing organic additives with or without quinone structure. *Journal of Power Sources* 158 (2): 1106–1109.

Hong, L. and Simon, J.D. (2007). Current understanding of the binding sites, capacity, affinity, and biological significance of metals in melanin. *The Journal of Physical Chemistry B* 111 (28): 7938–7947.

Huskinson, B., Marshak, M.P., Suh, C. et al. (2014). A metal-free organic–inorganic aqueous flow battery. *Nature* 505 (7482): 195–198.

Hwang, S.-W., Tao, H., Kim, D.-H. et al. (2012). A physically transient form of silicon electronics. *Science* 337 (6102): 1640–1644.

Hwang, S.-W., Huang, X., Seo, J.-H. et al. (2013). Materials for bioresorbable radio frequency electronics. *Advanced Materials* 25 (26): 3526–3531.

Hwang, S.-W., Park, G., Cheng, H. et al. (2014a). 25th anniversary article: materials for high-performance biodegradable semiconductor devices. *Advanced Materials* 26 (13): 1992–2000.

Hwang, S.-W., Park, G., Edwards, C. et al. (2014b). Dissolution chemistry and biocompatibility of single-crystalline silicon nanomembranes and associated materials for transient electronics. *ACS Nano* 8 (6): 5843–5851.

Hwang, S.W., Lee, C.H., Cheng, H. et al. (2015). Biodegradable elastomers and silicon nanomembranes/nanoribbons for stretchable, transient electronics, and biosensors. *Nano Letters* 15 (5): 2801–2808.

International Organization for Standardization (2013). ISO 14031, environmental management systems – environmental performance evaluation – guidelines. https://www.iso.org/obp/ui/#iso:std:iso:14031:ed-2:v1:en (accessed 23 March 2021).

International Organization for Standardization (2020). ISO 14006, environmental management systems – guidelines for incorporating ecodesign. https://www.iso.org/obp/ui/fr/#iso:std:iso:14006:ed-2:v1:en (accessed 23 March 2021).

Irimia-Vladu, M. (2014). "Green" electronics: biodegradable and biocompatible materials and devices for sustainable future. *Chemical Society Reviews* 43 (2): 588–610.

Irimia-Vladu, M., Głowacki, E.D., Troshin, P.A. et al. (2012). Indigo – a natural pigment for high performance ambipolar organic field effect transistors and circuits. *Advanced Materials* 24 (3): 375–380.

Irimia-Vladu, M., Kanbur, Y., Camaioni, F. et al. (2019). Stability of selected hydrogen bonded semiconductors in organic electronic devices. *Chemistry of Materials* 31 (17): 6315–6346.

Jo, K.Y., Wei, W., Sang-Eun, C. et al. (2014). Catechol-mediated reversible binding of multivalent cations in eumelanin half-cells. *Advanced Materials* 26 (38): 6572–6579.

Josée, D. (2006). Le compostage facilité : guide sur le compostage domestique.

Jung, Y.H., Chang, T.H., Zhang, H. et al. (2015). High-performance green flexible electronics based on biodegradable cellulose nanofibril paper. *Nature Communications* 6 (May): 1–11.

Kang, S.-K., Park, G., Kim, K. et al. (2015). Dissolution chemistry and biocompatibility of silicon- and germanium-based semiconductors for transient electronics. *ACS Applied Materials & Interfaces* 7 (17): 9297–9305.

Kang, S.K., Murphy, R.K.J., Hwang, S.W. et al. (2016). Bioresorbable silicon electronic sensors for the brain. *Nature* 530 (7588): 71–76.

Karnchanawong, S. and Nissaikla, S. (2014). Effects of microbial inoculation on composting of household organic waste using passive aeration bin. *International Journal of Recycling of Organic Waste in Agriculture* 3 (4): 113–119.

Kato, M., Cardona, T., Rutherford, A.W., and Reisner, E. (2012). Photoelectrochemical water oxidation with photosystem II integrated in a mesoporous indium–tin oxide electrode. *Journal of the American Chemical Society* 134 (20): 8332–8335.

Khanbabaee, K. and van Ree, T. (2001). Tannins: classification and definition. *Natural Product Reports* 18 (6): 641–649.

Kim, D.H., Lee, D.Y., Lee, H.S. et al. (2007). High-mobility organic transistors based on single-crystalline microribbons of triisopropylsilylethynyl pentacene via solution-phase self-assembly. *Advanced Materials* 19 (5): 678–682.

Kim, Y.J., Wu, W., Chun, S.-E. et al. (2013). Biologically derived melanin electrodes in aqueous sodium-ion energy storage devices. *Proceedings of the National Academy of Sciences of the United States of America* 110 (52): 20912–20917.

Kim, J.H., Lee, M., and Park, C.B. (2014). Polydopamine as a biomimetic electron gate for artificial photosynthesis. *Angewandte Chemie International Edition* 53 (25): 6364–6368.

Kim, K.C., Liu, T., Lee, S.W., and Jang, S.S. (2016). First-principles density functional theory modeling of Li binding: thermodynamics and redox properties of quinone derivatives for lithium-ion batteries. *Journal of the American Chemical Society* 138 (7): 2374–2382.

Kumar, P., Di Mauro, E., Zhang, S. et al. (2016). Melanin-based flexible supercapacitors. *Journal of Materials Chemistry C* 4 (40): 9516–9525.

Kymissis, I. (2009). Basic OFET fabrication and advanced OFET fabrication. In: *Organic Field Effect Transistors*, 60. Springer.

Laurichesse, S. and Avérous, L. (2014). Chemical modification of lignins: towards biobased polymers. *Progress in Polymer Science* 39 (7): 1266–1290.

Lee, G., Kang, S.-K., Won, S.M. et al. (2017). Fully biodegradable microsupercapacitor for power storage in transient electronics. *Advanced Energy Materials* 7 (18): 1700157.

Lei, T., Guan, M., Liu, J. et al. (2017). Biocompatible and totally disintegrable semiconducting polymer for ultrathin and ultralight weight transient electronics. *Proceedings of the National Academy of Sciences* 114 (20): 5107.

Lhopital, V. and Bordignon, M. (2018). *Overview of eco-design applications on various types of electronic product development*. In: *Eco-Design in Electrical Engineering* (ed. J.-L. Bessède), 3–8. Châteauvilain, FR: Springer.

Liang, Z., Tang, Q., Liu, J. et al. (2010). N-type organic semiconductors based on π-deficient pentacenequinones: synthesis, electronic structures, molecular packing, and thin film transistors. *Chemistry of Materials* 22 (23): 6438–6443.

Liang, Y., Jing, Y., Gheytani, S. et al. (2017). Universal quinone electrodes for long cycle life aqueous rechargeable batteries. *Nature Materials* 16 (8): 841.

Liedel, C., Wang, X., and Antonietti, M. (2018). Biobased polymer cathodes with enhanced charge storage. *Nano Energy* 53: 536–543.

Lin, K., Chen, Q., Gerhardt, M.R. et al. (2015). Alkaline quinone flow battery. *Science* 349 (6255): 1529–1532.

Lincke, G. (2002). On quinacridones and their supramolecular mesomerism within the crystal lattice. *Dyes and Pigments* 52 (3): 169–181.

Liu, T., Kim, K.C., Lee, B. et al. (2017). Self-polymerized dopamine as an organic cathode for Li-and Na-ion batteries. *Energy & Environmental Science* 10 (1): 205–215.

Liu, X.-x., Hu, X., Cao, Y. et al. (2019). Biodegradation of phenanthrene and heavy metal removal by acid-tolerant *Burkholderia fungorum* FM-2. *Frontiers in Microbiology* 10 (408): 1–13.

Luo, M., Martinez, A.W., Song, C. et al. (2013). A microfabricated wireless RF pressure sensor made completely of biodegradable materials. *Journal of Microelectromechanical Systems* 23 (1): 4–13.

Lydén, A., Larsson, B.S., and Lindquist, N.G. (1984). Melanin affinity of manganese. *Acta Pharmacologica et Toxicologica* 55 (2): 133–138.

Ma, T., Zhao, Q., Wang, J. et al. (2016). A sulfur heterocyclic quinone cathode and a multifunctional binder for a high-performance rechargeable Li-ion battery. *Angewandte Chemie Int'l Edition* 55 (22): 6428–6432.

Ma, Q., Zhang, X., and Qu, Y. (2018). Biodegradation and biotransformation of indole: advances and perspectives. *Frontiers in Microbiology* 9: 2625.

Maleki, H., Deng, G., Kerzhner-Haller, I. et al. (2000). Thermal stability studies of binder materials in anodes for lithium-ion batteries. *Journal of The Electrochemical Society* 147 (12): 4470.

Malliaras, G. and Friend, R. (2005). An organic electronics primer. *Physics Today* 58: 53–58.

Mamada, M., Nishida, J.-i., Tokito, S., and Yamashita, Y. (2009). Anthraquinone derivatives affording n-type organic thin film transistors. *Chemical Communications* 16: 2177–2179.

Mamada, M., Kumaki, D., Nishida, J.-i. et al. (2010). Novel semiconducting quinone for air-stable n-type organic field-effect transistors. *ACS Applied Materials & Interfaces* 2 (5): 1303–1307.

Manu, M.K., Kumar, R., and Garg, A. (2017). Performance assessment of improved composting system for food waste with varying aeration and use of microbial inoculum. *Bioresource Technology* 234: 167–177.

Mason, I.G. (2007). *A Study of Power, Kinetics and Modelling in the Composting Process.* Christchurch (NZ): University of Canterbury.

Matthews, C., Crispie, F., Lewis, E. et al. (2019). The rumen microbiome: a crucial consideration when optimising milk and meat production and nitrogen utilisation efficiency. *Gut Microbes* 10 (2): 115–132.

Mauger, A., Julien, C., Paolella, A. et al. (2019). Recent progress on organic electrodes materials for rechargeable batteries and supercapacitors. *Materials* 12 (11): 1770.

Melone, F., Saladino, R., Lange, H., and Crestini, C. (2013). Tannin structural elucidation and quantitative 31P NMR analysis. 2. Hydrolyzable tannins and proanthocyanidins. *Journal of Agricultural and Food Chemistry* 61 (39): 9316–9324.

Meredith, P. and Sarna, T. (2006). The physical and chemical properties of eumelanin. *Pigment Cell & Melanoma Research* 19 (6): 572–594.

Milczarek, G. and Inganäs, O. (2012). Renewable cathode materials from biopolymer/conjugated polymer interpenetrating networks. *Science* 335 (6075): 1468–1471.

Ministère du Développement Durable de l'Environnement et de la Lutte contre les Changements Climatiques (MDDELCC) (2018). *Lignes directrices pour l'encadrement des activités de compostage*. Québec: Direction des matières résiduelles.

Montazer, M. and Sadeghian Maryan, A. (2010). Influences of different enzymatic treatment on denim garment. *Applied Biochemistry and Biotechnology* 160 (7): 2114–2128.

Morgenstern, I., Klopman, S., and Hibbett, D.S. (2008). Molecular evolution and diversity of lignin degrading heme peroxidases in the Agaricomycetes. *Journal of Molecular Evolution* 66 (3): 243–257.

Morini, A.A., Ribeiro, M.J., and Hotza, D. (2019). Early-stage materials selection based on embodied energy and carbon footprint. *Materials & Design* 178: 107861.

Mostert, A.B., Powell, B.J., Pratt, F.L. et al. (2012). Role of semiconductivity and ion transport in the electrical conduction of melanin. *Proceedings of the National Academy of Sciences of the United States of America* 109 (23): 8943–8947.

Mostert, A.B., Rienecker, S.B., Noble, C. et al. (2018). The photoreactive free radical in eumelanin. *Science Advances* 4 (3): eaaq1293.

Mukhopadhyay, A., Jiao, Y., Katahira, R. et al. (2017). Heavy metal-free tannin from bark for sustainable energy storage. *Nano Letters* 17 (12): 7897–7907.

Nagaraju, D.H., Rebis, T., Gabrielsson, R. et al. (2014). Charge storage capacity of renewable biopolymer/conjugated polymer interpenetrating networks enhanced by electroactive dopants. *Advanced Energy Materials* 4 (1): 1300443–1300443.

Nam, S., Slopek, R., Wolf, D. et al. (2015). Comparison of biodegradation of low-weight hydroentangled raw cotton nonwoven fabric and that of commonly used disposable nonwoven fabrics in aerobic Captina silt loam soil. *Textile Research Journal* 86 (2): 155–166.

Navarro-Suárez, A.M., Casado, N., Carretero-González, J. et al. (2017). Full-cell quinone/hydroquinone supercapacitors based on partially reduced graphite oxide and lignin/PEDOT electrodes. *Journal of Materials Chemistry A* 5 (15): 7137–7143.

Osyczka, A., Moser, C.C., Daldal, F., and Dutton, P.L. (2004). Reversible redox energy coupling in electron transfer chains. *Nature* 427 (6975): 607.

Oudart, D. (2014), Modélisation de la stabilisation de la matière organique et des émissions gazeuses au cours du compostage d'effluents d'élevage, in *Ingénierie de l'environnement*, INSA de Toulouse.

Palaniveloo, K., Amran, M.A., Norhashim, N.A. et al. (2020). Food waste composting and microbial community structure profiling. *Processes* 8 (6): 723–753.

Park, M., Shin, D.S., Ryu, J. et al. (2015). Organic-catholyte-containing flexible rechargeable lithium batteries. *Advanced Materials* 27 (35): 5141–5146.

Perry, M.L. and Weber, A.Z. (2016). Advanced redox-flow batteries: a perspective. *Journal of The Electrochemical Society* 163 (1): A5064–A5067.

Petruzzelli, G. (1996). Heavy metals in compost and their effect on soil quality. In: *The Science of Composting* (eds. M. de Bertoldi, P. Sequi, B. Lemmes et al.), 213–223. Dordrecht: Springer Netherlands.

Pognani, M., Barrena, R., Font, X., and Sánchez, A. (2012). A complete mass balance of a complex combined anaerobic/aerobic municipal source-separated waste treatment plant. *Waste Management* 32 (5): 799–805.

Poizot, P., Gaubicher, J., Renault, S. et al. (2020). Opportunities and challenges for organic electrodes in electrochemical energy storage. *Chemical Reviews* 120 (14): 6490–6557.

Pradhan, R., Reddy, M., Diebel, W. et al. (2010). Comparative compostability and biodegradation studies of various components of green composites and their blends in simulated aerobic composting bioreactor. *International Journal of Plastics Technology* 14 (1): 45–50.

Pujol, A. (2012). *Modélisation du procédé de compostage - Impact du phénomène de séchage*. Institut National Polytechnique de Toulouse.

Ragni, R., Cicco, S., Vona, D. et al. (2017). Nanostructured silica from diatoms microalgae: smart materials for photonics and electronics. In: *Green Materials for Electronics* (eds. M. Irimia-Vladu, E.D. Glowacki, N.S. Sariciftci et al.), 287–313. Weinheim: Germany: Wiley-VCH.

Ramachandran, G.K., Tomfohr, J.K., Li, J. et al. (2003). Electron transport properties of a carotene molecule in a metal–(single molecule)–metal junction. *The Journal of Physical Chemistry B* 107 (25): 6162–6169.

Rastogi, M., Nandal, M., and Khosla, B. (2020). Microbes as vital additives for solid waste composting. *Heliyon* 6: e03343.

Robinson, J.M. (1990). Lignin, land plants, and fungi: biological evolution affecting phanerozoic oxygen balance. *Geology* 18 (7): 607–610.

Rodenberg, A., Orazietti, M., Mosberger, M. et al. (2016). Quinones as reversible electron relays in artificial photosynthesis. *ChemPhysChem* 17 (9): 1321–1328.

Root, S.E., Savagatrup, S., Printz, A.D. et al. (2017). Mechanical properties of organic semiconductors for stretchable, highly flexible, and mechanically robust electronics. *Chemical Reviews* 117 (9): 6467–6499.

S.C.O. Canada (2016). CAN/BNQ 0413–220/2016, *Organic soil conditioners – composts – determination of respiration rate – Part I: respirometric method — Part II: modified biochemical oxygen demand (BOD) respiration method.*

Salvatore, G.A., Sülzle, J., Dalla Valle, F. et al. (2017). Biodegradable and highly deformable temperature sensors for the internet of things. *Advanced Functional Materials* 27 (35): 1702390–1702390.

Sandrin, T.R. and Maier, R.M. (2003). Impact of metals on the biodegradation of organic pollutants. *Environmental Health Perspectives* 111 (8): 1093–1101.

Schön, J., Berg, S., Kloc, C., and Batlogg, B. (2000). Ambipolar pentacene field-effect transistors and inverters. *Science* 287 (5455): 1022–1023.

Segura-Aguilar, J., Paris, I., Muñoz, P. et al. (2014). Protective and toxic roles of dopamine in Parkinson's disease. *Journal of Neurochemistry* 129 (6): 898–915.

Shi, Z.-F., Black, H.T., Dadvand, A., and Perepichka, D.F. (2014). Pentacenobis (thiadiazole) dione, an n-type semiconductor for field-effect transistors. *The Journal of Organic Chemistry* 79 (12): 5858–5860.

Sieniawska, E. and Baj, T. (2017). *Tannins*. In: *Pharmacognosy* (eds. S. Badal and R. Delgoda), 199–232. Elsevier.

Singh, D.J. and Kalamdhad, A. (2013). Chemical speciation of heavy metals in compost and compost amended soil-a review. *International Journal of Environmental Engineering Research* 2: 27–37.

Solano, F. (2017). Melanin and melanin-related polymers as materials with biomedical and biotechnological applications – cuttlefish ink and mussel foot proteins as inspired biomolecules. *International Journal of Molecular Sciences* 18 (7): 1561–1578.

Söllinger, A., Tveit, A.T., Poulsen, M. et al. (2018). Holistic assessment of rumen microbiome dynamics through quantitative metatranscriptomics reveals multifunctional redundancy during key steps of anaerobic feed degradation. *mSystems* 3 (4): e00038–e00018.

Song, Z., Qian, Y., Liu, X. et al. (2014). A quinone-based oligomeric lithium salt for superior Li–organic batteries. *Energy & Environmental Science* 7 (12): 4077–4086.

Song, Z., Qian, Y., Gordin, M.L. et al. (2015). Polyanthraquinone as a reliable organic electrode for stable and fast lithium storage. *Angewandte Chemie* 127 (47): 14153–14157.

Stegenta-Dąbrowska, S., Sobieraj, K., Pilarski, G. et al. (2019). Analysis of the spatial and temporal distribution of process gases within municipal biowaste compost. *Sustainability* 11: 2340.

Tang, Q., Liang, Z., Liu, J. et al. (2010). N-heteroquinones: quadruple weak hydrogen bonds and n-channel transistors. *Chemical Communications* 46 (17): 2977–2979.

Temmink, J., Field, J.A., Van Haastrecht, J., and Merkelbach, R. (1989). Acute and sub-acute toxicity of bark tannins in carp (*Cyprinus carpio* L.). *Water Research* 23 (3): 341–344.

Thakur, V.K., Thakur, M.K., Raghavan, P., and Kessler, M.R. (2014). Progress in green polymer composites from lignin for multifunctional applications: a review. *ACS Sustainable Chemistry & Engineering* 2 (5): 1072–1092.

Tian, K. and Bilal, M. (2020). Chapter 15 – research progress of biodegradable materials in reducing environmental pollution. In: *Abatement of Environmental Pollutants* (eds. P. Singh, A. Kumar and A. Borthakur), 313–330. Elsevier.

Tran, H., Feig, V.R., Liu, K. et al. (2019). Stretchable and fully degradable semiconductors for transient electronics. *ACS Central Science* 5 (11): 1884–1891.

Trautmann, N. and Olynciw, E. *Compost Microorganisms. 1996* (2020). http://compost.css.cornell.edu/microorg.html (accessed 23 March 2021).

USDA Natural Resources Conservation Service (2010). *Environmental Engineering National Engineering Handbook*. USDA.

Valitova, I., Amato, M., Mahvash, F. et al. (2013). *Carbon nanotube electrodes in organic transistors. Nanoscale* 5 (11): 4638–4646.

Van den Bosch, S., Schutyser, W., Vanholme, R. et al. (2015). *Reductive lignocellulose fractionation into soluble lignin-derived phenolic monomers and dimers and processable carbohydrate pulps. Energy & Environmental Science* 8 (6): 1748–1763.

Vázquez, G., Fontenla, E., Santos, J. et al. (2008). Antioxidant activity and phenolic content of chestnut (*Castanea sativa*) shell and eucalyptus (*Eucalyptus globulus*) bark extracts. *Industrial Crops and Products* 28 (3): 279–285.

Vesala, R., Niskanen, T., Liimatainen, K. et al. (2017). Diversity of fungus-growing termites (*Macrotermes*) and their fungal symbionts (Termitomyces) in the semiarid Tsavo ecosystem, Kenya. *Biotropica* 49 (3): 402–412.

Vonlanthen, D., Lazarev, P., See, K.A. et al. (2014). A stable polyaniline-benzoquinone-hydroquinone supercapacitor. *Advanced Materials* 26 (30): 5095–5100.

Wang, H., Zhao, Y., Xie, Z. et al. (2014). The thermodynamic characteristics of organic crystal growth by physical vapor transport: towards high-quality and color-tunable crystal preparation. *CrystEngComm* 16 (21): 4539–4545.

Wang, T.-y., Li, Q., and Bi, K.-s. (2018). Bioactive flavonoids in medicinal plants: structure, activity and biological fate. *Asian Journal of Pharmaceutical Sciences* 13 (1): 12–23.

Warnock, M., Davis, K., Wolf, D., and Gbur, E. (2011). Soil burial effects on biodegradation and properties of three cellulosic fabrics. *American Association of Textile Chemists and Colorists Review* 11: 53–57.

Wei, Y., Wu, D., Wei, D. et al. (2019). Improved lignocellulose-degrading performance during straw composting from diverse sources with actinomycetes inoculation by regulating the key enzyme activities. *Bioresource Technology* 271: 66–74.

Wolfenden, R. and Snider, M.J. (2001). The depth of chemical time and the power of enzymes as catalysts. *Accounts of Chemical Research* 34 (12): 938–945.

Wong, L., H'ng, P., Wong, S. et al. (2014). Termite digestomes as a potential source of symbiotic microbiota for lignocelluloses degradation: a review. *Pakistan Journal of Biological Sciences* 17 (8): 956–963.

Wu, Y., Zeng, R., Nan, J. et al. (2017). Quinone electrode materials for rechargeable lithium/sodium ion batteries. *Advanced Energy Materials* 7 (24): 1700278.

Wuana, R. and Okieimen, F. (2011). Heavy metals in contaminated soils: a review of sources, chemistry, risks and best available strategies for remediation. *ISRN Ecology* 2011.

Xu, D., He, Y.B., Chu, X. et al. (2015). Synthesis of lithium iron phosphate/carbon microspheres by using polyacrylic acid coated iron phosphate nanoparticles derived from iron (III) acrylate. *ChemSusChem* 8 (6): 1009–1016.

Xu, R., Gouda, A., Caso, M.F. et al. (2019). Melanin: a greener route to enhance energy storage under solar light. *ACS Omega* 4 (7): 12244–12251.

Yanagisawa, H., Mizuguchi, J., Aramaki, S., and Sakai, Y. (2008). Organic field-effect transistor devices based on latent pigments of unsubstituted diketopyrrolopyrrole or quinacridone. *Japanese Journal of Applied Physics* 47 (6R): 4728.

Yang, L.-L., Wang, C.-C., Yen, K.-Y. et al. (1999). Antitumor activities of ellagitannins on tumor cell lines. In: *Plant Polyphenols*, vol. 2 (ed. G.G. Gross), 615–628. Springer.

Yilmaz, Y. and Toledo, R.T. (2004). Major flavonoids in grape seeds and skins: antioxidant capacity of catechin, epicatechin, and gallic acid. *Journal of Agricultural and Food Chemistry* 52 (2): 255–260.

Yin, L., Huang, X., Xu, H. et al. (2014). Materials, designs, and operational characteristics for fully biodegradable primary batteries. *Advanced Materials* 26 (23): 3879–3884.

Yin, L., Bozler, C., Harburg, D.V. et al. (2015a). Materials and fabrication sequences for water soluble silicon integrated circuits at the 90 nm node. *Applied Physics Letters* 106 (1): 14105–14105.

Yin, L., Farimani, A.B., Min, K. et al. (2015b). Mechanisms for hydrolysis of silicon nanomembranes as used in bioresorbable electronics. *Advanced Materials* 27 (11): 1857–1864.

Yokoji, T., Kameyama, Y., Maruyama, N., and Matsubara, H. (2016). High-capacity organic cathode active materials of 2,2′-bis-p-benzoquinone derivatives for rechargeable batteries. *Journal of Materials Chemistry A* 4 (15): 5457–5466.

Yoshida, S., Hiraga, K., Takehana, T. et al. (2016). A bacterium that degrades and assimilates poly(ethylene terephthalate). *Science* 351 (6278): 1196.

Yoshizawa, S., Horiuchi, T., Fujiki, H. et al. (1987). Antitumor promoting activity of (−)-epigallocatechin gallate, the main constituent of "Tannin" in green tea. *Phytotherapy Research* 1 (1): 44–47.

Yue, X., Liu, H., and Liu, P. (2019). Polymer grafted on carbon nanotubes as a flexible cathode for aqueous zinc ion batteries. *Chemical Communications* 55 (11): 1647–1650.

Zennaro, M., Cristofori, F., Formigoni, D. et al. (2005). Heavy metal contamination in compost. A possible solution. *Annali di Chimica* 95: 247–256.

Zhang, H. and Matsuto, T. (2011). Comparison of mass balance, energy consumption and cost of composting facilities for different types of organic waste. *Waste Management* 31 (3): 416–422.

Zhang, B., Lye, L., Kazemi, K., and Lin, W. (2012–2013), Development of advanced composting technologies for municipal organic waste treatment in small communities in Newfoundland and Labrador.

Zhang, X.Y., Bolag, A., Wu, Y., and Yamashita, Y. (2019). Preparation, characterization, and FET properties of 3,3′,5,5'-tetra-phenyldiphenoquinone. In: *Solid State Phenomena*, vol. 288, 44–50. Trans Tech Publications.

Zhao, M., Yang, X., Tsui, G.C., and Miao, Q. (2019) Trifluoromethylation of anthraquinones for n-type organic semiconductors in field effect transistors. The Journal of Organic Chemistry, Advanced Research in Materials Science II, 7th International Conference on Materials Science (ICMS 2018), pp 44–51.

Zhao, M., Yang, X., Tsui, G.C., and Miao, Q. (2020). Trifluoromethylation of anthraquinones for n-type organic semiconductors in field effect transistors. *The Journal of Organic Chemistry* 84 (1): 44–51.

Zheng, Y., Ashizawa, M., Zhang, S. et al. (2020). Tuning the mechanical properties of a polymer semiconductor by modulating hydrogen bonding interactions. *Chemistry of Materials* 32 (13): 5700–5714.

Zhou, Y., Wang, B., Liu, C. et al. (2015). Polyanthraquinone-based nanostructured electrode material capable of high-performance pseudocapacitive energy storage in aprotic electrolyte. *Nano Energy* 15: 654–661.

Zhu, M., Huang, Y., Huang, Y. et al. (2016a). Capacitance enhancement in a semiconductor nanostructure-based supercapacitor by solar light and a self-powered supercapacitor–photodetector system. *Advanced Functional Materials* 26 (25): 4481–4490.

Zhu, H., Luo, W., Ciesielski, P.N. et al. (2016b). Wood-derived materials for green electronics, biological devices, and energy applications. *Chemical Reviews* 116 (16): 9305–9374.

13

Circular Economy in Electronics and the Future of e-Waste

Nani Pajunen[1] *and Maria E. Holuszko*[2]

[1] *The Finnish Innovation Fund Sitra, Carbon Neutral Circular Economy, Sustainability Solutions, Itämerenkatu 11-13, PO Box 160, 00181 Helsinki, Finland*
[2] *University of British Columbia, Norman B. Keevil Institute of Mining Engineering, 6350 Stores Road, Vancouver, BC, V6T 1Z4, Canada*

13.1 Introduction

The world population's current standard of living increases the demand for the earth's finite natural resources. At the same time, population growth and urbanization are challenging these standards. We have been practicing a linear model economy since the Industrial Revolution. As a result of this behavior, we created a scenario, which is frequently referred to as a "take, make, dispose of" world, where we promote a single use of materials and products, creating a one-directional model of mass production, mass consumption, and finally, disposal after a single or limited time of use. This type of model is stretching our planet's physical limits (Esposito et al. 2018). Hence, it is vital for the planet's future to implement carbon-neutral and circular-economy solutions into our everyday life, improve strategies in the public and private industry sectors, and develop decision-making policies at the industrial scale. In other words, we need to rethink our business model entirely.

At present, only 9% of the materials already produced are recycled back to use (Statista 2021b). On average, close to 14% of municipal waste collected is recycled globally, with some countries reaching recycling rates close to 50%, like India and Australia. In the EU, the average recycling rate in 27 countries is at 43%, while it is 26% in Canada and 22% in China. However, some materials are recycled more than others. In the US, paper and paperboard are recycled at a rate of 64.7%, metals such as lead at 67%, magnesium at 54%, iron and steel at 52%, aluminum at 50%, glass at about 33%, and plastic 9.1% (Better Meets Reality 2021).

The global market economy is mainly based on the extraction of natural resources. It provides factories with the feedstock of materials that are then transformed into the products to be purchased and typically disposed of after a short time of use. Esposito et al. (2018) reported that currently, we are consuming goods at a 50% faster rate than replacing them.

Electronic Waste: Recycling and Reprocessing for a Sustainable Future, First Edition.
Edited by Maria E. Holuszko, Amit Kumar, and Denise C. R. Espinosa.

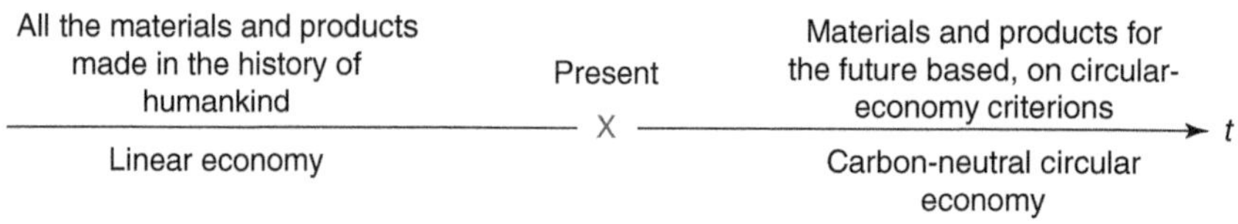

Figure 13.1 Recycling of materials – past and future.

In our everyday life, we use various mixed, hybrid, and composite materials, which are either bio-sourced or derived from metals or minerals, while more and more plastic and human-made types of materials are becoming prevalent. Many of these materials are still produced in a not-very-ecological, sustainable, nor efficient way. Most of these materials once produced, never used again after their end-of-life (EoL), while using virgin materials derived from natural resources. At the same time, the power generation is still mainly based on fossil fuels, hence, significantly contributing to global anthropogenic pollution and accelerating climate change. The major challenge for today's world will be to stop and make a difference – today – and start designing materials and products based on a circular-economy's principles and criteria, as indicated in Figure 13.1.

The circular-economy theme started to be debated in the 1990s, but it was not until the early 2000s when it began to be discoursed to achieve "true environmental sustainability." The definition of a circular economy was formulated by the European Commission (2020) and discussed further by Sillanpää and Ncibi (2019). The circular-economy model could address resource scarcity and issues with waste disposal and eventually provide a win–win situation from an economic and environmental perspective while creating a new value proposition (Homrich et al. 2018). The cradle-to-cradle flow concept was born out to replace the cradle-to-grave economic model, representing linear material flow. If the circular economy is carried out extensively, it could potentially reduce consumption of new raw materials by 32% within 15 years and by 53% by 2050 (Ellen MacArthur Foundation 2015; Esposito et al. 2018). The circular-economy concept has been promoted by the European Union and several other countries, including China, Japan, Canada, and the United Kingdom (Korhonen et al. 2018).

According to Araujo Galvão et al. (2018), many barriers were initially identified for moving from a linear to a circular economy, and these can be categorized as:

1) Technological
2) Financial and economic
3) Regulatory/policymaking
4) Managerial
5) Performance indicators
6) Lack of accurate information on environmental impacts
7) Customer-oriented and social barriers.

We must start looking at the "big picture" while handling all the products and materials coming to an EoL phase and learn from the life cycle of a product and the past experiences. The most challenging is the technological barrier encountered

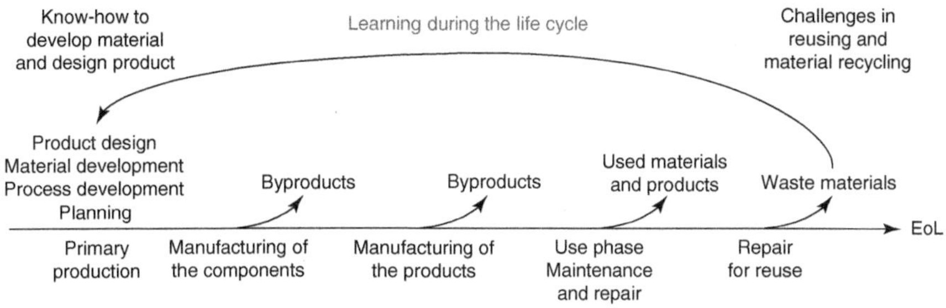

Figure 13.2 Learning during the life cycle.

in recycling, while the most significant is the decreased economic value of recycled materials and the recycling costs. In addition, there is a lack of understanding of environmental costs and an absence of pressure from societies to demand from the governments more stringent environmental policies and regulations around manufacturing and waste management, hence demonstrating significant social barriers.

According to Zhang et al. (2009), the main challenge facing societies is to find feasible strategies to disconnect economic growth from the environmental issues within the limits of the available resources. Rizos et al. (2016) indicated that the main barriers moving to the circular economy are lack of supply and demand for recyclable materials combined with the lack of capital and an absence of technical know-how in connecting material development, product design, and recycling with the proper government policies or supports. At present, a significant barrier to the transition to a more circular system and developing recyclable products (Figure 13.2) is a lack of demand or customers' willingness to pay more for recyclable products, both in business and in the general society at large.

This chapter aims to provide an overall picture of the challenges and discuss the way forward on how to change the industrial system from linear to circular. The approach is to assess this from the point of view of electrical and electronic equipment (EEE) devices, material design, and development. The following aspects will be addressed in this chapter:

- Digitalization and the need for electronics
- Perspective on recycling, life-cycle analysis in the material design phase
- Challenges for e-waste recycling and circular economy
- Drivers for change and complexity of the industrial system
- Barriers for change to move to a circular economy.

13.2 Digitalization and the Need for Electronic Devices

Electronics provide us with broad access to education, instant information, and nonstop entertainment while contributing to mass communication and improving our everyday life. Digitalization brought through electronics led to the advancement of industrial productivity, accessibility to information, and improved living standards in our present societies worldwide. It has also proven to allow us to continue functioning as a productive society during pandemic scenarios (COVID-19).

Every year, millions of new electronic devices are pouring into the market to facilitate a digital lifestyle for many consumers worldwide. According to Statista (2021a), by 2020, over 4 billion people have had a personal computer, while a total of 3.3 billion own a mobile smartphone. This translates into 41.5% users in the global population. Global sales reached a value of over 400 billion USD until the beginning of 2021, with over 150 billion USD worth of sales in China.

These numbers are staggering and provide us with insight into the scale of electronics production and its sales value worldwide. Digitalization is creating enormous demand for many different types of devices and, consequently, various raw materials. It is also affecting markets, businesses, and employment in a very significant way. Digitalization on its own requires a large amount of energy to power the computers and support internet transmission of the data to facilitate worldwide connectivity. Digitalization through the use of millions of electronic devices and their faith at the EoL becomes part of the problem in recycling complex materials used in the manufacturing of these modern electronics. At the same time, it can also be part of the solution for facilitating a circular economy.

Modern smartphones are an excellent example of the digital world's electronic products that everyone commonly uses; they are made of very complex materials to increase their digital performance. At the same time, they pose significant challenges in recycling. The digital–technical performance of smartphones has rapidly evolved from the perspective of the types of materials used in these products and their functionalities. Smartphones are made as complex devices, using close to 70 different elements with more than 60 metals in their composition. While the competition between smartphone producers has focused only on display size, resolution, or sensitivity, it has not concentrated on material efficiency, sustainability, or recyclability.

The number of smartphones sold to end users worldwide from 2007 to 2020 has increased from 122 million to 1560 million units (Statista 2021a). These products are in use approximately only for a maximum of one to two years (Forti et al. 2020; Kumar 2020). The take-back system and material recycling of most of the components and materials remain negligible. The large number of electronics continuously produced and owned globally illustrates the extent of the issue.

13.3 Recycling and Circular Economy

The pervasive expansion of electronics generates an unsustainable growth of waste of electrical and electronic equipment (WEEE). E-waste is the fastest-growing solid-waste stream, expected to reach 74.7 Mt by 2030 (Forti et al. 2020). Furthermore, electronics rely on the use of critical and scarce elements that are needed to make interminably newer and better-performing devices. Since we need these electronic devices to progress into the future, the recovery of these scarce and valuable materials should become our most significant driver for a circular economy. As a result, all of these products' life-cycle analyses – starting from the beginning to the end – should provide us with the incentive to reuse these critical materials. In addition to the scarcity of the necessary materials, there is an economic value in the circularity of materials recovered from electronic waste. Currently, the material value from electronics alone is US$ 62.5 billion, while there is 100 times more gold in a ton of mobile phones than in a ton of gold ore (World Economic Forum 2019). There is a significant incentive for the recovery of valuable materials from e-waste.

The challenge is also in the EoL phase of electrical devices since recycling is difficult, even impossible in some cases, especially when the waste is not homogeneous in terms of valuable metal content and with very little-known materials' composition. Electronic devices contain a wide variety of materials, including hazardous and toxic materials, for example, arsenic, lead, mercury, cadmium, and types of flame retardants that are harmful to human health and the environment (Chen et al. 2012; Guo et al. 2015; Kumar et al. 2018; Yu et al. 2017) if not disposed of properly. Meanwhile, some materials also represent a very high value, such as gold, silver, or copper (EPA 2012).

Suppose the e-waste goes to a landfill or is not treated in an environmentally responsible way at the end of a device's life; there is a high risk of environmentally harmful damage extending for the foreseeable future (Kumar 2020). In addition, the e-waste contains valuable resources that can be recovered and reused, reducing the need to use natural resources in the form of new virgin materials or metals (Hagelüken and Corti 2010). There is no functional recycling system for electrical devices to separate different metal alloys, mainly due to economic reasons, while recycling infrastructure for bulk metals, such as steel and aluminum, exists. The challenge is that the existing e-waste recycling systems cannot economically process nonhomogeneous e-waste scrap with many different substances in its composition.

Recycling in the traditional linear economy was usually designed to provide additional resources for primary production; it has been referred to as "an Aspirin" for the societal hangover due to overconsumption. It has been mainly designed to alleviate shortages of some raw materials while mostly providing an opportunity for segregated waste disposal in many jurisdictions. In developing countries, recycling is frequently carried out informally while providing a source of income for the most impoverished population engaged in the rudimentary recycling of valuable materials, mostly metals, for immediate monetary profit (Forti et al. 2020).

On the other hand, in the developed countries where there are more organized practices of municipal waste management, recycling can be considered an opportunity for resource recovery. This is even more so in the context of pressing issues of the criticality of some metals, critical elements, and raw materials to produce new electronics that we so heavily rely on in our everyday life. In the circular-economy model, the goal is to keep the materials and products in use as long as possible and increase their value during the life cycle. Material efficiency is part of the circular economy, and it promotes the efficient use of natural resources and the effective reuse of generated from waste byproducts. We need more research in recycling and the material-design and product-development phase, and the use of recycled byproducts, as this becomes an integral part of a material-efficiency strategy to facilitate a circular economy. In addition, we need to learn how to apply knowledge of materials to the beginning of the life cycle of the newly designed and manufactured products for better recyclability.

Alternatively, recycling and reusing materials must be justifiable from an economic, environmental, and social perspective. The transition must begin from acknowledging the challenges and opportunities in the "big picture" and from the systemic perspective; otherwise, we only partly optimize the system. Most of the research in material efficiency has always focused on how waste from one process can be used as a valuable raw material in another process elsewhere (Pajunen et al. 2016).

However, we have to start from the beginning of the process at the design stage, where we develop new materials from recovered secondary resources and use them in manufacturing new products. Also, we need to make changes in the whole value chain, where all of the components and materials have to go through the same design and development process (Pajunen et al. 2016). In other words, we have to start designing for recycling and focusing on maximizing what is already in use while considering the life cycle of the product from the sourcing stage and through the whole supply chain to consumption and including sourcing of secondary raw materials obtained from recycling. Moreover, recycling should also be aimed at producing the byproducts of a certain quality in terms of their physical characteristics and functionalities; hence, they can be diverted straight to developing the new materials and manufacturing stage.

Systemic change is needed in the EoL of products too. We have to create a global take-back system for all the different kinds of devices and products. This will allow a decrease in the complexity of recycling in industrial systems. The take-back approach can create a more homogeneous waste scheme by allowing for the EoL devices to come back directly to the original equipment manufacturer for recycling. This will incentivize the manufacturer to design for recycling material savings by using recovered secondary products from recycling, while these will become a source of secondary "raw" material for their new products. Simultaneously, it will lower the demand for virgin materials, allow for a closed-loop recycling system, and adapt eco-design approaches while creating economic benefits (Wagner et al. 2020). It will also ensure an adequate quality of secondary raw material obtained from recycling required for designing new materials by the same industrial set-up, the same manufacturing company.

13.4 Challenges for e-Waste Recycling and Circular Economy

New-age electronics are manufactured using newly designed materials, and functional requirements drive these new materials' designs. The development of new electronic devices and many novelty products requires new composite or hybrid materials to attain appropriate material characteristics, allowing them to be lightweight, have extreme strength, and a reduced size that is desirable for everyday use. Composites are materials prepared from two or more distinct complementary substances, for example, metals, ceramics, glass, and polymers. These materials produce another material with characteristics different from its components, often visible on a macroscopic scale. The hybrid materials are the ones in which the constituents are blended on the molecular or nanometer scale, and the individual components remain invisible on the macroscale (Pajunen et al. 2016). In the product design phase, many types of materials are combined; hence, these electronic products contain many metals, their alloys, and various chemical compounds, polymers, plastics, and flame retardants. The more advanced functions required, the more complex the materials need to be. For electronic products, these can include combining metal alloys with ceramics and fibers for creating composites, gluing honeycomb structures, making metal foams, depositing thin films, and creating nanoparticle structures (Reuter et al. 2013). Functionality is often further enhanced by coatings for improving wear, corrosion or fire resistance, safety, or improving aesthetic aspects. In conclusion, recycling such devices becomes very complex (Pajunen et al. 2016).

The economic drivers are the most important in the design phase, while the EoL phase aspects are usually not considered at all, except as depreciable value. In manufacturing, material efficiency should be considered from the economic and marketing point of view. Currently, there are few incentives or pressures to increase the recyclability of electronic products due to their technological complexity. Generally, only valuable metals or materials, which have a high monetary value, such as copper, gold, silver, palladium, rare-earth elements, or fiberglass, are being recycled from e-waste (Baldé et al. 2017; Forti et al. 2020; S. Zhang et al. 2017).

For example, in EU countries, recycling targets are set at the national level and are based on the percentage value of the material's total weight (The European Commission 2008). Incentives for recycling should include financial drivers for manufacturers to produce recyclable products and prefer to buy those products and help in their recycling collection. This can be enabled via taxation. For example, if the product is recyclable and the recycling system is available, it could have tax relief during its life cycle. Such tax relief or a price reduction should increase consumer demand for recyclable products. Monetary incentives for consumers and recycling companies are missing in the current linear system.

The main challenge also lies in the lack of appropriate recycling systems and efficient separation technologies for complex electronics. There is a lack of information about the EoL of an electronic product or its life cycle to understand opportunities for recycling electronics. As a result, it is essential to include EoL in the design phase for material efficiency. Designing for recyclability to recover high-value and low-value components for reuse in new products is the only way to solve recycling problems of structurally, naturally complex electronics.

The complexity of electronics will increase as the multifunctional hybrid materials, and advanced manufacturing technologies are used for more sophisticated electronics in terms of performance and practicality. Hence, there is a critical necessity for high-level multidisciplinary competencies and the development of new solutions, which can be crucial factors in renewing the global manufacturing industry. Manufacturing in the current industrial system is very complex, and most of the material cycles are multifaceted and interconnected in terms of material sourcing. The reuse of byproducts or metals from recycling to develop new products can become very complicated due to the already pre-existing linear-economy logistics. The change needs to be implemented right at the design and material-development stage to facilitate material circularity.

The main goal is to include life-cycle thinking and the recyclability of complex and hybrid materials in the material-development and product-design phases and integrate recycling and sustainability perspectives into decision making at strategic, management, and production levels and work toward increasing demand for byproducts from recycling and recycled materials as source material for new-age materials.

13.5 Drivers for Change – Circular Economy

A circular economy needs to be considered against the backdrop of modern societal needs and industrial logic. Industrial logic drives companies to strive for market share, increased sales, and improved profits by offering competitive products through innovations and improved technologies, while there are different drivers for different players-stakeholders. For example, both the general public and private sector organizations, with their industrial set-ups, are vital in any society. They need to be accommodated in terms of their needs and desires to transition comfortably and successfully toward a circular economy.

In addition to industrial set-ups and societal backgrounds, there are also urgent environmental concerns; for example, the levels of clean water and the availability of raw materials needed by the industrial set-ups are already reaching a critical point for the supply of critical raw materials (European Commission 2014). For example, if a single country dominates the market for a particular raw material, that country can directly influence its availability and price. This means that the availability of materials is also challenged by global trade as much as economic and policy issues, given the current situation with the worldwide supply of rare-earth elements. In the worst scenarios, problems with raw materials' availability may lead to difficulties in manufacturing and closure of industrial operations in certain jurisdictions.

The lack or shortage of natural resources and raw materials needed in the production phase can become a strong driver for promoting a circular economy. According to Esposito et al. (2018), the linear-economy model's inability to cope with the enormous demand for natural resources will result in a shortage of 8 billion tons of raw material supply. An approach in line with a circular-economy's principles is the best solution for the environment and the most practical way of operating from an economic sense. The circular-economy perspective is presented in Figure 13.3. In the transition toward a circular economy, there is a need for new solutions in every phase of the product's life cycle.

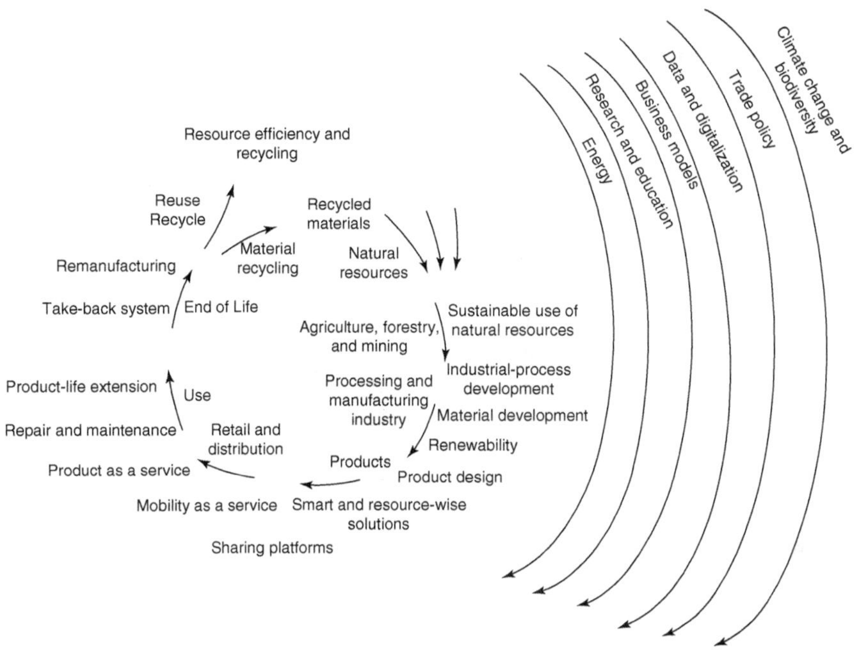

Figure 13.3 The circular-economy model.

In every part of a product's life cycle, there is a potential to reduce resource consumption and improve its environmental performance. From the design perspective, the six sustainability principles are the following (UNEP 2007):

1) Apply a "re philosophy," meaning to rethink the product and its functions
2) Make the product easy to repair
3) Replace harmful substances with safer, less hazardous alternatives
4) Design the product for disassembly so that the parts can be reused
5) Reduce energy, material consumption, and socioeconomic impacts throughout a product's life cycle
6) Select materials that can be recycled.

At present, the challenge from a recycling point of view is balancing the economy and environmental standpoint. The EoL phase of complex products creates nonhomogeneous waste. This waste may contain valuable materials that can be recovered and reused and parts that do not have any or have very little economic value when recycled, such as nonmetal fractions from e-waste, including various plastics with low monetary value.

In the industrial-design process, most of the environmental impacts are locked in at an early phase, and significant decisions are made when there is very little information about the details of the product's design. However, when these decisions are made earlier in the product's life cycle, their effects become crucial in the end. The opportunities to make changes in product design decrease with time, and the cost of making any changes increases at the same time. To make changes in sustainability

Figure 13.4 Transition toward circular economy step by step.

and recyclability, all of the important players in the supply chain have to be involved. They need to strive for the same goal to achieve the circularity of the materials used in the product that is being designed.

It is vital to take the recycling perspective into account both in the material-development and in the product-design phases when targeting circular economy and consider a take-back system for the components and materials to promote material circularity, as shown in Figure 13.4.

The first step of promoting a circular economy is extending the product's life cycle by using durable materials and making long-life products that can be repaired and reused at the end of their life cycles. In some cases, a product or component can be designed to be used for another purpose without chemical or mechanical modifications. Hence, further processing of the product does not require extra energy or new raw materials.

If a mechanical modification is not possible to recirculate the product, this must be considered in the design phase; i.e. the product's materials are developed to be reused later as a raw material for a new product. In addition, it is also essential to make sure that nonhazardous substances are used in composite materials as they impair recirculation and cause the materials to become hazardous waste or end up in incineration. The recirculation of materials can also be facilitated by chemical modification (chemically assisted); however, this is the most aggressive form of recirculation from a monetary-investment and energy-intensity point of view. In a chemically assisted modification, the material's chemical bonds are broken down so that the materials can be reprocessed and utilized as new raw materials in another industrial process.

The starting point for material scientists, product developers, and designers is to reduce waste and losses as these always incur additional monetary expenses. Waste and losses can be reduced in many ways. Everyone benefits when all the stakeholders participate in the product's life cycle, cooperate across corporate boundaries, and jointly provide feedback on its entire life cycle. The concept can be "to design out waste," meaning that we design to minimize the final waste as proposed in one of the five fundamentals traits by The Ellen MacArthur Foundation (2015).

An uninterrupted flow of information about the choices being made and cost-cutting opportunities have to flow through the life cycle, from the development phase right into the final project's execution. It is not enough that correct decisions and plans are made at the early stages – the plans must be executed and supervised throughout the production process and the product's life cycle until it reaches the customer. To succeed, there is a need to bring together expertise in all phases, from product design to reuse, remanufacturing, and recycling.

The novel circular-economy business model can be reshaped, for example, where a product can be shared as a part of service, creating a sharing-economy model and generating new business types for the industrial network (product-as-a-service) according to Lacy (2015) and as proposed in Figure 13.3. Some of these ideas were developed as a result of economic and social research on the operation and functionality of a circular economy as early as in the 1980s, 1990s, and the early 2000s (Hawken et al. 1999; Lifset and Graedel 2002; McDonough and Braungart 2002; Stahel 1986). This "product-as-service" business model can be adopted easily for electronic products such as cellular phones and computers. The customer buys services with the new product and can access software upgrades as part of the sharing service. However, this may be more difficult to apply to other electronics such as standard home appliances or specialty industrial electronics.

13.6 Demand for Recyclable Products

A vital part of redesigning and making electronic business greener is to have an economic driver toward sustainability. However, if the reward is somewhere else, a business is not going to choose an environment-friendly option without having a possibility of an economic benefit. As a result, business interest is nonexistent, and without consumers' demanding recyclable products in the manufacturing of new products, there will be no market forces to call for regulations in the use of recyclable materials in new products.

At present, the only driver in the smart/mobile phone market is to increase product sales. For manufacturing companies, the target is to be the global market share leader. All companies are developing new models with new technical features and applications; competition is fierce, and cost-efficiency targets are high. Accordingly, since 2012, the Samsung brand, the leading global smartphone vendor, has held a share of 20–30% in the smartphone market. The Apple brand is the second-largest, and Huawei is the third-largest vendor of smartphones worldwide, while Xiaomi and OPPO have recently increased their smartphone market share according to global market statistics (Statista 2021c).

The biggest electronic manufacturing companies and their retailers' marketing and advertising information is based only on their phones' technological features and new applications but is not focused on providing information regarding the device's composition, structure, or recyclability features. All companies provide similar information on the device's technological capabilities instead of information on their sustainability characteristics. It may still be a long way for a change in manufacturing companies' attitudes toward the concept of circularity and sustainability in the electronics industries unless the consumers strongly demand this shift.

This chapter aims to provide information on the issues related to electronic waste, which is linked with the design of electronics and their recyclability to promote responsible consumption with minimum waste production. Building a functional recycling system is crucial when targeting the transition toward a circular economy. However, without social pressure from the consumers, the manufacturers do not have any incentives to promote their electronic products' recyclability. The main driver of the manufacturing industry is the economy and for the customer is to acquire the best in terms of the latest technology electronic device for their everyday use and applications. As a result, there are no incentives for recycling, and there is a lack of recyclable products in the mobile phone marketplace. The marketing of new features with advanced technical applications and their capabilities are still the most significant challenges to sustainability through the device's life cycle.

There are many small- or medium-sized enterprises or companies (SME) in the mobile phone supply chain. However, large manufacturing companies' importance in the supply chain and their willingness to cooperate with smaller companies within the supply chain is paramount. If the larger manufacturer demands sustainability and environmental responsibility from their suppliers via procurement, smaller SME companies must comply out of necessity if they want to continue cooperating with the known manufacturer in the future; therefore, they have to respond to these market needs and conditions.

With the increasing costs of raw materials and waste disposal, finding new solutions to mitigate these issues in cost-effective ways will be essential. Thus, the most important driver will be to recycle electronics to secure critical and other, perhaps not-so-critical raw materials to protect the environment. However, from an economic perspective, there are currently no suitable or justifiable separation and refining processes for recovering some of the high- and low-value materials from e-waste at the same time.

For a transition toward a society with a circular economy, there is a need to promote recyclable materials and increase the demand from society to use such materials in electronics manufacturing. In other words, we need to educate the consumers' market – the more recyclable and good-quality products there are on the market, the more demand there will be for these materials. The move toward a circular economy requires perseverance and consistent progress in all aspects of the manufacturing process. The industrial organization needs to develop confidence in the circular economy and its associated strategies, such as a demand for suppliers' sustainability while continuing the advancement into efficient recycling technologies and developing high-quality recyclable products at the same time.

13.7 Summary

In this chapter, we have focused on the circular economy and the role of life-cycle analysis when making recyclable electronic devices. The transition toward a

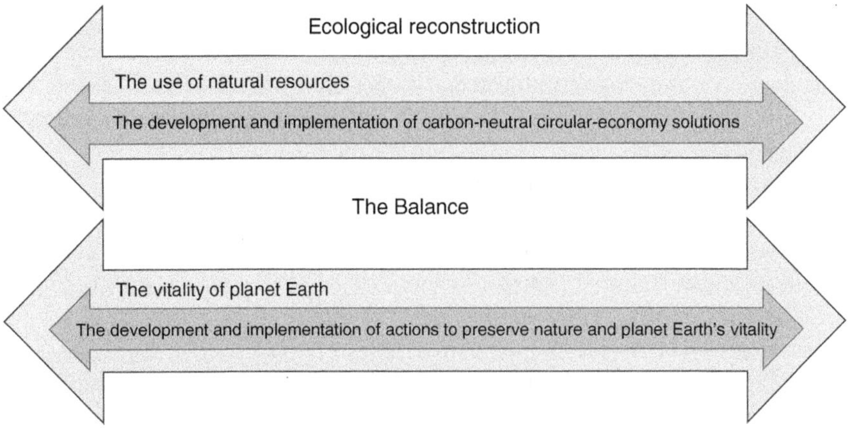

Figure 13.5 The balance between the use of natural resources and the planet's natural boundaries.

carbon-neutral circular economy and sustainability in small electronics such as smartphones might be possible due to the opportunities lying in the recovery of valuable components and the desire to build resilient manufacturing industries that will promote eco-design and a design for recyclability. For example, the "European Green Deal" is considered a roadmap for making the EU's economy sustainable. This can happen by turning climate change-related environmental challenges into opportunities across all the policy areas and sectors of society. There will be significant undertakings to integrate the circular economy and life-cycle thinking into the development and design processes.

Ideally, the goal is to live within the planet's natural boundaries and physical means, maintaining the planet's vitality (Figure 13.5) and keeping the extracted resources and products made from these natural resources in circularity as long as possible. When it comes to manufacturing products and materials, scientific and engineering expertise plays a significant role. However, no engineering field can tackle major challenges alone, such as transitioning from a linear economy to a circular economy. Sustainable solutions have to be carried forward while considering technological, economic, legal, administrative, social, and environmental factors. In this chapter, we attempted to present some ideas that could be implemented to improve the chances of recirculation of electronics by-products and e-waste to facilitate a circular economy. Nonetheless, to make this happen, there is a need for multisectoral, multidisciplinary, and global co-operation.

As for the future of e-waste, what is needed is designing new materials with an ecological design in mind, using recycled-from-e-waste components for the production of new electronics without toxic and hazardous chemicals, coupled with a strategy for future recycling. All this is necessary while mastering the recycling processes with minimization of final waste and utilizing all byproducts during the process; all these elements may eventually lead to near-zero e-waste in the future. This would be the most desirable effect in the context of a circular economy; it would be a win–win situation for the future of electronics in the new age of digitalization.

References

Araujo Galvão, G.D., de Nadae, J., Clemente, D.H. et al. (2018). Circular economy: overview of barriers. *Procedia CIRP* 73: 79–85. https://doi.org/https://doi.org/10.1016/j.procir.2018.04.011.

Baldé, C. P., Forti, V., Gray, V., et al. (2017). The Global E-waste Monitor–2017. Bonn/Geneva/Vienna. Available at: https://www.itu.int/en/ITU-D/Climate-Change/Documents/GEM 2017/Global-E-waste Monitor 2017.pdf (accessed 21 March 2018)

Better Meets Reality. (2021). *The Present & Future Of Recycling Metal – Better Meets Reality*. https://www.bettermeetsreality.com/the-present-future-of-recycling-metal/#comment-1161 (accessed 02 April 2021).

Chen, Y., Li, J., Chen, L. et al. (2012). Brominated flame retardants (BFRs) in waste electrical and electronic equipment (WEEE) plastics and printed circuit boards (PCBs). *Procedia Environmental Sciences* 16: 552–559.

Ellen MacArthur Foundation. (2015). *Growth Within: a circular economy vision for a competitive Europe*. https://www.ellenmacarthurfoundation.org/publications/growth-within-a-circular-economy-vision-for-a-competitive-europe (accessed 02 April 2021).

EPA. (2012). *Printed circuit board recycling methods*. https://www.epa.gov/sites/production/files/2014-05/documents/handout-10-circuitboards.pdf (accessed 02 April 2021).

Esposito, M., Tse, T., and Soufani, K. (2018). Introducing a circular economy: new thinking with new managerial and policy implications. *California Management Review* 60 (3): 5–19. https://doi.org/10.1177/0008125618764691.

European Commission. (2014). *Report on Critical raw materials for the EU*. https://ec.europa.eu/growth/sectors/raw-materials/specific-interest/critical_en (accessed 02 April 2021).

European Commission. (2020). *Communication from the commission to the European Parliament, The Council, The European Economic and Social Committee and The Committee of the regions: A new Circular Economy Action Plan. For a cleaner and more competitive Europe*. https://eur-lex.europa.eu/legal-content/EN/TXT/HTML/?uri=CELEX:52020DC0098&from=EN (accessed 02 April 2021).

Forti, V., Baldé, C. P., Kuehr, R., & Bel, G. (2020). *Global E-waste Monitor 2020: Quantities, flows and the circular economy potential*. https://www.itu.int/en/ITU-D/Environment/Documents/Toolbox/GEM_2020_def.pdf (accessed 02 April 2021).

Guo, J., Zhang, R., and Xu, Z. (2015). PBDEs emission from waste printed wiring boards during thermal process. *Environmental Science & Technology* 49 (5): 2716–2723.

Hagelüken, C. and Corti, C.W. (2010). Recycling of gold from electronics: cost-effective use through 'Design for Recycling'. *Gold Bulletin* 43 (3): 209–220. https://doi.org/10.1007/BF03214988.

Hawken, P., Hunter, L., & Amory, L. (1999). *Natural capitalism: creating the next industrial revolution*. http://www.environmentandsociety.org/mml/natural-capitalism-creating-next-industrial-revolution (accessed 02 April 2021).

Homrich, A.S., Galvão, G., Abadia, L.G., and Carvalho, M.M. (2018). The circular economy umbrella: trends and gaps on integrating pathways. *Journal of Cleaner*

Production 175: 525–543. https://doi.org/https://doi.org/10.1016/j.jclepro.2017.11.064.

Korhonen, J., Honkasalo, A., and Seppälä, J. (2018). Circular economy: the concept and its limitations. *Ecological Economics* 143: 37–46. https://doi.org/https://doi.org/10.1016/j.ecolecon.2017.06.041.

Kumar, A. (2020). *Characterization of non-metal fraction of waste printed circuit boards to promote recycling* [University of British Columbia]. https://open.library.ubc.ca/collections/24/items/1.0394917 (accessed 02 April 2021).

Kumar, A., Holuszko, M.E., and Janke, T. (2018). Characterization of the non-metal fraction of the processed waste printed circuit boards. *Waste Management* 75: 94–102. https://doi.org/10.1016/J.WASMAN.2018.02.010.

Lacy, P. (2015). *The Circular Economy. Great Idea, But Can It Work?* https://www.forbes.com/sites/valleyvoices/2015/01/20/the-circular-economy-great-idea-but-can-it-work/?sh=549cad732e28 (accessed 02 April 2021).

Lifset, R. and Graedel, T.E. (2002). Industrial ecology: goals and definitions. In: *A Handbook of Industrial Ecology* (eds. R. Ayres and L. Ayres). Edward Elgar Publishing https://doi.org/10.4337/9781843765479.00009.

McDonough, W. and Braungart, M. (2002). *Cradle to Cradle: Remaking the Way We Make Things*, 1e. North Point Press https://books.google.ca/books?id=KFX5RprPGQ0C&pg=PP10&source=gbs_selected_pages&cad=2#v=onepage&q&f=false (accessed 02 April 2021).

Pajunen, N., Rintala, L., Aromaa, J., and Heiskanen, K. (2016). Recycling – the importance of understanding the complexity of the issue. *International Journal of Sustainable Engineering* 9 (2): 93–106. https://doi.org/10.1080/19397038.2015.1069416.

Reuter, M. A., Hudson, C., van Schaik, A., Heiskanen, K., Meskers, C., & Hagelüken, C. (2013). UNEP (2013) Metal Recycling: Opportunities, limits, infrastructure. A report of the Working Group on the Global Metal Flows to the International Resource Panel.

Rizos, V., Behrens, A., Van der Gaast, W. et al. (2016). Implementation of circular economy business models by small and medium-sized enterprises (SMEs): barriers and enablers. *Sustainability* 8 (11): 1–18. https://doi.org/10.3390/su8111212.

Sillanpää, M. and Ncibi, C. (2019). Getting hold of the circular economy concept. In: *The Circular Economy* (eds. M. Sillanpää and C.B.T.-T.C.E. Ncibi), 1–35. Academic Press https://doi.org/https://doi.org/10.1016/B978-0-12-815267-6.00001-3.

Stahel, W.R. (1986). Product life as a variable: the notion of utilization. *Science and Public Policy* 13 (4): 185–193. https://doi.org/10.1093/spp/13.4.185.

Statista. (2021a). *Cell phone sales worldwide 2007–2020*. https://www.statista.com/statistics/263437/global-smartphone-sales-to-end-users-since-2007/ (accessed 02 April 2021).

Statista. (2021b). *Consumer Market Outlook – market directory*. https://www.statista.com/outlook/consumer-markets/market-directory (accessed 02 April 2021).

Statista. (2021c). *Global smartphone market share from 4th quarter 2009 to 4th quarter 2020*. https://www.statista.com/statistics/271496/global-market-share-held-by-smartphone-vendors-since-4th-quarter-2009/ (accessed 02 April 2021).

The European Commission. (2008). Directive 2008/98/EC of the European Parliament and of the Council of 19 November 2008 on waste and repealing certain Directives. In *Official Journal of the European Union*. https://eur-lex.europa.eu/legal-content/EN/TXT/PDF/?uri=CELEX:32008L0098&from=EN (accessed 02 April 2021).

UNEP (2007). *E-waste Volume 1: Inventory Assessment Manual*, 1e). United Nations. https://wedocs.unep.org/bitstream/handle/20.500.11822/7857/EWasteManual_Vol1.pdf?sequence=3&isAllowed=y. (accessed 02 April 2021).

Wagner, F., Peeters, J.R., De Keyzer, J. et al. (2020). Closed loop recycling of WEEE plastics: a case study for payment terminals. *Procedia CIRP* 90: 416–420. https://doi.org/https://doi.org/10.1016/j.procir.2020.01.084.

World Economic Forum. (2019). *A New Circular Vision for Electronics – Time for a Global Reboot*. www.weforum.org (accessed 02 April 2021).

Yu, D., Duan, H., Song, Q. et al. (2017). Characterization of brominated flame retardants from e-waste components in China. *Waste Management* 68: 498–507.

Zhang, H., Hara, K., Yabar, H. et al. (2009). Comparative analysis of socio-economic and environmental performances for Chinese EIPs: case studies in Baotou, Suzhou and Shanghai. *Sustainability Science* 4 (2): 263–279. https://doi.org/10.1007/s11625-009-0078-0.

Zhang, S., Ding, Y., Liu, B., and Chang, C. (2017). Supply and demand of some critical metals and present status of their recycling in WEEE. *Waste Management* 65: 113–127. https://doi.org/10.1016/j.wasman.2017.04.003.

Index

Electronic Waste: Recycling and Reprocessing for a Sustainable Future, First Edition.
Edited by Maria E. Holuszko, Amit Kumar, and Denise C. R. Espinosa.

d

e

f